Author's Note

DEATH FACTORY U. S. A. is a story of when the paths of 'Right' & 'Wrong' crossed. Larry lost his parents long before their time, lost a child, lost $1.5 million in livestock within days. He and his family were terribly injured, his brother and family were horribly injured, they have been paying the ultimate price now for 30 years. Yet with courage and inspiration for a better world for everyone, he wrote a book "DEATH FACTORY U. S. A. then it became Part I and Part II. Larry Encourages every American to read both of them, join him in challenging the many polluters, & encourage them (big or little) to clan it up. Larry says he knows That Silent Spring by Rachel Carson stunned the world, but he Believes that DEATH FACTORY U. S. A. Part I and Part II will Not only leave the world stunned but as you read it you will be Wiping away the tears running down your cheeks. Each time you wipe a tear away, "think of a friend or relative you can tell Them to get a copy or maybe as a gift of DEATH FACTORY U. S. A. Part I and II. It's a book you will never be able to pack it away.

It is the secret war gasses, agents, and a Plethora of man-made chemicals, Biological, Virological, Radioactive pollutants, such as Cesium, Plutonium, Radon, Beryllium, radioactive salts, Iodine, Tritium, then Pesticides, and Herbicides, (they are all The "Silent Killers"),(Known Carcinogens), (Destroyers of DNA), (Known Mutants). (Known for a plethora of birth defects). They are Strangling, Suffocating, and will bury us all in the debris of intentionally created pollutants left from our technologically, inventive and Creative irresponsibility.

Larry blew the whistle on technological inventive and creative Irresponsibility by our government, and Corporate giants alike. Larry likes to think he helped put teeth in at least some of the environmental laws. We cannot allow the government or corporate giants to go on polluting and contaminating everything and everyone everywhere they go. All we want and expect is for them to do what They must but

do it safely. There is no room for irresponsibility in the Handling of Ultra toxic and Ultra hazardous materials.

Larry had the key ingredient to success, "DESIRE" is what he had. In 1972 he desired intensely, he acted upon it, 1988 his bill passed the Full House of Congress giving him the right to sue the United States of America. The first complaint was filed in the Court of Claims in 1988 by 1997 the Court process was over although he lost any further legal claims against the United States he won an award for his family's illnesses, and his livestock claims against the United States. House Bill H. R. 816 was sent back to Congress for them to determine and appropriate an award, based on the facts present in the record before the Court of Claims and its review board. This "House Bill" went back to the Full House of Congress for action February 1997, Congressman Bob Schaffer, Congresswoman Diane DEGETTE, Congressman Scott McGinnis Senator Ben Night Horse Campbell. And Senator Wayne Allard, have not taken the time to do their Part as required by Law, acting on the Bill for the Land's.

BEST WISHES FROM THE AUTHOR
LARRY D. LAND

Please read Part I, Part II, and for a finale of a World dilemma Part III, to be out sometime in 2004.

"If Man loses the capacity to foresee and to forestall what he has begun. It will all **End** by destroying the earth and us all."

LARRY D. LAND

Death Factory U.S.A Part II

Death Factory's in our Midst

By

Larry D. Land

This book is based on facts, News accounts and information from the courts, and the U.S. Senate. A documented story of the life and times of the Author Larry Land. Certain names and places were changed to protect the innocent.

ISBN: 1-4033-9354-0 (e-book)
ISBN: 1-4033-9355-9 (Paperback)
ISBN: 1-4107-1084-X (Dust Jacket)

Library of Congress Control Number: 2003090676

This book is printed on acid free paper.

Printed in the United States of America
Bloomington, IN

1stBooks - rev. 08/06/03

About the Book

"DEATH FACTORY U. S. A." Is about man made pollutants that have contaminated our environment. The pristine image of nature is slipping away. Deadly chemicals, germ warfare and Atomic radioactive weapons, testing and materials have killed or Shortened, & taken the lives of millions since the 1950's.

The Government, Corporate giants, the Army, the Pentagon, the Department of Energy, have waged war on their own people. They have intentionally used people without their knowledge as guinea pigs testing of chemicals, biological, virological, and radioactive products, and deadly warfare agents. Weapons of mass destruction Have been tested & used on the American people in the Environment, in the schools, at their work, & without their knowledge.

Forty years since Rachel Carson's "Silent Spring" A great part of my Inspiration to let everyone know "Silent Spring" was the Truth and a point of knowledge, on human carelessness, irresponsibility, & greed that has devastated our Environment. It may be the epitaph of a world not very far beyond us in time. DEATH FACTORY U.S.A. shows the Reckless Negligence, and how it went on unchecked, Larry closed them up. He went further to help close up many more polluters or at least made them clean it up. What began in 1972 as the paths of the Rocky Mountain Arsenal, & Shell Oil Co crossed, with Larry Land it started a battle not easily won. By 1983 Shell Oil Co. & the Rocky Mountain Arsenal were stopped in their tracks and closed forever. Larry Won the battle in getting them and others to clean it up, before suffocating us and Burying us in the chemical and radioactive debris.

Larry sometimes feels he didn't win because it left him totally disabled, but the Courage and inspiration came when he hears how many he may have helped and asks everyone to read the books part I and Part II. DEATH FACTORY U.S.A. is but another attack on human carelessness, greed, and irresponsibility of our Government and Corporate giants everywhere. "It's time they pay the Piper" or, be shut down, & they pay the price to clean it up. It's now costing them billions $$to clean up what should not have happened. Millions of Americans, and their children have paid the ultimate price, and will for decades to come they were unaware of what really happened. Read Part I, and Part II, then watch for the "Finale" Part III, The shocking truth of what has been done to clean it up, or "Cover it Up".

BEST WISHES FROM THE AUTHOR
LARRY D. LAND

CHAPTER ELEVEN

Science is triumphant with far-ranging success, but it's triumph is somehow clouded by growing difficulties in providing for the simple necessities of human life on the earth.

Gerald Pierce, 1993

*** On monday, August 30, 1993 at 10:00 a.m. Larry's attorneys, and attorneys for the United States Department of Justice, met at the Federal Courts Building, 717 Madison Place, NW, Washington DC. This was the United States Court of Federal Claims. Purpose here today is the Defendants Motion in Limine, for a protection order, and Motion for Summary Judgment.

*** The proceedings are as follows:

The Clerk: "All Rise. The United States Court of Federal Claims is now in session, the Honorable Judge Robinson presiding, Case No. 1-88.

Judge: "Who appears for the Defendant"?

"I do your Honor Mr. Reagan, and Mr. Lavoy.

Judge: "Who appears for the Plaintiff in this case?"

"I do your Honor, Denise Callen."

Judge: I will allow thirty minutes for the argument from both sides, with a ten minute rebuttal period, if it is necessary.

*** Def: "Thank you. Your Honor, this action is brought pursuant to Congressional Reference No. 1-88. Plaintiffs brought this action alleging legal and equitable claims against the United States allegedly arising from the operation of the Rocky Mountain Arsenal just north of Denver, in Colorado.

*** By previous order dated September 13, 1990, this court granted the United States Motion in Limine. (limits) and defined the scope of equitable claims in Congressional Reference Actions. In that order, this Court held that, "absent fault of the common law negligence variety, an award would amount to a mere gratuity." The court thereby put plaintiffs on notice, that to establish their equitable claim, pursuant to the reference. they must establish that their alleged injuries were caused by some wrongful act or omission by the United States.

*** As for the Motion by the Defendant for a Protective Order, This court finds that Defendant, under this court's rules, has failed to show good cause for the issuance of a protective order relating to the general subject of airborne contamination. Count III q 48 of plaintiff,s amended complaint specifically alleges that defendant had trespassed by spilling or otherwise releasing into the environment toxins which would migrate and enter upon and under plaintiffs property. That allegation is sufficiently broad to cover airborne contamination.

*** "Your Honor, the complaint specifically alleges that the Arsenal is negligent in manufacturing, designing, researching, testing, constructing, and operating the chemical treatment systems and disposaal basins at the Arsenal.

*** Plaintiff's allege that as a result of that negligence, toxic chemicals from the Rocky Mountain Arsenal contaminated their ground water and air. That sometime in the spring and summer of 1972, the plaintiff's suffered personal injuries and injuries to Mr. Land's Livestock."

*** Judge: "Is the issue of personal injuries to lievestock and the persons involved, the plaintiff? Are those the only issues now before this court?"

*** Def: "No. Your Honor. Those issues are not before the court. The only issue before the court is whether or not the United States committed any wrongful act or omission. The issue of causation is the only issue before the court at this time.

*** Judge: "I didn't mean in your motion. I meant what is the Congressional Reference limited to?"

*** Def: "Yes, your Honor. It is limited to personal injuries and injuries to the cattle operation. By previous stipulation and order of this court, This issue of actual property damage to real property is not in question here. However in question here is because of the Courts bifurcation of the Trial, which places Causation first then personal damages at a second trial.

*** Judge: "and there has been no amendment of the Congressional Reference to encompass injury to the Land."

*** Def: "that is correct, Your Honor. There has been no amendment of the Reference."

*** Judge "all right, please continue.

*** Def: "Your Honor. Plaintiff's amended complaint seeks recovery pursuant to five separate plead counts. Those counts are Neglignece, Strict Liability for conducting an Ultra-Hazardous Activity, Trespass, Nuisance, and Deceit based on Concealment. Each of the issues addressed here, will be discussed specifically.

*** "I would like to recap briefly the procedure of this case and how we've gotten to this stage. This action was brought pursuant to the Reference. Soon after the case was filed, the United States filed it's motion in limine and defined the scope of the equitable claim. That motion was briefed and argued before this court in 1990. In the order that came out in 1990, this court applied the standard as set out in Bee Amusement Company and Subsequent cases which required a showing of wrongful conduct in order to establish an equitable claim and rejected the notion that equitable claim was based on some broad moral duty to pay.

*** "Following that order in September of 1990, Plaintiff's filed their amended complaint. On the heels of the amended complaint, they filed approximately one-hundred pages of written discovery. The United States responded to that written discovery in January 1991, and supplemented the response to June of 1991.

*** "This Court issued a scheduling order which required, among other things, that plaintiff's produce their expert reports by October 31, 1991."

*** "Their response was a promise to provide expert evidence to establish the standard of care by which the acts or omissions of the United States could be compared. Plaintiff's did not file their report in October, but by means of an agreed extension filed their final expert reports in January of 1992."

*** "When those reports were filed there were five expert reports. There was an expert report by an economist on the issue of damages, that has nothing to do with the issue of wrongful conduct. There was a report by a medical doctor who is a toxicologist. His opinions go only to the issue of causation of injury. Again, that has nothing to do with the issue before the court today."

*** "There were to opinions by veterinarians, obviously, neither of whom could give opinions regarding the nature of the negligence or wrongful conduct in the disposal systems at the Arsenal. The fifth was a report from a hydrologist, who stated no opinions regarding the operations of any system at the Arsenal. The only opinion stated in the hydrologist report were that the chemical DIMP, which was found in the plaintiff's wells, this is a stipulated fact that it could have come from the Arsenal."

*** "That is not in dispute. The United States has stipulated that the DIMP ultimately found in the plaintiff's wells must have come from the Arsenal. Your Honor, those reports are completely devoid of any opinion regarding the United States activities at the Arsenal, and evaluating whether or not they were conducted wrongfully or negligently."

*** Judge: "How about the issue of liability without fault? If you've got a dangerous instrumentality that escapes from your property and invades the property of someone else, even though you take every reasonable precaution to prevent that, what's the law in Colorado with respect to that issue?"

*** Def. "Well, your Honor, I'd like to address that. The law in Colorado at the time of these acts or omissions probably would not have established strict liability in those circumstances."

*** Def, "The Plaintiff's cite in their brief the City of North Glen's case which was decided sometime in the 1980's. That case specifically makes reference to the fact that prior to the adoption of Chapter 21 of the Restatement, Strict Liability under Colorado Law was highly limited."

*** Def, "Second, no strict liability will attach to any abnormally dangerous activity that is conducted pursuant to a public duty. Clearly, the Arsenal was operated in all of its activities pursuant to public duty. It was established in the 1940's to manufacture chemical warfare agents for the purpose of the common defense."

*** Def, "Section 521 of the Restatement 2nd on Ultra-Hazardous activity, sets out that there shall be no strict liability under those circumstances. Comment A to Section 521 is directly to the point. Comment A says, 'There shall be no strict liability to the Government for the manufacturing or storage of high explosives.'"

*** Def, "And for those reasons, I don't believe strict liability would be within the meaning of the equitable claim, and I think the prior order is consistent with that ruling. It would be a great expansion to allow strict liability against the Government. It's not allowed in FTCA actions, and there is nothing in the Reference that hints that there was a broadening of the scope of liability.

*** Def, "Turning back to the negligence count, Your Honor, plaintiff's allege that the United States breached a duty of reasonable care in the design and the conduct of the activities of the disposal basins at the Arsenal. It's clear that it's the Plaintiff's burden of proof to establish negligence."

*** Def, "It's also clear that in order to establish a breach of duty to use reasonable care, one must set out what the standard of care is for any specific activity of which one complains. The activities complained of in this lawsuit all involve complex engineering systems and engineering and scientific decissions. Those things, the standard of care for those activities, are clearly beyond the competence of ordinary persons to define.

*** Def, "Under Colorado Law, in the case of Palmer V. Robbins, which is cited both in our reply brief and in the plaintiff's response brief, states that, 'Where the standard of care to be applied is beyond the competence of ordinary persons, expert evidence is required to establish that standard."

*** Def, "Plaintiff's evidently acknowledged that requirement when they responded to interrogatory No. 38 and indicated that they intended to prove negligence by way of expert evidence. yet, when they filed thier expert reports, none of them even addresses the question of standard of care or states any opinion regarding whether the United States conduct would have breached any such standard of care."

*** "Under Celotex, having had plenty of opportunity to discover the facts that would support expert opinion, and having failed to do so, despite over two years of litigation and despite the fact that sixteen months have run since this court put plaintiff's on notice that they must establish wrongul conduct, Summary Judgment should be granted on the negligence count.

*** "To the extent that either the trespass or the nuisance count is predicated on the negligent conduct of the United Sates, they must fail for the same reason. To the extent that neither the trespass or the nuisance counts are predicated on intentional wrongdoing, they must also fail. There is no record of evidence that would establish that the United States intended to cause a trespass or nuisance on the plaintiff'ss property.

*** Def, "Under Section 158 of the restatement, as it applies to trespass, when one alleges an intentional trespass, he must prove intent. Intent can be the volitional act that actually constitutes a trespass. That is, if I step over the proprty line in my neighborhood, the act itself constitutes a trespass. Plaintiff's are not alleging that type of trespass here, your Honor. They are alleging that the United states intentionally caused a thing to enter their property. Specifically, the thing entering their property is contaminates in their ground water.

*** Def, "Under Section 158 of the Restatement, when one alleges that type of intentional trespass, the plaintiff must show that the United States intended for the thing to enter the property. One can meet that test by meeting the so-called substantial certainty test set out in comment I. That is, Plaintiff must show at the time of the act or omission complained of, that the defendant knew or should have known to a substantial certainty that the invasion would result.

*** Def. "The facts of this case would not support that finding There is no record of actual intent. Nor is there any recorded evidence by testimony or report that at the time of any specific act of omission, a trespass of neighboring property was a substantial certainty. In fact, the recorded evidence suggests otherwise."

*** Def, "The original unlined basins were built pursuant to the recommendations of an expert. His recommendations include the assumption upon which the Army acted, that the basins would form an impermeable blanket on the bottom. Ultimately, that may have proved incorrect. Nevertheless, it would not support a finding of inevitability of substantial certainty as required by the Restatement, and as required by Colorado Law."

*** Def, "Each of the trespass cases that the plaintiff's cite in their brief are intentional trespass cases. Plaintiff's response brief sets out three elements that make a prima facie case for Trespass in Colorado. That list of elements is incomplete. There is no such list cited in the case they cite, the Burt case.

*** Def, "There is a fourth element in addition to those three, and that is the need to prove intent. Burt makes that clear, both by the cite to the restatement, and by the use of the word 'inevitable' which precedes the cite to the Restatement.

*** Def, "The other cases involved similar facts where, for instance, one, there was a two-story sloped roof in an area of high snowfall, constructed only seven feet from the neighboring one-story building. It was inevitable that the snow was going to fall on the neighboring property. You don't have that here. These basins are miles from the plaintiff's property, (in reality the distance is barely at

1 mile.) They are miles inside the border of the Arsenal, (another fib, Irondale Trailer Park is one mile from Basin F.) When they were constructed, the assumption was they would seal. There is nothing in the record that would support a finding of inevitability or substantial certainty as required by Colorado Law and the Restatemnt on trespass.

*** Def; As to nuisance, the same substantial certainty test is incorporated into the definition of intentional nuisance in sections 822 and 825 of the Restatement on tort. For that reason, Your Honor, both the trespass and nuisance claims must fail. There is no evidence of an essential element that intent is the wrongful conduct required for those torts. And because of that lack of evidence, summary judgment would be proper. The Def. Is asking."

*** "Def; final attack, "the final count of the plaintiff's amended complaint is a count for fraud and deceit based on concealment. That is a Tort that is defined by Section 550 and 5551 of the Restatement 2nd. It arises from the duty to disclose material information in a transactional relationship between the parties. That is, where one party induces another to enter into a contract or some sort of business relationship by withholding material information."

*** Def; "There is no evidence in this record nor is there, in fact any allegation of a transactional relationship between the Plaintiff's and the United States. For that reason, the count for fraud base on concealment also lacks evidence of an essential element of the claim and should be dismissed.

*** Def; Summation, "In summation, Your Honor, each count in the plaintiff's complaint fails for there is no evidence on an essential element of each count of the complaint. For that reason, the United States respectfully requests that summary judgment be entered in his favor in each and every count. Thank You."

*** Judge; "All right, thank you to the Defendant, the Plaintiff are you ready."

*** Pltff; "Yes your Honor".

*** Judge; “Very Well, Lets proceed.”

*** Pltff; “Your Honor, we are here today on defendant’s motion for summary judgment in this case. This court is obviously familiar with the standards for the granting or denying of a motion for summary Judgment, so I won’t spend a lot of time on that issue. The rules of the Claims Court do not differ from those of the United States Federal District Court rules, nor do they differ from most of the states. Which is that a motion for summary judgment must be denied if there is a material issue of fact presented. Our brief, Your Honor, clearly sets forth the appendices that were attached to a clear issue of fact with respect to each of the claims that are being brought in this action. That includes negligence, Trespass, Nuisance, Ultra hazardous activity, and Fraud by Concealment.

*** Pltff; “Before I begin to review each of those claims and the facts that we have before you, I would like to address this argument that the defense is making concerning that to prove an equitable claim in a Congressional Reference Case, we are limited to negligence or an intentional tort. When I read this Court’s order dated September 13, 1990, Which stated that, ’In order for plaintiff’s to recover under an equitable claim, they show that: (1) the Government committed a wrongful act; and (2) that this act caused damage to the claimants.’”

*** Pltff; “That order appeared to be clear to me, Your Honor. But it seems as though we are rehashing an old issue. I do not interpret your order as limiting an equitable claim in a Congressional Reference case to an intentional wrongdoing or to negligence. In fact, in the case of Spaulding, which was recently decided by this Court in May of 1993, that can be found in 28 Federal C.T. 242 1993, the court in that case reaffirmed your prior September 13, 1990 order. In that case the court held that, ’Absent a finding of negligence or wrongdoing on the part of the government or employee, any award would be a gratuity.’”

*** Pltff; “If the court’s intent was to limit a Congressional Reference equitable claim to negligence, why does that court consistently in their opinion use the word ‘wrongful act.’ I cannot find a case that does not define 'Wrongful Act,' Your Honor. But no case has held that it's limited to negligence. Since the court does consistently use in the recent line of cases, where they've gotten away from the moral standard, 'Wrongful Act,' is used consistently. What does that mean, Your Honor?

*** Judge; "If you go to Black's Law Dictionary, it is broadly defined as follows: 'Any act which in the ordinary course will infringe upon the rights of another to his damage, unless it is done in the exercise of an equal or superior right. Term is occasionally equated to term negligent, but generally has been considered more comprehensive term, including criminal, willful, wanton, reckless and all other acts.'"

*** Pltff. "Shoot I lost the rest of the definition."

*** Judge; "just take your time."

*** Pltff, "It's here somewhere, ...got it. 'Which in the ordinary course will infringe upon the rights of another to his damage.' That is the key language, Your Honor. 'And all other acts which will infringe upon the rights of another to his damage.'

*** Pltff; "Reasonable interpretation, not only under Black's Law Dictionary's but also under the line of cases this court has interpreted in order to prove an equitable claim, would be that a 'wrongul act' should be included in encompass tortuous conduct. That's a broad spectrum from intentional to negligent. Trespass, nuisance sits in there somewhere, to strict liability-type tortuous conduct. An interpretation that 'wrongful act' includes all types of tortuous conduct, is reasonable and it will not interfere with the intent of the claims court, which is to avoid gratuitous awards of damages in Congressional Reference Cases.

*** Pltff; "Claimants under an equitable claim theory in a Congressional Reference case will still have to prove a tort, whether it's intentional, negligent, or the strict liability nature, or of the trespass and nuisance nature, Your Honor. It would not be a gratuity, as this court is aware".

*** Pltff; "You have issues of proving that the chemicals and the by-products of those chemical warfare weapons actually arrived on the Larry Land Farm, Either via the ground water or the air emissions, from when they were burning the Mustard gas, and the nerve agents."

*** Judge; "Is the point that counsel is making, is that we look at the evidence that has been developed thus far in discovery, because they are saying that you can't point at an expert's testimony to anything that the government has done that would constitute, even under the broader concept of wrongdoing or negligence. That if you look to the evidence that you have adduced thus far, that's not going to be sufficient in the aggregate to prove causation and damage. If you had an expert that, based upon the deposition testimony, that you've taken, who could say that, 'yes, in my opinion as an expert in this area, there is no question that the defendant should have known that the seal was not working and that in the normal course of events, there would be an escape of toxic substances into the ground water, so that cause could be shown."

*** Judge; "But you don't have that evidence. And while I could…I may be able to accept the theory that there is some kind of overall concept of wrongdoing that supersedes, just merely what's stated in the restatement."

*** Judge; "if I grant the motion… excuse me… I mean, if I deny the motion and we go to trial, if you are limited to just what you have adduced thus far, you are going to be faced with a motion at the end of your case for a directed verdict. And right now I'm looking at that evidence and seeing, this doesn't prove enough to the Court. This is what's troubling me with this case, not the broaader concept of there being some wrongdoing that could be encompassed in a Congressional Reference. That's what I need you to talk about."

*** Pltff; "Clearly, Your Honor, cause and damages in this case have been a hotly contested issue. We do have proof, Your Honor, that toxic substances did get into the Land family's well water in the early 1970's. These reports produced concerning DIMP and other toxic and Carcinogeni compounds were clearly established there in 1972."

*** Judge; "But the defendant says that they could be there without there being any wrong doing, you see. Obviously, they escaped. There is no question about that, because they have admitted that there's quantities there enough to be detectable, and the only place it could have come from is that retention facilitys' Basins A, C, and F. If we knock that out, then I'm looking for some evidence of wrongdoing.

*** Pltff; "Your Honor, the defendant is trying to graft onto the trespass and nuisance claims an element of intent that is not required in Colorado, under Colorado Law.

*** "Pltff; "In Colorado, trespass and nuisance are viable, long-time claims in Colorado. It's developed primarily through our water rights law, Your Honor. Colorado has separate water courts because of the shortage of water, and people in Colorado take their water extremely seriously."

*** Pltff;" In the case of Burt V. Beautiful Savior Luthern Church, it was set forth that to prove a trespass claim, it only must be established that there was a physical intrusion on the property of another without the permission of the person lawfully entitled to possession of the real estate, which caused damage to the person or property of another."

*** Pltff;" There is no requirement, Your Honor, that the person intended to trespass. What has to be intended is an intent to do the act. In this case it is the manufacturing and disposing of chemical warfare weapons. It is not required that the Government intended for these chemicals and by-products to escape off the Arsenal and get into the underground aquifers and contaminate the surrounding landowner's ground water and wells."

*** Pltff;" The Burt case made this abundantly clear. It said, 'In Colorado, liability for tespass requires only an intent to do the act intself, that constitutes or inevitably causes the intrusion."

*** Judge "So, let me see if I've got this right. You are saying the act is the intent to store the contaminants, is that right?"

*** Pltff; "That is correct. and to dispose of it into the various basins. The Basins A through F were utilized over long time periods, and at different time periods, Your Honor. Also the method of disposing of the Lewisite, mustard gas, Sarin and GB nerve agents, which was released or emanated into the air, which we are alleging caused damage to the Lands which Dr. Teitlebaum discussed in his report.

*** Pltff; "Therefore, Your Honor, it's the fact that a landowner sets in motion a force, which is the act, which causes damage ulitimately to the property of another which constitutes the trespass. Colorado does not require the level of intent that the defense is arguing. The defendant is trying to make the same argument, Your Honor, with nuisance.

*** Pltff: "Your Honor, water is the most precious, limited natural resource we have in this country. But because water belongs to no one, except the people. Though special interests, including Government, and Corporate polluter's, who use it as their private sewers."

*** Pltff; "Nuisance and trespass are similar type claims, although the only thing that needs to be shown in Colorado to establish a nuisance is that the defendant unreasonalby and substantially interfered with the use and enjoyment of his property by enjoinment (Forbid/prohibit). A defendant's liability in Colorado under Allison V. Smith, which held that defendant's liability rests on the duty he owes as an occupier of land, to prevent conditions on his land from deteriorating, particularly those he has created.

*** Pltff; "Here they have created the chemical warfare weapons, and they have created the conditions where they needed to dispose, and in fact, disposed of them themselves. There is even a stipulation on the record from Mr. Delk, and we cited that in our brief, where they are not even contesting the fact. That, 'yes we made the chemicals, and yes DIMP is a by-product, and yes it came from the Arsenal. It couldn't have come from anywhere else.' Let's face it, most people in America do not have a book in their house on how t make nerve gas."

*** Pltff; "It is important to recognize that it's the interference under nuisance, just like it is with the act itself under trespass, which must be unreasonable, Your Honor. Here we have alleged numerous questions of fact in our brief. They are cited on page 26 which created a material issue of fact for purposes of the motion for summary judgment."

*** Pltff; "It's clear that toxic chemicals entered the plaintiff's ground water making the water unusable. The entry of Nerve Gas from the Rocky Mountain Arsenal also got to the Plaintiff's property in the mid 1970's when the Arsenal decided to burn some Mustard gas, and GB, and sarin nerve agents and dispose of it that way."

*** Pltff; "With respect too the Ultra-Hazardous activity argument, Your Honor, I would take issue with the statements made by opposing counsel. Under Colorado Law, Strict Liability is imposed on parties who engage in Ultra-Hazardous activities.

*** Pltff; "In Garden of the Gods Village V. Hellman, which is at 294 Pacific 2d 597. which is a 1956 case, the Court of Appeals held the defendants strictly liable for blasting operations. In the City of North Glenn V. Chevron, which was decided by the Federal District Court in Colorado in 1991, they did find that the Supreme Court of Colorado would hold that in this particular case, wher large quantities of gas being stored and other petroleum products that leaked into the underground water, poisoning the wells of many residents in the City of North Glenn, not dissimilar to allegatons in this case. And they did find that it was an Ultra-Hazardouls activity".

*** Pltff; "More importantly, Your Honor, a recent Tenth Circuit case called Daigle V. Shell Oil Company, which can be found at 972 Fed. 2d 1527, involving Shell Oil Company and the Rocky Mountain Arsenal."

*** Judge; “Are these cases cited in your brief?”

*** Pltff; “Daigle is not, Your Honor That came out in August of 1992, I believe, which was subsequent to the filing of our briefs. However, the City of North Glenn and Garden of the Gods are. In the Daigle case, Your Honor, Residents that lived to the northwest of the Arsenal in a trailer home park brought an action against both the Rocky Mountain Arsenal and Shell Oil Company for clean up activities concerning Basin F that were occurring in the late 1980’s.

*** Pltff; “At that time, Your Honor, part of the method by which they were trying to clean up Basin F was by burning some of the Sludge, mud and everything else. The residents there had brought an action against the Arsenal and Shell Oil Company, although the Arsenal was dismissed on its FTCA case, found for and on personal injury damages, and property damages.

*** Pltff; “In the Daigle case, the Tenth Circuit reversed Judge Carrigan’s decision finding that it was not an Ultra-Hazardous activity, and held that, ’Looking to the factors to be weighed under Section 520 of the Restatement, 2nd section of torts, this clean up activity of the Arsenal was an Ultra-Hazardous activity.’”

*** Judge; “When did the burning take place? What was the year?”

*** Pltff; “There were two separate time periods, Your Honor. 1972 through approximately, August of 1974, it was called project Eagle. The Arsenal was disposing of Mustard Gas bombs and also GB Nerve Gas bombs. The method utilized at the time by incineration.

*** Pltff; “The second time period that is discussed in Daigle, was the 1988 clean up of Basin F where a lot of this chemical waste had been disposed of since 1954. Part of that on going process also involved burning of the chemical warfare by-products and pesticide by-product that Shell Oil Company dumped into that basin.”

*** Pltff; “With respect to the fraud and concealment action that we have brought, Your Honor, it is simply incorrect that a fraud and concealment action is limited to a transactional type activity. The elements for a fraudulent concealment action in Colorado are set forth in Carson v. Garrison, and that case is cited on page 29 of our brief in opposition. Nowhere in the elements that are required to establish fraud and concealment is it stated that it is limited to a transactional type activity. We all know of product liability situations where manufacturers of, may say, for instance, all terrain vehicles hid a lot of the knowledge and information they had concerning the dangers of those vehicles."

*** Pltff; "If a third-party non-purchaser was just using it, and was injured, you could still bring an action for fraud and concealment of the knowledge that Honda, Kawasaki, and Yamaha had concerning the dangers of those vehicles."

*** Pltff; "It's similar to this situation here, Your Honor. The defendant knew in February of 1953 that DIMP was created during the GB nerve gas manufacturing process. There is also evidence, that the defendant knew by 1958 that ground water contamination was resulting from their activities not only on, but off the Rocky Mountain Arsenal as well, it extended to an area of several square miles to the North and Northwest, off the Arsenal."

*** Pltff; "Indeed, Your Honor, in the late 1950's there were farmers that were off site of the Arsenal that had crop damage, and that the Government paid off several of those claims."

*** Pltff; "Defendant's response to this contamination, which they knew or suspected was emanating from the Rocky Mountain Arsaenal, was to no longer use Basin A, and to build Basin F, which was completed in 1955. Defendant knew that Basin F was leaking contaminants by at least 1965, and 1969. Defendant did not disclose this material fact to anyone until 1988, when it was required to do so as part of the Cercla Action taken by the State of Colorado against the Arsenal by the Colorado State Health Department and the Attorney General's Office as well.

*** Pltff; "George Donnely who was in charge of the Arsenal, their facilities department from 1967 through 1976, testified in a lawsuit that was brought by Shell Oil Company against various insurers, trying to force its insurance carriers to cough up their insurance proceeds, so to speak, to help pay for the clean up costs at the Arsenal."" Larry reminded them "What Shell had done at the Arsenal was not an accident, nor was what the Army done considered reasonable or proper. It was purposefully and willful mishandling of Ultra-Hazardous materials, causing pollution in both the air and the water."

*** Pltff; "One of the defenses was that the insurance companies were not required to do so because Shell had defrauded them, and the Arsenal U. S. Army had defrauded them at the time these policies were entered into by not disclosing that they knew that contamination had, in fact, occurred, off post."

*** Pltff; "Your Honor, from pages 31 through 33, in our brief, is George Donnelly's testimony from that trial, and it clearly establishes fact, that the Arsenal knew, at least by sometime in the mid-seventies, and probably earlier than that. Because at the end of 1969, they had brought in outside consultants to examine the Basin F Liner and discovered that it had indeed deteriorated. The United states Army Hygiene and environmental agency, in hand with the United States Health, Education, and Welfare agencies, (HEW) and two General's from the Pentagon, joined that inspection of basin F. Much to their surprise a great part of the liner had been dissolved, and disintegrated over the years, It was missing, and they strongly suspected (knew) that there was off-post contamination occurring, and had been for quite some time, at that particular point in time they estimated 25 square miles off the Arsenal had already been polluted. The Army was warned at that time by all, that any further use of Basin F would be with the Knowledge it was leaking, and on the premise it is contaminating underground water off the Arsenal property."

*** Pltff; "At no time despite that knowledge did they warn any of the surrounding landowners, which would have been a very simple step."""The Army chose to ignore common sense, and the warnings and memo's from their experts, and continued to cotaminate even a far larger area, with deadly toxic warfare agents, off the Arsenal for an additional 15 years under the premise it was leaking. What they done here was willful, and wanton Negligence."

*** Pltff; "There is no excuse for what the Army did at the Arsenal. "It's true, Your Honor, that we do not have an engineer that's going to come forward and say that it as negligent in 1955 to dispose of these materials in this manner, etcetera. Our claim, Your Honor, is based upon failure to warn, while operating and using unreasonable conduct which it does not require expert testimony, and the Army did not."

*** Pltff; "It is also our contention, Your Honor, that there is a question of material fact concerning the Rocky Mountain Arsenal's off-post and on-post water monitoring, The Rocky Mountain Arsenal, and the Testimony is cited in our brief and in our appendix, that the Arsenal stopped off-post and on-post water monitoring despite the fact that they had done so in the late 1950's and early 1960's. and all testing was void until 1974, when Jim Land, began complaining that his wheat field crop had sustained damage. It wasn't until approximately that point in time that the Arsenal once again began its off-post water monitoring program, as mandated by EPA. and Colorado State Health Department, Thank you Your Honor."

*** Judge; "All right, thank you to the Plaintiff, then ordered the defendant to use the remainder of his time for rebuttal."

*** Def; "Your Honor, a few points on rebuttal. For plaintiff's to say they did not interpret your September order as requiring proof of some kind, I think is false. Their own brief in the response to our motion in liimine reccognize that if this court adopted the B amusement standard, they would be required to prove negligence.

*** Def; "Secondly, I must again address the Burt case and plaintiff's recitation of the elements for trespass under Colorado law, and I urge you specifically to read Burt. Nowhere does it set out three and only three elements of a trespass case in Colorado. Again I must point out that Burt and the other cases cited by plaintiff specifically cite to section 158 of the Restatement, which is the section that governs intentional trespass." Larry says, "Intentional, The Army had several basins including basin F, they were and had been operating since the early 1950's knowing they were infact leaking into the under ground water stratus off the Arsenal." "Plaintiff states that's considered unreasonable conduct."

*** Def; "It also includes comment I. which sets out the substantial certainty test. Similarly, as to nuisance, Plaintiff's contention is that under Colorado Law, the focus on the work 'unreasonable' only applies to the consequence is simply incorrect. That is the case where you have an intentional nuisance. When you have an intentionald nuisance alleged, it's clear you must meet the same substantial certainty test, and again, I refer you to section 822 and 825 of the restatement."

***** Larry says, "Stack records that had been placed into evidence clearly shows that far more air particulate is being released than was considered safe, for Mustard agents, and the GB agents, and any other chemical warfare agents." "they could not even maintin the stacks in the safe range, that was a nuisance and a trespass both, Then lets talk of the basins they knew were leaking yet they continued using them with the premise they were leaking deadly pollution off post." "that is intentional."

*** Def; "The Allison case that she quoted and relied upon makes that clear, and I'll quote directly: 'The unreasonable and substantial interference test enunciated in Lowder and Miller necessarily include a consideration whether the questioned activity is reasonable.' That is a focus on the defendant's conduct under all surrounding circumstances."

***** Larry says; "Let's focus on the Army's activity at the Rocky Mountain Arsenal, Basin F as designed and planned to hold 180 million gallons, any thing else would be overfill, of toxic waste. With a maximum life expectancy of ten (10) years. The Army increased the capacity from 180 million gallons to 250 million gallons, this was after the fact they had been told the basin was leaking. It was constructed in 1955, 1965 was the end of the life expectancy, but the Army increased the gallonage, by 70 million gallons, to the point of overfill, knowing full well it was leaking, they continued using it through 1982.""This sounds as if it as intentional and unreasonable and willfully Negligent. Shell Chemical Company had doubled, tripled, and quadrupled their production of some of the deadliest chemicals in the world, that had already been banned in the United States. Yet the United States Army continued allowing them to use basin F. knowing full well it was leaking those deadly chemicals into the water auquifers off the Arsenal."

*** DEF; "Thus, the legitimate but unsightly activity may become a private nuisance if it is unreasonably operated. Again, That's a focus on whether the defendant's conduct is unreasonable or not. It has nothing to do with the ultimate consequences."

***** Larry says, "The defendants conduct was a nuisance, it was unsightly and was unreasonably operated. It was Unconscionable, on the part of Shell Chemical Company and that of the United States Army."

*** Def; "Your Honor, Plaintiff made reference to the injuries alleged from the D-milling process in the early seventies. Dr. Teitlebaum's opinion, which they attack, attempts to link those up to the D-milling process specifically for the chemical GB Nerve Gas. It does not mention Mustard. The Mustard D-mill did occur in 1972, and the plaintiff's injuries were in 1972, but the GB D-mill did not begin even until October 1973. (Not true, he's telling stories again. It'a somewhat of a red herring, I believe. See I thought something smelled fishy and its him.)"

***** "Larry says that is simply not true, we have records that the GB D-milling began in 1972, and the Army at that time was having trouble because it was leaking GB through the stacks and into the air where a good southerly breeze would blow it directly to the Land property. This is just another lie by the Justice Department to cover up the toxic tracks the Army had made."

*** Def; "With regard to the strict liability question, a couple of points. No strict liability for something done pursuant to a public duty. Nothing under Colorado law, what ever the state of the law might have been at any particular time, obviates that requirement."

*** Def; “It is clear that up until the time of the North Glenn case strict liability in Colorado was limited to blasting cases. This is not a blasting case and doesn’t fit into the scenario. This case specifically mentions that strict liability for these types of actions could not have occurred under Colorado law until such time as the Restatement on abnormally dangerous activities was adopted in 1977, three or four years past any act or omission that could have anything to do with plaintiff’s injuries.”

*** DEF; “There is a lot of extraneous things being brought in here for the first time. The Daigle case, just to set the record straight, had nothing to do with incineration. There were allegations of air-borne contamination kicked up as dirt was moved around. But there was no incineration on the site. And again,, that has nothing to do with the plaintiff’s complaint.”

*** Def; “As to the failure to warn that the plaintiff makes, there are some points that are clear in the record. First of all, there were contaminants reported off-post in 1954. That was an area northwest of the Arsenal. The ground water studies that were conducted define the plume, and it came nowhere near the plaintiff’s property. (Then what killed my parents?)

*** Def; “Those complaints were public knowledge, they were not concealed.

*** Def; “As to the issue of whether it was negligent to stop testing in 1964, contrary to plaintiff’s complaints,, that would require expert testimony. You would need an expert to say why that breached any standard of care, given the facts of the case. The complaints came out in 1954, and the Army immediately started testing with the state of Colorado, as would be reasonable. They identified the cause of the crop failures, and they identified the possible source as the unlined Basin A. The Army immediately remedied that by building Basin F, Which was, if not the first, one of the first lined basins used anywhere for industrial waste storage.

***** Larry says, "That is very true the Army did begin a testing program aligned with the Colorado Department of Health. However they also did quit all testing in 1964, and would not allow the State Health Department to do their job either, because they knew Basin A had already leaked, so they stopped use of it and began using basing F in 1955, but allowed Shell Oil Company to continue using basin A for their waste. The Army was aware that Basin F was leaking and refused to do any further testing, and continued using Basin F on the premise it was leaking into the water stratus off the Arsenal property. They also Classified the facts that basin F was leaking upon the advice of the pentagon, Then never disclosed that material fact until 1988, when it was demanded of them by Cercla, Colorado Health Department, and the Colorado Attorney General's Office. Basin F. was the first and the last lined basin in the World, The liner consisted of 3/8's inch blown asphalt."

*** Def; "In addition to that, they continued to test for more that eight years after the time Basin F went into use. To say that the cessation of testing, after ten years, somehow is negligent would, I think, be beyond... and that breaches some sort of standard of care....is clearly beyond the competence of ordinary persons. And you would require some expert in environmental science to say that it was unreasonable to do so, given those facts.

***** Larry says, "Given those facts it would seem unreasonable, and a breach of that duty, after knowing Basin F was Leaking in 1965, and 1969 and to continue it's use on the premise it was leaking, then discontinue all testing, and classified all information on basin F, was beyond all comprehension and unconscionable, and willful trend of Negligence. Keeping everyone completely in dark for over 25 years."

*** Def; "Finally, one last thing. The testimony of Mr. Donnelly indicated only that the Arsenal suspected the Basin F liner was breached, not... there is nothing in that record that suggests they had any knowledge that the breach, which was at the surface line, had, in fact, impacted the ground water aquifer below. They immediately started testing, and the testing in the viscinity, and you worked from the basin out. They did that testing immediately. There is nothing in the record that indicates that, was what caused any off-post migration of contaminates.

***** Larry says; "This is all after the fact, in 1965, and 1969, after a thorough, and extensive investigation by The United States Army Hygiene Department, and the United States Department of Health along with the Pentagon officials official reports stated it was leaking and had already contaminated at least 25,000 acres to the north and northwest off of the Arsenal. That report contained written warnings to continue using basin F would be on the premise it was leaking. THAT IS WHEN THEY DISCONTINUED ALL TESTING ON AND OFF THE ARSENAL, AND AT THAT EXACT TIME IS WHEN ALL INFORMATION ON BASIN "F" WAS CLASSIFIED TO EVERYONE. IT WAS AN ARMY GENERAL FROM THE PENTAGON WHO FIRST SUGGESTED ALL INFORMATION ABOUT BASIN F BE CLASSIFIED BECAUSE IT WAS LEAKING."

*** Def; "It is plaintiff's burden to produce evidence, not allegations, now that we are three years into the lawsuit. There is no evidence that we had any knowledge of off-post migration from any breach of the liner Basin F, and that any duty to warn arose from that."

*** Judge; "all right, sir. Thank you. Anything further from the plaintiff?"

*** Def; "Nothing further, Your Honor."

***** Larry says, "The United states Army failed to warn, and withheld information knowingly and negligently, from the Land's that would have eliminated or mitigated the damages to the Lands. The Army knew full well the spray raft they operated in basin F began in 1963. An Army observer reported the odor of Basiin F was as strong on the northern boundary outside the Arsenal, where the Land Farm's was located was as if one were on the banks of Basin F. The height of the spray was over 100 feet, it was a fact that aerosol droplets were carried for some distance. Up until 1977 a report reflects was contaminating the Plaintiff's air for many years. The Army was aware as early as 1965, and was confirmed in 1969 that basin F was in fact leaking and had already polluted 25,000 acres off the Arsenal to the north and North west. The air contamination to the Land's during the 60's and 70's was deliberate, willful negligence on the part of the Army."

***** Larry says;" information supplied in the breifs. Reservior Flushing: Clean irrigation water was placed into Basin C beginning in 1957. The purpose of such water was to "flush", and "dilute", the polluted aquifer. "The very aquifer they knew full well was Polluted". Such waters also had the effect of driving pollutants beyond the boundary of the Arsenal and increasing the height of the water table. The record does not reflect when or whether this practice ceased at all but does reflect that water was in evidence in Basin C as late as 1974. A leak in the south plant area provided a mound of fresh water which served the same purpose."

***** Larry says; "Blose Deposition p. 119, 11-3-23 States there are other well documented sources of the pollution which affected the plaintiffs even if Basin F had not been leaking. Basin F is one of the better illustrations of how the Arsenal solved difficult problems. See letter dated 5, Feb 70 from A. N. Hardman, Maintenance Engineer, entitled report of a trip to Rocky Mountain Arsenal, 21-22 Jan 70. 'Plaintiffs' Exhibit 19, for comments concerning classification of adverse information. Just like they did in 1969 when they knew Basin F was leaking they classified everything from then on.

*** Judge; "All right. Since there are some cases that have been cited, the court wants to take a closer look at the first time by plaintiff, the court will not make a ruling from the bench today. I would like to make several observations."

*** Judge; "While the Court can certainly accept the possibility that there may be a broader area of wrongdoing that's not identified within specific allegations of negligence and tort principles. Nevertheless, the Court would expect the plaintiff, if it intends to pursue this case to trial, assuming that I deny the motion, to adduce concrete evidence of wrongdoing, either by expert testimony and the like.

*** Judge; “And if you don’t have that plan to adduce it, then I think you need to tell the Court that now, that you’re going to rely solely on the evidence that has already been presented in your briefs with respect to wrongdoing. And the expert testimony that you have already indicated you have and plan to adduce.

*** Judge; “So, with that, I would like to ask plaintiffs’ counsel, is there any other expert evidence or testimony that they would expect to adduce at trial, other than what the Court has already been made aware of?”

*** Pltff; “I guess I don’t know how to respond to that, Your Honor. There is a lot more evidence, obviously, we would produce at trial.” Injury to others.” “Looking back to evidence presented, you to be Judge:” “To recover under the negligence theory, a plaintiff must show that the defendant breached a duty of care owed to the plaintiff and thereby caused the plaintiff's damages.” “The defendant claims entitlement to summary judgment on plaintiff's negligence claims on the basis of defendants erroneous assertions that there is no ‘specific’ evidence of negligence on the defendants part.”

***** Larry says;” the Plaintiffs’ attorney’s state, plaintiff's have Substantive, Competent Evidence which is Sufficient to Establish Every Essential Element of Plaintiffs’ claims of negligence against the United States Army.”

*** Judge;” But you’ve concluded Discovery?”

*** Pltff; “Yes, Your Honor.”

*** Judge; “Discovery is complete.”

*** Pltff;” Yes, Your Honor.”

*** Judge;” And you haven’t listed any other experts other than those that defendant’s counsel has referred to?”

*** Pltff;” That is correct.”

*** Judge; “All right. Those are the experts you plan to make your case on?”

*** Pltff; “That is correct Your Honor.”

*** Judge; “All right.”

*** Pltff; “Include Geotran, which is an engineering firm. They were endorsed in our expert witness endorsement.”

*** Judge; “And have they been deposed?”

*** Def; “Your Honor, may I address that?”

*** Judge; “Yes.”

*** Def; “The geotrans is the hydrologist that I refer to in my argument. Nowhere in the report is there any stated opinion that goes to any engineering system. They merely opine that the chemical could have reached the Land’s well from the Arsenal. And, as I mentioned, that is not in dispute.”

*** Def; “I would like too remind the Court that we had a disputed history on expert reports in the extension. And this court, by its order of January 22, 1992, has held that the plaintiffs, because of that dispute, that the plaintiffs cannot either add new experts or supplement the reports as issued.”

*** Judge; “I appreciate that reminder. So we are limited then by ruling of the Court already…”

*** Def; “Yes, Your Honor.”

*** Pltff; “If I may, Your Honor. When defense filed their motion for summary judgment, we entered into a stipulation to stay any further discovery, including any depositions they may want to take of our experts. And, in fact, there was a stay of their duty to endorse their experts, pending the outcome of this, just to save costs. So, the answer to your question is, no no experts have been deposed."

*** Judge; "Well, if I reopen discovery, you wouldn't intend to identfy any other experts and make them available for deposition. You are not being denied any opportunity for discovery by virtue of the stay. In other words, you are not being prejudiced by virtue of the Court's stay order to adduce evidence of wrongdoing in this case?"

*** Pltff; "No, Your Honor."

*** Judge; "All right. Thank you. All right. The Court will take this matter under advisement. There being nothing further, this Court is now adjourned."

*** Judge; "(Where upon, at 11:00 a.m. the hearing was concluded.)

With the official end of Court another long wait would start as the Judge began the sometimes long process of weighing the facts and testimony as given by both sides of counsel and examining Colorado law, upon which he must base his ruling. The Judge has also ruled there is no room in a Court of claims for Federal Tort Claims Act, FTCA."

In his heart, Larry thought that after all the deaths, and all the Material Fact on the Army's negligence, and total disregard for human life, under no circumstance would the Judge do anything but dismiss the Defendants Motion for Summary Judgment. The instincts he had learned long ago not to trust these people when so much was at stake. He knew from past experience that the Army had a way of suppressing the truth, and keeping it from public knowledge through classification. The intentional and conscious exclusion from consciousness certain thoughts or feelings.

"THE REPORT OF JUDGE ROBINSON"

***** "The plaintiff's in this case challenged the motion for summary judgment, urging that the record does reflect genuine issues of material fact precluding summary judgment." "For the reasons stated below, defendant's motion is denied."

***** "In 1942, defendant, through the United States Army, established the Rocky Mountain Arsenal (RMA or Arsenal), located near Denver, Colorado, to manufacture toxic chemical and incendiary muntitions. Shortly thereafter, one of the RMA's primary objective became the demilitarization of obsolete hazardous and toxic munitions. Portions of the Arsenal were leased to private Corporations, such as Shell Oil Company for the manufacturing of what became the deadliest pesticides, and Herbicides known to mankind. What is called the dirty dozen that have been banned from manufacturing or the use of, world wide."

"THE MAJESTIC BALD EAGLE"

**** Judge; "the RMA halted monitoring in 1960. It refused to resume testing until 1974-75 despite the fact that a report on groundwater contamination in the relevant area by (HEW).

***** Larry says, "That was because they were told Basin F. was leaking, and the Army was advised to stop its use. The Army merely classified all information, stopped all testing, and resumed using

Basin F. through 1975 for RMA's waste, but continued allowing Shell Oil Company to use Basin F thru 1982."

**** Judge; "Contamination of the aquifer is likely to persist for many decades. In 1965, the Army Environmental hygiene Agency recommended that steps be taken "to eliminate Lake 'F' as soon as possible, the therby remove much of the present environmental hazard of exposed surface storage of toxic wastes. By 1969 there was evidence that Basin F was leaking; indeed, a physical inspection reflected that sections of the protective membrane were absent. Thus, by the spring of 1970, the Army understood that it was operating Basin F on the premise that it is leaking."

**** Judge; "Of the 520 yearlings Larry Land had purchased,, one-half died or had to be destroyed, the remainder were sold for slaughter because of their condition. Plaintiffs rely on a report by Dr. Frederick W. Oehme, D.V.M., ph.D., which opines that the cause of the death or eventual destruction of all but three of the calves was the presence in the well water of the man-made compound DIMP and it's breakdown product IMPA.

**** Judge; "During this same period of time in 1972, and afterwards, all sixteen plaintiffs bathed in and consumed water from the wells and, obviously exposed to the air. Each asserts that he or she has endured abnormal medical histories, citing symptoms such as: lethargy, headaches, airway irritation, eye irritation, bronchial irritation, nausea, diarrhea, and mental and emotional changes.

***** Judge; "They rely on a recent reports by Dr. Daniel T. Teitelbaum, M.D., concluding that the cause of the health problems suffered by plaintiffs and Larry Land's calves was DIMP, and IMPA contamination of the Land well water and the air. Moreover, Dr. Teitelbaum concludes that; Review of the records which were supplied to me establishes unequivocally that the Land Family's groundwater was contaminated with materials from the Rocky Mountain Arsenal. In particular, the finding of DIMP….establishes beyond any question that the source of the material in the groundwater under the Land's property and in their drinking water and hygiene water wells was the Rocky Mountain Arsenal. No other source of this material is conceivable in that area. THis finger-printed chemical establishes without any question that the toxic material which reached the Land family had, as its source, the Rocky Mountain Arsenal GB manufacturing facitlity. In addition, the finding of IMPA, a metabolite of DIMP, further identifies the material as having a source in the Rocky Mountain Arsenal."

**** Judge; "In December 1974, the Colorado Department of Health detected DIMP in a well near the city of brighton, located seven miles north of the RMA and six miles north of the Land farm. On April 7, 1975 This resulted in the Colorado Department of Health issuing three (3) Cease and desist orders against SCC and RMA."

***** Judge; "Activities on the Arsenal have resulted in one of the worst hazardous waste pollution sites in the Country., and Basin F is only a small portion of the problem. Army Officials has estimated that the twenty-seven square mile Arsenal has 120 contamination sites which contain huge quantities of liquid and solid wastes, some of the which are unique because of the mixture of private herbicide and pesticide manufacturing activities with Army Munitions manufacturing activities."

***** Judge; "The Arsenal is among the federal sites grouped as the number two priority's on the National Priorities List, and EPA compiled list which prioritizes CERCLA sites for cleanup based on the relative risk of danger to public health or welfare or the environment. There was fourteen specific sites,, that required intrim response".

***** Judge; "The House of Representatives referred this case to this court on September 22, 1988. Plaintiffs filed an amended complaint on October 15, 1990, requesting legal or equitable relief. Plaintiffs allege five counts in the amended complaint: (1) Negligence; (2) Strict liability, Ultrahazardous activity and inherently dangerous activity; (3) Trespass; (4) Nuisance; and (5) Deceit based upon concealment".

"H.R. 816 states:"

"A BILL"

"For the releif of Larry Land, Marie Land, et.al."

"Be it enacted by the Senate and House of Representatives of the United States of America in Congress assembled, That the Secretary of the Treasury is authorized and directed to pay, out of any money in the treasury not otherwise appropriated, the sum of $____________________to Larry Land, Marie Land, et.al." "This sum shall be in full and complete satisfaction of all their claims against the United States based upon health problems and other related injuries resulting from the operations and activities at the Rocky Mountain Arsenal, Denver, Colorado, conducted under the auspices and control of the United States Army."

"H. Res. 61 states:

RESOLVED, That H. R. 816 entitled "a bill for the relief of Larry Land, Marie Land, and others," together with all the accompanying papers, is hereby referred to the Chief Commissioner of the Court Claims pursuant to section 1492 and 2509 of title 28, United States Code, for further proceedings in accordance with applicable law."

**** Judge; "Under section 2509, the hearing officer is charged with proceeding "in accordance with the applicable rules to determine the facts… [The Hearing Officer] shall append to his findings of facts conclusions sufficient to inform Congress whether the demand is a legal or equitable claim or a gratuity. The Court rules require that "[the hearing officer shall find the facts specially.]" The applicable rules are "such rules of the United States [Court of Federal Claims] as may be required to determine the facts."

**** Judge; "The Defendant moved for Summary Judgment in this case, arguing there is no genuine issues as to any material fact. Plaintiffs counter, arguing that the reference in this case allows a broader basis for recovery than negligence. Further, Plaintiffs argue that the record currently before this court reflects genuine issues of material fact. RCFC.56, which follows Rule 56 of the Federal Rules of Civil Procedure very closely, imposes an initial burden on the moving party, (def) to inform the court of the basis for its motion and to demonstrate through it's filings that no genuine issue of material fact exists as to its claim for relief."

***** Judge; "in viewing the movants, (Def) evidence, the rule also requires that all factual inferences drawn from such evidence be viewed in the light most favorable to the non-movent i.e., the Plaintiffs in this instant matter....Any doubt as to whether a genuine issue of material fact EXISTS, and will be resolved in favor of the non-moving party, (plaintiffs). Thus "It is axiomatic that, if the moving party fails to meet it's initial burden under Rule 56(c), (Def.) It has no entitlement to Summary Judgment."

***** Judge; "Although not raised by the plainntiff and although no discussion is necessary here as it does not change the outcome of this motion, the Court questions the propriety of a Summary Judgment motion in a Congressional Reference case where the hearing officer is charged with finding facts specifically and the moving party only concedes, rather than stipulates to, facts for a limited purpose."

***** Judge; "The plaintiff must present specific and credible facts that either tend to establish all the necessary elements of its case or it position, or raise a genuine issue of material fact. Material facts are those which the substantive law identifies as affecting the outcome of a suit. A dispute is genuine only if, on the entirety of the record, a reasonable finder of facts could resolve a factual matter in favor of the non-movant, (plaintiff)."

***** Judge; "In a Congressional reference, Plaintiffs may assert a legal or equitable claim against the Government. LEGAL CLAIM: A legal claim arises from substantive law, i.e. the Constitution, a statute, a regulation, or some principle of common law,, and is based on the invasion of a legal right. Plaintiffs are barred from legal relief. The time period for this court's general statute of limitations is a period of six years."

***** Larry says;" The Courts general statute of limitations should not bar a legal claim in this court, since there is ample evidence the Army had lied, denied, and classified information by deceit and fraud that would have opened the doors for this case in the early 1970's."

***** Judge; “Equitable claims. The principles of right, justice and morality create equitable claims.” “The term equity in this context is used in the sense of broad moral responsibility, what the Government ought to do as a matter of good conscience.”

***** Judge; “In its broadest and most general signification, equity denotes the sprit and habit of fariness, justness, and right dealing which would regulate the course of mankind.” “Do unto others as we desire them to do to us: It is therefore the snynonym of natural right or justice.... It is grounded in the precepts of the conscience, not in any sanction of positive law. The Nation, speaking broadly, ‘owes a debt’ to a [party when its] claim grows out of general principles of right and justice.”

***** Judge; “The Colorado Department of Health informed Larry Land, in correspondence date June 11, 1975, of the presence of chemicals in the the house well. P. App. 16. The Tri-County District Health Department informed Larry Land of chemicals in the well water on July, 7 1975. It indicated the RMA as the source of contamination.”

***** Judge; “In any event, “[a]n equitable claim in a Congressional reference must rest on some unjustified Governmental Act that caused damage to the claimants. Absent a finding of negligence [or wrong doing] on the part of Governmental employees, any award here in would be a gratuity.”

***** Larry says; “The Judge seems to be leaning toward the FTCA Federal Tort Claims Act, because that is where a negligence act must be found by the employees. Judge Robinson himself in his own report stated there is no room in the Court of Claims for a Tort claim. When the Judge indicated we had no legal claim because of the time period lapsed under the FTCA, Federal Tort Claims Act.

***** Judge; “Equity, thus contemplates a remedy to a claim in a variety of circumstances where relief would be other wise unavailable in a traditional juridicial setting, such as when there is no remedy under existing law or the remedy is time barred by a statute of limitations. See Spalding, 28 Fed. Cl. At 250. While such a claim, like a legal claim, may be founded on the Constitution, a statute, a regulation, or a common law principe, such bases are not necessary to warrant relief in equity. Nevertheless, most equitable claims are based on traditional principles, in which the standard to be applied are those that govern the subject area of the law generally bearing on the claim. Seee, e.g., Kochendorferr v. United States, 193 Ct.Cl. at 1045, 1056 (1970). In judging whether there is an equitable claim, however, these legal guidelines need not be rigidly applied. Rather, the particular facts may be examined to determine whether the plaintiffs suffered an invasion of rights for which compensation is due them. White Sand Ranchers, 14 Cl.Ct. at 570."

***** Larry Says;" The Judge had already cleared, and ordered in his report that Colorado Circumstantial Law will Rule. The plaintiffs relied on that order in preparing the case for trial. The Plaintiff did not under any circumstance prepare it nor were they prepared for the Tort Claims Act. FTCA. for trial. Judge Robinson had specifically ruled and ordered it to be heard under Colorado Law, and agreed it would be heard under circumstantial evidence to rule. He stated specifically there is no room in the Court of Claims for the Tort Claims Act. The Plaintiffs at this point relied on this information, and the law of Equity. I believe the only reason for them barring the Legal claims was to alleviate the Government of being found Criminal.

***** Judge; "In any event," "[a]ny equitable claim in a Congressional reference must rest on some unjustified governmental act that caused damage to the claimants. Absent a finding of negligence [or wrong doing] on the part of governmental employees, any award herein would be a gratuity." "To obtain a favorable recommendation from this court on any equitable claim, plaintiffs must show: (1) that defendant, acting through the United States Army, committed a negligent or wrongful act; and (2) this act caused damage to the plaintiffs." "Defendant first contends that plaintiffs' recovery on any of their claims is dependent on showing of negligence by the Army. The reference to this court is not so limited, nor is the applicable law.

***** Judge; "In addition to negligence, plaintiffs may recover upon a showing of "Wrongful" Conduct. When this word was added to the FTCA., it provided for liability for "wrongful" as well as "negligent" acts." "The treatment of "wrongful" action under the FTCA is persuasive authority for similar treatment, by this court of "wrongful" action in a congressional reference."

***** Judge; "This court finds it conceivable that there is an area on non-intentional, non-negligent. "wrongful" action, for which the government may be held liable in equity. in a congressional reference. Indeed, as plaintiffs noted at oral argument, "wrongful act" has been defined as "[a]ny act which in the ordinary course will infringe upon the rights of another to his damage, unless it is done in the exercise of an equal or superior right. [The t]erm is occasionally equated to [the] term 'negligence,' but has been considered [a] more comprehensive term, including criminal, willful, wanton, reckless and all other acts which in ordinary course will infringe upon rights of another to his damage." Black's Law Dictionary 1446 (5th ed. 1979)." "Moreover, even though this court accepts the more recent legal precedents that equitable relief in a congressional reference necessarily must rest on the existence of some type of an unjustified act or omission."

***** Judge; "Despite earlier authority finding that an equitable claim may arise without fault, this court is not convinced that a technical demonstration of negligence is required. Indeed, the reference in the instant matter is not so limited. Therefore, this court shall not bind plaintiffs to a preliminary showing of all elements required to support a strict negligence claim. This courts' order of September 13, 1990, also does not limit plaintiff to a strict showing of negligence." "Wrongful" within the meaning of a congressional reference. As stated above the court conceives that there may be an area on non-intentional, non-negligent, "wrongful" action, for which the government may be held liable in equity in a congressional reference.

***** Judge; "In regard to whether the activity at the RMA constitutes ultrahazardous or abnormally dangerous activity, defendant has failed to convince the court that Colorado would not consider it such. Defendant has been disposing of toxic chemicals and incendiary munitions since 1942 in a facility located a few miles from the city of Denver. Seven miles from the city of Brighton, and within a few miles of farmland such as plaintiffs.'" "Only one mile from RMA's operations." "Defendant was aware as early as 1951 that its chemical disposal techniques posed a risk to neighboring farmers. In the 1950's, the Army began receiving studies and reports finding that contaminants from the RMA were polluting land to the northwest of the arsenal. The Army also received reports outlining recommendations to stop the polllluting. The Army ignored many of the recommendations, however, including one in 1965 to discontinue use of Basin F. The record reveals further that the Army knew, as early as 1969, that Basin F was leaking, yet it took little or no immediate action to resolve the problem. Furthermore, deposition testimony indicates a "tracer" could have been used to detect off-post contamination, but was not. In addition, the record indicates that defendant conducted off-post tests after complaints in the 1950's, but discontinued this activity in 1960"s." "After learning Basin F was in fact leaking and also polluting the air, It had already polluted 25,000 acres to the north and northwest of the arsenal. The Pentagon Offically pointed out to the army Basin F was leaking, and the Army was advised to "classify" it by none other than the pentagon, after making a statement we sure don't want any one to find out about that." "Pentagon."

***** Larry says, "It was added to the evidence, showing that the Army and the pentagon were involved, and co-conspired to hide the truth of what had been taking place at the Arsenal, and the very reason the Army stopped all ground water testing. With deceit by classification they were able to continue using Basin F while increasing the volume from 180 million to 250 million gallons by the 1970's and 80's to the point of overfull far beyond the liners protective barrier. With the knowledge that it was polluting the water aquifers off the Arsenal and polluting the air as well. This was evidenced to Judge Robinson, and it was in fact declassified in 1988, very quietly, over 20 years after the fact." "Judge Robinson was very much aware of the deceit, fraud, nuisance, trespassing, with ultra hazardous materials, The Judge was very much aware that arsenal chemicals had been found in the Land wells. That had allowed them to pollute the country side near Denver. Judge Robinson allowed the

Army to stand before the Court and admit, and stipulate into evidence that," "The Army was aware and admits that Arsenal Chemicals were found in the Land wells."

***** Judge; "This Court accepts that the manufacture and disposal of toxic munitions is ultrahazardous or abnormally dangerous activity. Daigle v. Shell Oil Co.," "Stated that Colorado could consider such activity ultrahazardous or abnormally dangerous." This Court accepts, the above activity is ultrahazardous or abnormally dangerous activity. Defendant contends plaintiff's claims based on ultrahazardous or abnormally dangerous activity must be denied as outside the scope of this court's September 13, 1990 Order, and because the activity at the RMA was carried on pursuant to a public duty imposed upon them."

***** Larry says; "This is almost exactly the same "tune" played by several nazi administration at the Hague when ask why did you carry out these orders to kill., But then the Judge went on to say. According to defendant, memorandum of Approval No. 438 dated May 12, 1942, and prepared under the name of Robert P. Patterson, Under Secretary of War. "establishment of a new arsenal at Denver Colorado, for the production of certain gases and loading facilities. The Army contends that legislation established the RMA, it was operating pursuant to a public duty imposed by the legislature, and thus, should not be liable for it's activities. For this the most obvious one is that of a sanction given by statutory authority, or by a well defined local law....This statute has been interpreted as condoning the consequences in advance, and have refused to hold defendant liable." "The tendency in the later cases has been to avoid [concluding a party is not liable], If it was not sanctioned work," "In this case there shall be liability, for the manner of doing the work, that resulted in damage itself, are not authorized or necessary consequences of the sanctioned work."

***** Judge; "In this case the defendant has failed to show any special circumstances evidencing an intent by the legislature to relieve the government of liability. Generally, authorization to engage in abnormally dangerous activity does not in and of itself demonstrate an intention to relieve liability." "Strict construction of the memorandum does not evidence an intent to relieve liability for either the production or destruction of toxic munitions. Moreover, defendant has not shown that its manner of storing and disposing of toxic muntitions was authorized, or that any resulting damages were a necessary consequence of that activity. Even to assume that the defendant was operating the RMA pursuant to a public duty, this defense may not be available to defendant in a congressional reference."

***** Judge; "The judge confirms that FTCA reserves sovereign immunity for negligent misrepresentations. After reviewing the legislative history of the referencing legislation, this court concludes referencing legislation constituted a waiver of sovereign immunity which defendant would otherwise have been able to claim under the FTCA. With the sovereign bar lifted jurisdiction is hereby conferred on this court, to Hear, determine, and render judgment on plaintiffs' claims." "To hold otherwise would be to presume that Congress passed the bill in question merely to send plaintiffs to this forum for a ritual confirmation of the neustadt rule [that the FTCA barred such claims] and concomitant dismissal of their petition."

***** Judge; "similarly, in Sperling & Schartz inc. v. United States, (1978), the court found that the defense of absolute privilege which shields a government agency from liability for an employee's statements made in his official capacity was tantamount to sovereign immunity, but was not allowed in a Congressional Reference."

***** Judge; "The referencing statute in the present contains no concession of liability. However, Congress would not have needed to refer this case to the Court of Claims against the United States based upon health problems and other related injuries resulting from the operation and activities at the Rocky Mountain Arsenal, merely to affirm that the FTCA of RMA enabling legislation barred any recovery for liability arising out of the operation of the RMA.

***** Larry sees; "What the Judge has just said above, is the Congressman did not have to refer this case to the Court of Claims, they would have been able to of handled this in Congress. To think of this at the time in the 1970's and 80's and 90's as this has drug on for ever, (30) years. It has been a most weary and difficult issue to get any congress person to even listen.... In 1972 –74 area of time it seemed impossible to get any congress person, or senator to listen at all. Senatory Gary Harts office did listen, then they lost what was given to them and they listened again but nothing ever got done, and everyone in the World could see what his problem was, In his bid for the presidency, Congresswoman Pat Schroeder did take an interest but had to refer to our congressman Jim Johnson who worked very hard to get it before the House judiciary Committee but failed and it then went on to his successor where it got the utmost of attention by Congressman Hank Brown, and his aide Mac Mcgraw. Mac was also the aide to congressman Johnson, He had come and seen first hand several times, to see the dilemma first hand. who quickly got it to the House Judiciary Committee. Congressman Hank Brown is the man who in 1988 got the bill presented to the Judiciary Committee, where we were given a personal welcome to the House Judiciary Committee by Congressman Barny Frank who made a personal comment about the Army and why they had not come to the hearing. It quickly was passed by the Sub-committee and then to the full House of Congress where the Reference was past by the full house October 1988, giving the Land's chance in the Court where it spent the next ten years being debated. Congressman Schroeder had sent several letters to the Army trying to attain medical help for the Land's and their dilemma. Congresswoman Schroeder not only made time to meet in her office, she came out to the Farm to see first hand just what was happening. In D.C. I always felt welcome to stop and talk, and get some ideas of what could be done. The army refused to correspond, and if they did it would be classified information unable to give it out. So we were poisoned by chemicals from the Army and they refused to work with any of the treating physicians, leaving us with no help at all. Then if they do listen for a moment the next thing they say is we don't do private legislation. When I started out it was difficult to even get anything as to what was happening the RMA just classified every thing. The information alone on Basin F, that the RMA had been classified back in the 1960's wasn't even released until sometime in 1990 to the public."

***** Judge; "The defendant has failed to convince the court that the reference precludes recovery based on ultrahazardous or abnormally dangerous activity. This court reiterates that it is not opening the door to liability of the government without showing of fault whatever. Rather defendant has failed to convince the court that operation of an ultrahazardous or abnormally dangerous activity in the instance and under the attendant circumstances, i.e., a few miles from working farmland, only one mile from the Land farms in the instant case, with knowledge that basin F was leaking, is not a "wrongful" act and that compensation for injuries resulting from such activity might be considered "wrongful", The Defendants conduct falls within the scope of a congressional reference.

***** Judge; "A plaintiff must show the following to prove common law negligence in Colorado, (1) The existence of a duty owed by the defendant; (2) A breach of that duty; (3) The breach actually and proximately caused the injuries; (4) That the plaintiff actually suffered injuries." "For a determination of negligence, or any other substantive claim, the law of Colorado Governs., Benson v. United States (1958). Erie R.R. v. Tompkins, (1938)." The question of whether a defendant owes a plaintiff a duty to act to avoid injury is a question of law to be determined by the court." Smith v. Denver, (Colo. 1986); Metropolitan Gas Repair Serv., inc. v. Kulik, (Colo. 1980). The court determines, as a matter of law, the existence and scope of the duty that is, whether the plaintiff's interest that has been infringed by the conduct of the defendant is entitled to legal protection.

***** Judge; "It is axiomatic that in engaging in a particular activity every person is "bound to exercise that reasonable prudent and cautious person under the same or similar circumstances." Denver Consol. Elec. Co. v. Simpson, (Colo. 1895). Under this reasonable person standard "the greater the amount of care required to avoid injury to others." Blueflame Gas, Inc. V. Van Hoose, (Colo. 1984). Thus the degree of care varies according to the degree of risk associated with a particular activity." "Judge Robinson compared RMA to one requiring the highest degree of care which human ingenuity can practically exercise, and that, "Judge Robinson stated as a matter of Law." He compared RMA to the highest degree of care with the utmost care and diligence in protecting the public from the dangers necessarily incident to the carrying on of a hazardous business."

***** Larry says;" For the enhanced standard of care theory holds that the reasonable or ordinary care in cases such as the Rocky Mountain Arsenal dealing with the manufacturing and destruction of the "deadliest" and "most dangerous" chemicals known to mankind. Which all must acknowledge to pose a high risk of injury or death to others, is "the highest care which human ingenuity can practically exercise, and that, as a matter of law."

***** Judge; "As a matter of law, courts will hold every reasonably prudent and careful [person] to exercise of the utmost care and diligence in protecting the public from the dangers necessarily incident to the carrying on of a hazardous business." "There are instances where due to the gravity of the risk created by them an enhanced measure of care on the part of the defendant is required.

***** Judge; "Plaintiffs briefly address the existence of a duty in their opposition brief. Plaintiffs argue that the Army's conduct on the RMA, manufacturing and demilitarizing of chemical weapons and disposing of chemical wastes, created a substantial risk." This Court agrees". "This Judge would like to point out to the plaintiffs on the duty of care, the point is made that the Army has, from the 1940's been engaged in extremely risky activities that demand the utmost care and diligence to protect the public."

***** Judge; "This court is satisfied, after resolving all doubts as to the facts, presumptions, and inferences in favor of plaintiffs, that genuine issues of material facts, as to plaintiffs' claims including those sounding in negligence. see Auxier v. Auxier, (Colo. 1992); Webb v. United States, (1970); U.S. Fidelity & Guaranty v. Salida Gas Serv. (Colo. 1989);. "Defendant has failed to convince the court that no issues of material fact exist as to defendant's negligence or wrong doing." "This court is concerned that plaitiffs have not produced the expert testimony on the issue of defendant's duty of care and defendant's breach of that duty. Expert testimony is normally required to establish a standard of care only when it involves questions beyond the competency of ordinary persons." "However, defendant, this far has failed to convince the court that such testimony is required in this case."

***** Judge;" For all the reasons stated above, defendants motion for Summary Judgment, is ["denied"]

***** Judge; "In 1965 the Army Environmental Hygiene Agency recommended that steps be taken [to eliminate Basin F], as soon as possible, and thereby remove much of the present environmental hazard with air pollution because of exposed surface storage of toxic wastes. P. App. 10 at 3-4. By 1969, there was evidence that Basin F was leaking; indeed, a physical inspection reflected that sections of the protective membrane were absent. Id. At 11-12. Thus, by the spring of 1970, the Army understood that it was operating Basin F on the premise that it was leaking. Id. At 13." "It was ordered classified by the pentagon]," And continue using.

***** Judge; "In its broadest and most general signification, equity denotes the spirit and habit of fairness, justness, and right dealing which would regulate course of men the rule of doing to all others as we desire them to do to us; to live honestly, to harm nobody, to render every man his due. It is the synonym of natural right or justice. It is grounded in the precepts of the conscience, not in any sanction of positive law."

Coyote

The adaptable coyote has expanded its range throughout the United States into urban and semi-urban areas, such as the Arsenal. Approximately 50 made the Arsenal their home. These intelligent animals usually hunt alone, preying on pheasants, insects, prairie dogs and other small mammals. Coyotes are well-known for their melodious howl, often hard.

Western Tanager

The western tanager migrates through the Arsenal each spring and fall. These colorful songbirds next in Colorado's coniferous forests and migrate to the Central American mountains for the winter. Insects are the western tanager's primary food source.

CHAPTER TWELVE

Over increasingly large areas of the United States, spring now comes unheralded by the return of the birds, and the early mornings are strangely silent where once they were filled with the beauty of bird song.

We have been massively intervening in the environment without being aware of many of the harmful consequences of our acts until they have been performed and the effects, which are difficult to understand and sometimes irreversible, are upon us. Like a sorcerer's apprentice, we are acting upon dangerously incomplete knowledge. We are, in effect, conducting a huge experiment on ourselves. Yet, because we depend on so many detailed and subtle aspects of the environment, any change imposed on it for the sake of some economic benefit has a price. Sooner or later, wittingly or unwittingly, we must pay for every intrusion.

Of all the life forms, both past and present, that has ever lived on this planet, only the most civilized, and the most intelligent, feels it is necessary to destroy everything it comes into contact with or touches, including ourselves.

This was yet another low point in Larry's life. He was both ashamed and hostile at the Army, the Pentagon, and the Arsenal and its employees for what they had willingly and knowingly done over the years. He was ashamed of his government for trying to take the coward's way out by denying anything that had happened, and he was ashamed that many of the people living close to the Death Factory, (Arsenal) (Shell Oil Co. had been unknowingly murdered or put to death by their own fellow countrymen through their craving for World Power, Shell Oil for the almighty buck. He is most ashamed after hearing only part of the truth of the Persian Gulf War Syndrome, but wants to share with everyone with at least all he knows. He has dedicated chapter 10 to the Gulf War Vets their families, and their quest to know the reasons it back fired on the United States, while Saddam Hussein sat in his chair and called it the Mother of all wars. His own Armageddon, and the soldiers who fought that war were his victims.

He knew that the price for liberty is strength, and no man enjoyed his freedom as much as he did, but the question that kept coming to his mind is why the governments of the world felt it necessary to create and use such dangerous chemicals. Where was the humanity in the souls of those people when they made these decisions. No one wants war and only a fool or a depraved person such as Adolph Hitler his henchmen, and Saddam Hussein and his henchmen enjoy war, Death, and Power. If we must have wars on occasion why can't they be kept as conventional warfare, with no thought of ever using Nuclear, Chemical, Biological warfare, or Virological warfare. They also use Bio-chemical and Viro-chemical war-fare such as what was used in the Persian Gulf war, and most will never know what was used on them, nor did they know most of it was produced in the United States of America. They tried to cover their own A__ by denying it. See chapter 10 it will send chills right to your toes. I say what I say because it has all been documented classified and declassified Proof, and should not be hidden from public opinion any longer.

The Persian Gulf war is the war that had backfired on the United States. They conspired with Iraq to lay waste to Iran. Through out the 1980's they had supplied Iraq with technology, and products for Atomic, Chemical, Biological, and Virological warfare agents. Then Russia supplied them with the technology for both Bio-Chem and Viro-Chem war fare agents.

By Pauline Jelinek Associated press October, 2000, It was discovered the Pentagon had informed the wrong vets of nerve gas exposure. Based on a review of weather at various locations Gulf War Troops will get letters amending their status;

The pentagon reversed itself about who was exposed to nerve gas during the Persian Gulf War, and who wasn't.

Some 30,000 Gulf War Veterans are to be notified in the coming weeks that they probably came in contact with low levels of sarin nerve gas after being told in 1997 that they had escaped exposure. and 30,000 believed to have been exposed will get letters saying they probably weren't exposed to sarin.

"These are very low levels low enough that there is no known expectation of a health risk," Austin Camacho, spokesman for the Pentagon's Office on Gulf War Illnesses, said Friday.

This revision will again "raise questions of credibility" among critics of the government's decade-long effort to answer health questions of vets who served, said Dr. Vinh Cam an immunologist and member of the Special Oversight Board for the Department of Defense Investigation of Gulf War Chemical and Biological Incidents. The board's chairman, former Sen. Warren Rudman, said the revision proved the opposite. "It shows (pentagon Officials) have tried very hard, with all the evidence, to come back and re-evaluate," he said.

"Does it mean any more people are ill? So far there is no evidence of this. The revision in who was exposed is based on better information on such details as to weather, and troop locations," said oversight board spokesman Roger Kaplan.

The issue in question for veterans is what was the health effect and how many people were exposed to it. When U.S. Forces demolished Iraqi munitions and rockets at the khamisiyah weapons depot in March 1991 it turned out some of the rockets contained the highly toxic sarin and the even more lethal cyclosarin chemical warfare agents that paralyze the nerves, shutting down the lungs and other organs.

The report is to be announced in mid-to late November, and letters will be sent the day of the announcement to some 110,000 troops who in and around the area of Khamisiyah, officials said.

Of that number, some 99,000 were told in 1997 letters that they were believed to have been exposed and 11,000 were told that they were not.

In a review of work done and still under way on a number of Gulf War studies, Bernard Rostker, Pentagon undersecretary for personnel, told the oversight board in its last public session Friday that officials in the coming weeks will release a report revising figures on which of the troops deployed in the depot area might have been exposed. If there might have been low level exoposures to Sarin, you can bet there was exposure to several of the Biological agents as well. Saddam Hussein Played the game as the Russians had taught him. Bio-Chem. Using chemicals to do the initial injury and the Germ warfare would be there causing secondary infections, it would take over and finish the job the deadly chemicals began. This is the way all his weapons were set up.

***** Larry "questions the governments credibility, on how they tested for any other chemical agents that may have been used. How did they detect all the various Biological agents. As I understand there is no technology available that would warn any one of a biological attack. The technology used is not sufficiant in low level attacks even for sarin."

"How has any credibility or common sense been made, of how the United States supplied Iraq (A rogue nation) with atomic weapons, deadly chemical weapons and biological weapons.

What went so wrong in the Persian Gulf that caused Iraq to back fire on the United States, with their own chemical and biological war fare."

How can we trust the United States as being the Worlds policeman, when they won't follow their own laws, or rules for living together in this world. They actually believe they are above that. The last person known that sat him self and others above others as the perfect race was Adolph Hitler, and henchmen.

***** "A treaty praised as a tool for fighting pollution; Nations agree to ban 12 toxins, "The World's Dirty Dozen" Twelve of the the most Ultra Toxic, Ultra Hazardous Chemicals known as the most persistent and deadly pollutants in the World. I will list them nine of them Pesticides three are industrial chemicals and give a brief scenario on its use, and the major health effects of each.

Aldrin; A pesticide used to control termites, and insects, Aldrin is very toxic, and is carcinogenic.

Dieldrin; A pesticide, similar use as Aldrin, It is as toxic as Aldrin and harms the reproductive system.

Endrin; A pesticide, used to control pests in grain, cotton, sugar cane, rice, also used to control mice.

Chlordane; A pesticide used to control termites, ants, the rhinoceros beetle, Will harm the immune system.

DDT; A pesticide used to control malaria carrying mosquitoes were probably getting alot more back than we even thought.

In September of 1993 the most disturbing news came to Larry's home.

Larry and Marie were returning home from dinner one evening, he saw Jennifer's car sitting in the driveway. After parking his car and explaining to Marie who she was, he walked up to her.

"Hi Jen, its been a while. How have your been?" He asked.

"Pretty good," she said. "I found a new job a lot closer to home, so at least I don't have as far to drive, Larry I brought something out that I think you should read, and please don't ask how I got it. Everything we have done all these years has been for nothing. They are going to do whatever they want to and nothing is going to stop them, not us or anybody."

"C'mon Jen, don't feel so down, we're winning, just look at all the things we have accomplished, the Arsenal's shut down, Shell Chemical Company at the Arsenal has been shut down, and they were forced to move elsewhere for good they are cleaning things up a bit. It can only get better."

"Larry, I have to be going, Read this after I'm gone. As hard as you have fought over the years to shut that hell hole down, I couldn't stand to look at the pain in your eyes after you read this You'll have pain and anger in your eyes. See ya later."

"All right, bye Jen."

Walking inside Larry sat down at the kitchen table and opened the envelope. What he read made him physically ill.

A report that stated that the Army was planning to build a massive hazardous waste dump in the heart of the supposedly safe wildlife refuge where the public would be free to wander.

State health officials say the Army's $2.1 billion plan violates numerous federal, and state environmental laws.

The 1.5 square mile dump would be built on top of the old Basin A site that often fills with ground water. Health officials have said the ground water will always slosh through the highly toxic contaminants from pesticides and deadly warfare agents from chemical weapons and travel underground forever polluting and destroying our environment and our lives.

"It's a terrible idea" said one official. "We must stop producing these ultra hazardous chemicals, warfare items, pesticides, and herbicides period, not make more room for it. What has already been created here at the Arsenal will be here forever, It can never be cleaned up. The only thing left now is to encapsulate it, and maintain with vigil eyes so as not to let it pollute any worse than it already has. We truly have exactly what the EPA. Environmental Protection Agency called it the most Contaminated piece of land on earth. 18,000 acres plus 10's of thousands of acres off the Arsenal have been contaminated. We now have the technology available to us to begin making a completely biodegradable material that is harmless to everyone. We just don't need that kind of toxic ultra hazardous gunk fouling or destroying everything in nature, inherently endangering every person on the face of the planet in some way, with these terrible terrible, toxic and ultrahazardous chemicals. tsetse flies, illegal control of crop pests. DDT harms the reproductive system.

Heptachlor; A pesticide used to control termites and ants, controls cut worms. Heptachlor is carcinogenic, and harms the reproductive system.

Mirex; A pesticide used to control termites, ants, used as a fire retardent. Mirex is carcinogenic.

Toxaphene; A pesticide used in agriculture for mosquito control. Toxaphene, can cause thyroid tumors and cancer.

Hexachlorobenzene; A pesticide and industrial chemical fungicide. Hexachlorobenzene harms the immune and reproductive systems.

Polychlorinated Biphenyls (PCBs); Industrial chemicals used for electric transformers and capacitors, and as a paint additive. PCBs linked to reproductive failure and suppression of the immune system.

Dioxins; and industrial by product. Dioxins are carcinogenic and harm the reproductive system and the immune system.

Furans; Toxic byproducts of waste burning and industrial production. Furans is a suspected carcinogen, and effects the immune system.

U. N. officials announced Sunday that 122 countries have agreed on this treaty banning the "Dirty Dozen" Several of the chemicals on the dirty dozen list are known as persistent organic pollutants. They have been linked to cancer, birth defects, a whole list of genetic abnormalities. the toxic poisons can be transmitted to infants through breast-feeding.

This treaty is to be signed in May of 2001 Stockholm, Sweden, Then it will take up to five years before the treaty will take effect, and the Dirty Dozen will finally be banned. Electrical equipment using PCB's would be allowed until 2025, and they have agreed not to have any leaking equipment. There is talk of more chemicals that may be added to the treaty. The treaty is set up to err on the side of safety.

Most of these chemicals have already been banned in the industrialized countries, but let us remember it was the industrialized countries that had only banned their use in their countries, yet continued to manufacture, and pollute with the same Ultra Hazardous Chemicals by manufacturing them and shipping them off to the non industrialized countries. Those non industrialized counties such as South America who ships millions of tons of food supplies grown in their countries using the banned pesticides that are known to stay in food chain. Then we set this out on the tables for our families. Many of the dirty dozen are "systemic" pesticides, that are absorbed by the plants or trees, making them toxic, all the way through fruition. Then the fruits, vegetables, and nuts, back on our tables. We have actually gotten back what we had thought we got rid of. "Every time you purchase some type of produce imagine where did it come from because if it came from some other country, you can bet they are packaging up the systemic pesticides they used, and sending it back." Many have been killed by misuse of pesticides, using too much, they are untrained in the proper use, in many cases.

The old basins are little more than a large swamps or seep hole in the ground. They were low places, diked and used for 40 years as unlined waste basins, A, B, C, D, E, which has leaked millions of gallons of toxic poisons into the underground water, there is nothing to stop the pollutants from draining and leaching through the sandy ground which is very permeable straight into the under ground auquifers. Basin C known as straw basin, the army pumped water coming onto the Arsenal which was clean untainted irrigation water, for years they pumped it into basic C using it to flush contaminants from under the Arsenal. They say they are going to top it with a protective roof, that is not going to stop if from leaking for hundreds of years into the underground water.

Army officials say they're consolidating waste from throughout the polluted twenty-seven square mile Arsenal into a central location that is already contaminated. The Army would have us believe that they have struggled to get rid of wastes at the Arsenal for years, now they want to consolidate ultra toxic material right in the same place that is known to be leaking into the underground water, there going to put a protective roof on it but they didn't' talk about how what is under that roof is not just going to leak and leach into the under ground water stratus.

It sound as if what they are trying to do is put enough band aids on everything so they can quickly turn it over to the State of Colorado to worry about, and believe me they will be changing band aids here for 1,000's of years, and the tax payers of Colorado will be paying for it. The life expectancy of somthing like this is not just a year or two, or ten or twenty years. The chemicals and toxic wastes are not biodegradeable. They are a man made Ultrahazardous toxic horror that will be with us for eternity. They can't even be incinerated safely, they already learned they cannot be buried safely. Man should by law not be allowed to manufacture toxic chemicals unless they have the method for its safe destruction filed in a government office, like EPA, or the State Health Department.

In the draft plan for final clean up, the dump area was to be capped with an impenetrable layer, then landscaped so it looks like the adjacent prairie.

Let's remember back for a moment when they planned Basin F, in the planning stage it sounded good, looked good. but the operational part flopped, when they knew it was leaking they continued it's use for an additional 20 years, It's capacity was a maximum of 180, million gallons. But after they knew it was leaking in 1965, they increased that capacity to over fill for and additional 20 years beyond it's life expectancy. When basin F was planned and constructed the life expectancy given was ten years, that ten years was up in 1965 when they first learned that the basin was leaking. They were advised to stop using basin F in 1969 pointing out to the Army, they are using Basin F on the premise that it is leaking, and polluting the countryside.

Again, it would appear that the Army had learned nothing from the hard lessons of the past. Again they had every intention of building an unlined pit in which they could dispose of their contaminants. Their plan to "cap and landscape the basin" so that it would look like the adjacent prairie, right smack "out of sight", "out of mind" philosophy. It was ludicrous to think that in the midst of a Superfund clean up, they would even conceive such an idea. "Again history was about to repeat itself."

The Army says the solution is practical and will save money. If all the wastes were shipped off the Arsenal, they would fill, every landfill in the nation and could cost taxpayers as much as $6 billion. Suddenly they would have us believe that concerns were for us, rather than for their own expediency, and negligent creation, of the most contaminated place on earth.

This almost sounds as if the Army is trying to blame their own Wilfull, Reckless, Negligence and misuse of their Ultra Hazardous activity on the public. Instead they should be taking full blame for what they have done that created the "Most Contaminated Piece of Land on Earth". "Next paragraph you will see the Army has admitted to exactly what I've just said, at the Rocky Mountain Arsenal, and at the same time not mentioning the "Blame". Just trying to make it sound as if they have a most horrendous job without saying who created it.

According to the technical director at the Arsenal, "Body is being exposed to any kind of hazard and we are excavating 1.5 million tons of soil, That's one of the largest cleanup operations in the country. Right now, it's the biggest and the toughest in the country."

The Army has already spent over $750 million just studying contamination and doing only temporary clean up at the Arsenal. The Army and Shell Oil Company have paid the first $500 million of the clean-up cost. The Army will pay 65% and Shell 35% of the rest of the tab, up to 1.7 Billion, then Shell is out of the loop for clean up costs per agreement. The rest will fall on the Army and the taxpayers for the billions more it will cost just to maintain the most contaminated piece of Land on earth.

Ultimately, the Environmental Protection Agency and State Health Oficials will, and are overseeing the clean up, and maintenance program at the RMA.

"We're taking a realistic overview" Shelton said, who works for the Colorado Health Department. "We don't expect all the toxic waste out there to miraculously disappear. But the Army needs to find a safer solution."

Army engineers counter that similar clean up plans have been perfectly acceptable at Superfund sites even closer to human populations. "We feel you have to take everything, including money, into consideration." explained Brian Anderson, the Arsenal's chief of remedial planing.

LarryCounter's "There is no other Superfund Clean Up site in the United States, or any other Country that would even begin to compare to what the Army, and Shell Oil have done here at the RMA." Mr Anderson, blowing a smoke screen for the Army trying to appease the public with more mistruths, not facts." With some engineers who obviously don't have any idea or scope of the Ultra Hazardous toxicity that the Army and Shell Oil have created at the RMA." "Without taking everything into consideration." Did the Army take money into consideration when they used Basin F. twenty (20) years past it's life expectancy. Twenty years past the time they were informed, Basin F was leaking, and contaminating the Country side off the Arsenal. The Army used Basin F twenty (20) years past its life expectancy to dispose of deadly Ultrahazardous, and toxic materials on the premise Basin F was unsafe and not to be used because of it's contamination of the air and the water.

Since the beginning of this temporary clean up the Army has identified 181 sites polluted with 80 different chemicals. Their currnt plan is to clean up 118 sites and leave the rest. The toxic liquids from Basin F are being incinerated, but a pile of toxic solids that were dredged from the basin has been capped. The Army plans to recap the wastes and leave the 40 acre hill in place.

Most of the arsenal's 900 buildings, including a six (6) story reinforced concrete Nerve Gas Factory, would be blown up or torn down and dumped in the Basin A area. Some buildings that are more highly contaminated with Nerve-Gas, Mustard agents and/or pesticides would be incinerated.

The Army plans to indefinitely treat ground water before it leaves the Arsenal and flows toward the South Platte River. There are currently five water-treatment facilities at the arsenal and the Army plans to build at least one more near Basin A. The Army would excavate soil from dozens of toxic sites, and the worst soil, about one (1) million cubic yards, would be incinerated at low temperature and placed into the Basin A, dump site.

Waste from around the Arsenal would be placed on top of the old contamination in Basin A, which would include potentially explosive chemical weapons and barrels of smelly pesticides placed there from the mid-1950's. "Note here it says smelly leaving out the part of it being Ultra Hazardous and Tocic."

Larry Says, "Let us remember this Basin A specifically, and B, C and D are all sitting on top of the large underground water table at between Ten (10) to twenty (20) feet that flows off the Arsenal, and has polluted all the deeper water tables off the Arsenal North and Northwest clear to the South Platte River where the water flows into Nebraska. This does include the Brighton and Ft. Lupton areas. This area was once called straw basin, because the Chemicals would quickly disappear, and basin C is the one that Government documents explain exactly how the Army was using it with clean irrigation water to flush contaminants from under the Arsenal and into the Country side to the North and Northwest of the Arsenal.

Basin A is a natural drain into the auquifers yet they are building a site right on the top of Basin A where it will leech and leak into the underground water auquifers forever, and a day.

Former nerve gas production building in South Plants area which must be torn down and disposed of along with all former manufacturing buildings at Rocky Mountain Arsenal.

Just North of Basin A, the Army would build a relatively small 40 to 80 acre hazardous waste landfill for incineration ash, and other hazardous items. That landfill would supposedly comply with environmental laws and have two liners and a cap.

"Making the Arsenal pristine would destroy plant life and leave wildlife habitat devastated," said Anderson.

"Pristine may equate to a moonscape," he said. "The moon is sterile, it has no contamination on it. When you incinerate, the further you go with these aggressive technologies, the greater sterility you have, the greater reclamation you have to do," said Anderson. At least the moon doesn't have intrusive man made toxins on it.

Environmental regulators including the EPA and the State Health Department have reviewed and commented on the draft document, along with Shell Oil Company, and the U. S. Fish and Wildlife Service. They will inherit the Arsenal when it becomes a national wildlife refuge. It will be far from clean and pristine.

Larry says, "I have somthing to add here about the wildlife and how they got there at the arsenal. It's not because they were born there, nor is it because they just happen to be passing through and decided to make it their home. The mule deer have been brought in and introduced to the arsenal by the government. along with the prairie dogs, bald eagles, hawks, coyotes, badgers, and anything else they may think of. Just like the fish they brought in, they died. The animals, birds, fish, at the arsenal remain in peril, should any die, they are buried immediately. The animals that make up what they are trying to call it's pristine wildlife refuge, are completely fenced in with an eight to ten foot fence the prairie dogs probably would have a hard time squeezing through."

"The wildlife refuge has been nothing more than a smoke screen they are blowing, trying to lessen the blame on them for what they have done."

The Army plans to present the documents to the public in workshops and public meetings. A final version is due out in mid November. The final decision on the Arsenal clean up came out in September 1994, the Army has estimated that it will complete its clean up plan by the year 2015.

"But," replied Anderson, "it will never be clean." It will never be clean, he said, but the Army insists that it will be safe for the general public. That's wisdom and common sense for you.

LETS'S TAKE A LOOK AT WHAT REMAINS PERMANENTLY BURIED IN BASIN 'A', AND THEIR SAYING IT'S SAFE.

*World war II and Korean War vintage chemical weapons that may or [probably are] still be full of Nerve or Mustard gas.

*Left overs from conventional weapons, grenades, mortar and artillery rounds many which still carry explosive charges.

*Crushed storage barrels and tanks, along with contaminated metal, wood, and concrete debris that were disgarded in a cheap inexpensive wilfully negligent and reckless way, hopeful that no one would find out.

*Remnants from pesticides, and insecticide manufacturing by Shell Oil Company, including millions upon millions of gallons of ultrahazardous and ultratoxic waste byproduct.

*Contaminated machinery, hardware and protective suits that were used in everyday operations at the Arsenal.

How dare the military or certain portion of government, or for that matter large corporations decide for the American public how we are to live our lives! Are we not a democratic society, where the majority of the populace make the laws that governs us all? Why do we do nothing when we see our country being raped, our environment plundered, and polluted and poisoned by people who scoff at the Federal Superfund law and find or make loop holes in the law so they can continue their polluting ways, and leaving behind the devistation that can never be cleaned up, all for the almighty dollar?

But what of the Superfund? Is the Federal Law and EPA. doing what it was designed to do?

The news out now is the Superfund will be out of money by the end of 2001. Then what happens for all those hazardous waste sites, that are on the Environmental Protection Agencies Super Fund Priorities list? These I will list shortly for each State.

Surprisingly, the consensus of most people is that the Superfund is not working, and you know something, their right. For example: South Bend, Texas stands like a ghost town on the outide of Houston. In what once was a thriving community there lies a toxic nightmare. The Brio Superfund site, is an abandoned refinery that has polluted the air and water with chemicals leaking from unlined storage pits.

Its been a Federal priority site for almost a decade. Hundreds of millions of taxpayer dollars have been spent on lawsuits and planning, but virtually nothing has been done to clean it up. Eight years of studying, talking, and negotiations, and in anyone's "book" that's a long time for doing nothing.

The sad fact is, that what's happened outside of Houston is all too common in Superfund clean up sites. A classic case of a well intentioned government program, that simply hasn't worked.

Created in 1980, The Superfund was suppose to deal with the toxic legacy of the past forty years. It was a kind of environmental S.W.A.T. team. The government would spend a few billion taxpayer dollars on a few hundred of the nations worst toxic sites, clean them up and the problem would be taken care once and for all. But the Toxic devistation kept growing at a pace so rapidly, it surpassed the sites being clean up, and so did the cost. Now, more than 1300 toxic sites are on the priority list. With only about 217 sites have been completely cleaned up, at a cost of fifteen billion dollars, and people are fed up.

This Federal law has to be changed, but it needs more than just cosmetic changes, it needs major changes, in fact it needs a major overhaul. At issue is, how much clean up should be done, and who decides.

Larry believes that this issue should not even be open for discussion. The land must be put back to its original state and if these large Corporation, Companies, and the Army are unwilling to do this, then they should forfeit their right to be able to work in this country. If the financial responsibility for this clean up is to great and they are forced into closing their doors, then so be it. One must remember that many of these corporations and large companies are not even American owned. Like the Shell Oil Company, a Delaware Corporation (foreign owned). They have ravaged and raped the United States, all for profit. That almighty buck. The time has come for the American people to stand up and stop playing Mr. Nice Guy, and take back our own destiny. We must stop watching our tax dollars, and our lives, being flushed down the superfund toilet.

By the time the Superfund, and our government clean up what Shell Oil Company was allowed to do alone at the Rocky Mountain Arsenal, Shell will have gotten by with about 1 cent on the dollar from them for the clean up. Shell Oil Company a mere coyote has out foxed the fox in the clean up effort at the Rocky Mountain Arsenal just north of Denver Colorado. Where it will remain and is known as the single most toxic piece of Land on earth. They will go on somewhere else making promises to the community, and jobs for the people, and when they pull out what remains will be another multi billion dollar Super fund toilet where taxpayers can watch their money, and their lives being flushed down the drain. They can live right next to huge deadly ultra hazardous toxic waste pits that will remain unchecked for 100's or even 1,000's of years to come. I'm about to show you some of the shocking truth in your State, maybe in your back yard, some people are finding it right in their own homes, or their neighbor's homes.

Fully one third of the money spent in Superfund so far has gone to lawyers, as polluters faced with the bill try too put the blame elsewhere blaming anyone who may have used the site. Much like the Army did with Shell Oil Company the allowed them to contract dumping millions of gallons of deadly waste into basin A, B, C, D, E and into Basin F, and the Army was totally aware all the basins were leaking deadly toxins into the underground water aquifers.

The legal battles often take years, as do the decision's on how to clean up the toxic devistation left behind in the first place. Under the Superfund law, there are no national standards, it seems too be managed by EPA, and the State Health Organizations. EPA, seems to work hand in hand with the large Corporations, and the Government, especially the United States Army. At every site throughout the Country, it's as if people are remaking the wheel. Making decisions over and over again for the very first time, as if by the seat of their pants. All because of the Ultra Hazardous materials, and the methods that were used to store the waste, or hide the waste. Many of them are large pits, and there are no records as to what is in there but they know it is hazardous and toxic.

Much like one site somewhere near Brighton Colorado, but no one knows where. There are hundreds of thousands of tons of phosgiene nerve agent buried in the 1940's no record was kept, the man in charge of the Army Corps of engineers at the time wrote a letter to Congresswoman Patricia Schroeder explaining how worried he was that the barrels would be rusted out, that letter was written in 1969 from California. It can turn into a very toxic gas, and is a highly flammable liquid. Brighton sits just north of Denver Colorado.

The Environmental Protection Agency makes the assumption at most sites that someone is going to build a house on the property, that is just simply silly. It's like a rail yard at most of these sites, no one is going to build a house in the middle of a rail yard. One out of four Americans now lives within four miles of a Superfund site. Yet, many feel left out as the fate of their community is decided by bureaucrats in Washington D.C., while they pay the price for delay with their lives, and that of their families.

The Hazardous Waste Sites are only a few that are named to the Superfund list. Each site listed may have upwards of two hundred large individual dump locations within each one listed, for instance Colorado, Federal states 3, the Army alone had nearly 150 sites for the priority list.

HISTORY OF THE ROCKY MOUNTAIN ARSENAL IN BOTCHED OPPORTUNITIES.

Patriotic team of workers needed just six months in World War II to build the Rocky Mountain Arsenal into one of the Worlds biggest factories of "DEATH".

The U. S. Army and Shell Oil Co. spent the greater part of the next four decades turning it into the most poisoned piece of land in the World. In the course of the arsenal's history, Army and Shell Officials often were warned that their waste disposal policies were endangering the environment. Those warnings for the most part went unheeded for years, or just ignored, or even classified, in order to keep the truth from the public, keeping them in the dark much like a mushroom, "keeping them in the dark."

Hazardous Waste Sites
Source: Environmental Protection Agency, Priorities List, Feb 1992

State/Terr	General	Federal	Total*
California	67	20	87
Pennsylvania	90	4	94
Florida	47	4	51
Arkansas	10	0	10
Virginia	19	1	20
Colorado	13	3	16
Indiana	32	0	32
Kansas	9	1	10

Louisiana	10	1	11
Massachusetts	32	3	25
Missouri	19	3	22
North Carolina	21	1	22
Oregon	7	1	8
South Carolina	22	1	23
Utah	7	4	11
Virgin Islands	0	0	0
Wisconsin	39	0	39
Guam	1	0	1
Tennessee	12	2	14
Alabama	10	2	12
Alaska	2	4	6
American Samoa	0	0	0
Arizona	7	3	10
Comm Marianas	0	0	0
Connecticut	14	1	15

Delaware	19	1	20
Dist Columbia	0	0	0
Georgia	11	2	13
Hawaii	0	1	1
Idaho	7	2	9
Illinois	32	4	36
Iowa	19	1	20
Kentucky	17	0	17
Maine	7	2	9
Maryland	8	2	10
Michigan	77	0	77
Minnesota	39	2	41
Mississippi	2	0	2
Montana	8	0	8
Nebraska	5	1	6
Nevada	1	0	1
New Hampshire	15	1	16
New Jersey	102	6	108
New Mexico	8	2	10

New York	79	4	83
North Dakota	2	0	2
Ohio	30	3	33
Oklahoma	9	1	10
Puerto Rico	8	1	9
Rhode Island	9	2	11
South Dakota	2	1	3
Texas	25	3	28
Trust Territories	0	0	0
Vermont	8	0	8
Washington	31	14	45
West Virginia	5	0	5
Wyoming	2	1	3
Total	1067	116	1183

Feeding them a lot of sh__." A pattern of destruction, that killed ducks and geese in the thousands, destroyed farmer's crops, contaminated 10's of thousands of acres off the Arsenal. They polluted the wells and drinking water killing livesotck and people. to the North Henderson, Brighton, Ft. Lupton, endangering peoples lives everywhere.

Former nerve gas production building in South Plants area which must be torn down and disposed of along with all former manufacturing buildings at Rocky Mountain Arsenal.

What the Army and Shell Oil Company left behind was a dangerous trail of foul-ups, and cover-ups that threatened nearby residents. Years of leaking waste basins, broken sewers, faulty industrial pipes and careless disposal policies created so much environmental devastation that the federal government has already spent billions in tax money, and havn't yet totally outlined the magnitude of the problem. Shell only paid a tenth of a cent for each gallon of hazardous waste they dumped at the Arsenal.

In 1990, the nation's manufacturing companies reported releasing a grand total of 4.8 billion pounds of toxic chemicals. The ten companies that released the most toxic chemicals accounted for 1.1 billion pounds, or 23% of all reported toxic chemicals released.

These chemicals were released to the air, water and land, injected underground, discharged to public sewage treatment plants and sent off-site to treatment, storage disposal operations.

Hazardous substances in the environment are one of the most insidious and personal of all environmental threats because they could strike anywhere, anytime striking families or neighborhoods in the places where people least expect them and where they feel the safest; in the home, at work, children at school, day care, or even at the dinner table. Hazardous substances have disrupted and separated many communities. As neighbors, People have been forced to scatter after learning their homes, and schools have been built over or near toxic waste sites. These toxic substances have also generated feelings of betrayal in the companies that provided them with jobs, or in large corporations or in the State and Federal Government itself.

The problems caused by toxic chemicals are twofold. First, they cause a wide range of harmful effects on human health, from significant health maladies, many forms of cancer, It alters the chromosomes that contain most or all the DNA, the genes of the individual, causing a wide range of birth defects, children born prone to seizures' terrible learning disabilitie's and even spontaneous abortions.

"Mutations" Significant and basic alteration, (a). a relatively permanent change in hereditary material involving either a physical change in chromosome, or a bio-chemical change in the codons that make up genes; the process of producing a mutation. (b). an individual or strain resulting from mutation.

Second, these chemicals can cause long-term or permanent damage to the ecosystem. Since manufactured toxic chemicals can accumulate in the environment over time, the long term danger is especially great. The use of some chemicals, such as toxic synthetic organic compounds, has increased about fifteen fold since World War II.

The number of chemicals in existence is also increasing, with 74,000 chemicals now used regularly, and 500 to 1000 new ones added each year. Much is understood about the damage caused by those chemicals, but the greatest harm of all may come from what we do not understand about them, nor can they be taken back apart and made harmless. "no chemical should ever be developed, or used unless its destruction can be shown and documented."

Dioxin for example is a highly toxic chemical. A common contaminant of herbicides and pesticides, it is one of the most toxic substances known to man. A single drop of Dioxin can cause cancer, immune system depression, and birth defects in laboratory animals.

The use of many chemicals is forbidden here in the United States, but not there production. These same chemicals are sold to many third world countries that do not have current information telling them of the dangers or how to correctly mix these very toxic agents, proper storage or disposal.

Consequently many chemicals are used far stronger than is necessary. Many of the crops grown by these farmers, are then imported back into the United States, and we serve this to our families at the table, unaware they may be saturated with harmful chemicals, many being systemic. (Absorbing into the plant making it and the fruation toxic to anyone).

The overpowering evidence of corporate greed and irresponsible decision making policy is before us, the cards have been placed on the table for all to see. What shall the punishment be for those who have willfully, and woefully committed these crimes against nature and the humanity of the world and how do they pay for their crimes?

When these destructive acts are committed by large corporations, because of their corporate power and governmental influence, it seems as if a coating of newly fallen show has descended upon the earth, covering all tracks of those involved.

Surely with a warm breath brought by a new spring, those tracks will be seen, and leave less chance of snow fall to cover those ever so obvious tracks.

While we are on the subject of numbers, government and corporate irresponsibility, it is as good a time as any to talk about another sensitive issue, the storage of toxic chemicals and chemical waste.

Steel tanks have long been used for the storage of just about everything from inedible objects such as gasoline, to edible material like molasses. Not all owners, however, have recognized the need to protect their storage facilities against corrosion. as a result, the EPA, some eight years ago, estimated that of the three of five million existing underground tanks, some 100,000 or more were leaking, exposing our underground environment and water supplies to serious Hazardous and toxic contamination. Many of these chemicals are known carcinogen's (cancer Causing). This doesn't take into account the above ground facilities that are constantly exposed to the elements. Most of the toxic wastes stored, in itself are not only toxic but corrosive and acidic, able to actually eat away at the storage tanks from the inside out. Corrosive acids (more at gnawing).

Recognizing that the technology existed, and had long been used in other parts of industry, Congress enacted Legislation in 1984 to initiate the regulation of underground storage tanks and piping. This led the EPA to issue the current rules known as 40 CFR part 280. The purpose of the rules is to minimize the release of product from underground storage tanks and piping (UST Systems as they have become known). The rules, for some strange reason, that is only conprehensible to someone with the I. Q. of a politician, It finally went into effect December, 1988 and covered both new and existing UST systems.

The EPA found that corrosion was an important cause of chemical leakage, so a large portion of the rules deal with corrosion control. Owners and operators of UST ststems, therfore, need to understand the effects of corrosion in order to meet the requirements of the regulations and to protect the population and the environment from harm.

When a steel structure is placed on or buried in the ground it usuallly corrodes. The resulting pitting causes penetration on weakness in the tank which permits the product to leak into the soil. Corrosion takes place because of a nattural electrochemical reaction between the steel and the soil, or because of the presence of stray direct current. In either case, direct current is discharged from the surface of the steel which then destroys the metal and causes leaks.

There are several methods used to prevent corrosion of UST systems and thus prevent release of products into the soil. Cathodic protection, which is nearly always combined with a coating, is a common and practicable method of preventing corrosion. This method consists of introducing direct current into the earth which is picked up on the surface of the steel and overcomes the corrosion current which is trying to leave.

Non-metallic material, notably fiberglass, is also used for tanks and piping. This material is immune to electrochemical corrosion, but can be damaged due to the incompatibility between the stored chemical and the chemical makeup of the fiberglass.

The current laws and storage tank specifications are most admirable. Unfortunately, if the bottom line is that these safeguards are on the honor system and very rarely enforced because the government does not want to spend the money, then we have gained nothing. The government and corporate America have repeatedly shown their true colors. They cannot be entrusted to do what is right, so why give them any more 100 Billion dollar chances?, with the taxpayers money.

"Quick reminder The Rocky Mountain Arsenal, United States Army and Corporate Shell Oil Company. In 1965, and 1969 after being warned by the United States Department of Health, and the Armie's own Hygiene Agency warned the United States Army that Basin "F" was leaking Parts of the protective liner are absent. Further use of Basin "F" will be on the premise it is in fact leaking." ""The Pentagon iladvised the Army we cannot let the public find this out [classify everything] from here on out. They continued using it for nearly an additional 18 years overriding the volume limit maximum of 180 million gallons increasing it to overfill of 250 million gallons, for 18 years, giving Shell Oil Company free access. This remained classified until late 1990"".

Webster's New World Dictionary Second College Edition defines Treason as: 1. betrayal of trust, or faith; Treachery 2. Violation of the allegiance owed to one's sovereign or state; betrayal of one's country, specifically, in the United States.

Webster's New World Dictionary Second College Edition defines Treason as: 1. betrayal of turst, or faith; Treachery 2. Violation of the allegiance owed to one's sovereign or state; betrayal of one's country, specifically, in the United States.

By allowing the Army to contaminate the soil, air, and water in and around the land that makes up this counrty which in most cases is leased to the government, it would seem that by definitions 1 and 2, the government is guilty of treason. By allowing large corporations to commit the same crimes against the country and it's people, does it not stand to reason that the United States Army and these large corporations, including the people who issued the actual orders, are guilty of treason as well?

Again lets look at what the Army done when they knew first hand basin "F" was leaking. They had it classified, then allowed Shell Oil Company to do what was never before done. "TREASON" This was an irresponsible decission that allowed them to continue contaminating the earth and the air for 30 years. Then when it was released they stated had we known, "They knew full well the consequences for what they had done." Yet it is the taxpayer's who pay for it with not only money but their lives and the very lives of their children, and the childrens children. It is a horrible monster out of control, will it ever again be controled.

Just before Christmas, 1993, Jennifer delivered two reports to Larry's home both dated April, 1993. Both reports were from the Colorado Department of Health.

In an attempt to hide something, someone had deliberately blacked out much of the information contained in one of the reports, only one small paragraph remained and it stated: "The Army does not plan to use state standards in the offpost water clean up, saying there 'inconsistent application and ambiguous language.' These standards, however, are enforceed at all other superfund sites in Colorado, and have been used by the Army itself for earlier ground water clean up at the Rocky Mountain Arsenal. The State of Colorado wants the Army to recognize these standards for clean up area's offpost.

Larry say's, "The State Department of Health is obligated to force the Army to use Colorado's safe drinking water standards, and will woefully have to force EPA to enforce Colorado Standards. EPA cannot set a standard, for instance DIMP, Colorado standard was set at .008 ppm. Then EPA set a standard of .500 ppm as an enforcement standard then tried to force Colorado to ablige. Colorado Health Stated Most certainly not what you have is not a safe drinking water standard, further more EPA cannot set a safe drinking water standard for Colorado and must ablige Colorado's Standard also for the clean up standard in Colorado."

The second report was issued by the Colorado Department of Health, entitled: State concernes on the Rocky Mountain Arsenal offpost proposed plan.

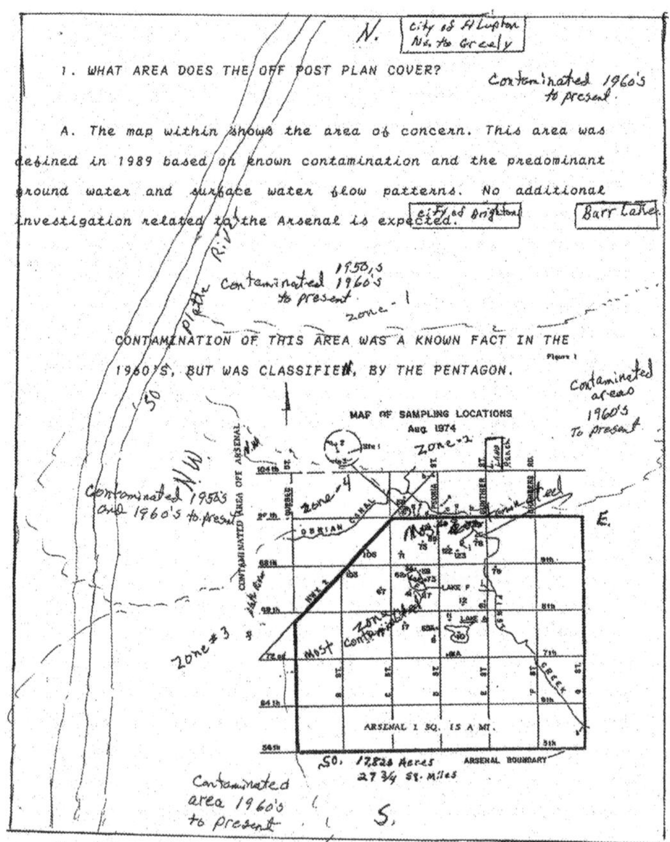
1. WHAT AREA DOES THE OFF POST PLAN COVER?

A. The map within shows the area of concern. This area was defined in 1989 based on known contamination and the predominant ground water and surface water flow patterns. No additional investigation related to the Arsenal is expected.

CONTAMINATION OF THIS AREA WAS A KNOWN FACT IN THE 1960'S, BUT WAS CLASSIFIED, BY THE PENTAGON.

DOES THE STATE SUPPORT THE PLAN? WHY OR WHY NOT?

According to federal law,, appropriate clean up levels are based on existing state and federal environmental standards, if they exist. In addition, risk assessment that estimate cancer and non-cancer risks are used to determine clean up levels when environmental laws either do not exist or are not considered to be protective at a particular site. A risk assessment compares the levels of contamination to EPA established numbers to determine hazard indices for non-cancer risk. Cancer risk is established through excess cancer risk predictions. An "excess" cancer means a cancer in addition to the predicted cancer risk. According to the American Cancer Society, one in three of us will develop a cancer sometime in our lives. The state has concerns with what the Army considers acceptable risk. These concerns are explained below:

CANCER RISK

Zone 2, 3, and 4 as depicted in the illustration, are the most highly contaminated areas of the offpost study [MY FARM] is located at the very bottom of zone 2. Contamination has been found in the ground water, soil, and surface water. At current concentrations such contamination, according to the Army studies, could pose excess cancer risks of aproximately 3 in 10,000. The Army states that potential risks as high as 1 to 2,000 are acceptable However, the state believes that federal law requires Superfund clean up to aim for an excess cancer risk of not more than 1 in 1,000,000, unless that number can not be achieved.

NON-CANCER RISK

Federal law states that hazard indices reflecting non-cancer risk should not exceed one. The proposed plan indicates that the hazard index exceeds one, in zone 2, 3, and 4, and a portion of Zone 1. This means the people exposed to existing contamination in those areas could suffer adverse health effects other than cancer, ranging from short term affects such as eye and skin irritation to long term effects such as asthma, liver or kidney damage. The state believes that the risk should be reduced at least to the hazard index of one.

ACCESS AND USE

Zone 3 and 4 are owned now by Shell Oil Company, Zone 2 is mostly privately owned. The Proposed Plan does not include active clean up of soil in these three zones. In addition, ground water contamination will remain there for decades while it is gradually flushed by water treated at the North Boundary System. The proposed plan does not provide any mechanism for preventing people (or animals) from drinking contaminated ground water while it is being flushed. Nor is there a commitment to provide access and use controls (like deed and well restrictions) to prevent exposure to water or soils. Therefore, the state would like the Army to evaluate active remediation of the soil and at the very least initiate measures which would prevent exposures to ground water until it is cleaned up.

Does the Army think it has the right to play Almighty God where the lives of Men, Women, and Children are concerned? These Agency's of the federal government use this ridiculous "acceptable risk" argument as if they were dealing with pencil's instead of people. The Army is in the business of protecting people, but it seems that they have become so inhuman non-caring, and non-feeling to the value of a human life that we are lowered to the status of a number. A percentage of the American population are classified as an acceptable risk. "Don't Worry, we've only lost 30,000 this year, we're still well within our guidelines for aceptable risk." "It's called expendable".

Larry Says, "Stop and think for just a moment about the Armed forces. They are not out there to protect us, people are considered expendable Indices they can choose from. The armed forces are to protect our government and their ideals. The government is supposed to govern and protect it's people, under a constitution, but doesn't it make you wonder how they have used the courts to strip away our constitutional rights locally, and nationally, without most people even being aware of it." ""it's really scarey if you stop to think for a moment how Adolph Hitler and his military machine stood up against his own people his own neighbors, and then the world. There were already 15 to 20 million people dead before the word got out what was going on. He toppled France, Poland, Belgium, sweeden, Hungary, Austria just to name a few. He was well on his way with the Russian invasion being completed, and England was right on the edge. With Benito Mussolini holding a tight grip on Adolph. On the other hand Japan had devistated China, the Phillipines and Pearl Harbor. Very few people actually realize that Japan had bombed the United States, and in at least one effort struck within 2.5 miles where Sarin nerve agents, Mustard and Lewisite were stored on the Rocky Mountain Arsenal Denver, Colorado where all the Nerve agents were manufactured."

"Larry knows this first hand as he was the one who first seen the balloon carrying many bombs. Larry's Father had told him to always watch for any balloons landing, coming down out of the sky or even going over. The Army had put out an alert warning of the balloons carrying bombs, and to call the Army quickly, and not to go near them. Larry was about 5 years old at the time he became the Armie's eye to the sky, he lived just one mile north of the Arsenal. He'd be up before sunrise until long after dark. one day he noticed this huge balloon coming down, watched where it landed and ran told his Father about the balloon and where it was. His Father called the Army, telling them where it was, and within minutes the whole area was covered with military vehicles closing off roads and any access. The Army did call later briefly said thanks, and said to keep an eye to the sky."

"Japan at that time was so sure they were going to be here invading the United States, that they printed money for the invasion and occupation of the Untied States."

Then on the other hand we had Bonito Mussolini in Italy who was working a deal for complete control of Europe, and possibly World take over with Adolph Hitler. Benito Mussolini was removed from his government post twice, each time he was helped to get a handle on it, but in the end he was located and shot to death. As Hitler was working toward the seizure of South America, and the Panama Canal. Hitler for Germany, Mussolini from Italy, and Yamamoto for Japan, They were Orchastrating a one World Power, with a one World Government. Japan didn't want anything to do with that. They wanted Asia, North America, through Canada and Alaska, and all the islands between, they were thinking more in the area of a two government World, and between the two of them they had discussed how the America's was going to be divided, they argued breifly as to how it was going to be divided. However they agreed Japan would rule North America, and the Germans was to rule South America. The panama canal would be used unilaterally, but under the control of Germany.

Germany and Japan both had briefly attacked the Panama canal. Below is a copy of the Japanese currency that was to be used when Japan Occupied the Untied States. "Larry was able to get some for his paper money collection. He wants to share it with you so you can see it and believe it, and it doesn't say Yen, It is Japanese invasion money for the United States. $1.00.

Hitler's government had made some deals with several different countries in South America, and they were for walk in controls of the governments. The Panama canal was definatly taken care of like socks in a shoe. They were ready to do what ever necessary to assist.

One life is to great of a price to pay for something that was done so carelessly and irresponsible to begin with.

When Larry was growing up, his parents gave him a strong feeling for 'love of life', and being like most American's he would develop over the years of his youth an equally strong 'love of nature.' "Larry became an expert in animal husbandry". Larry can not and will not give up these beliefs just to quietly sit back and become a number on some desk bound paper pusher's fact sheet, so the Army can have its way. He and others like him are made of stronger stuff than that. He pushes back, makes them stop and think, and for safety and the environment helps change their ways of thinking and planning.

DIMP IN THE GROUNDWATER

In 1990, the State requested that the Water Quality Control Commission set a ground water standard for Dimp, and for IMPA (Diisopropylmethylphosphonate) & (Isopropylmethylphosphonate). Both by products of two types of Nerve Gas production at the Arsenal. A current EPA Health Advisory level of 500 ppb. (parts per billion) or .500 ppm (parts per million). This has been used by the Army and EPA to determine what areas of ground water should be cleaned up. The Army will consider clean up only in those areas where Dimp level are greater than the EPA Advisory Level. The State believes that a more conservative figure should be used.

The Army has asserted that the part of the ground water plume with DIMP, and IMPA concentrations above 600 ppb has not moved past their offpost intercept treatment system. The most recent testing done by the Colorado Department of Health has found DIMPat 800 ppb in a private well at least ½ mile past the intercept system, indication that DIMP well above EPA's Health Advisory Level is already in private drinking water supplies. The well owner was already receiving bottled water from the State. The State believes the Army should address the significantly elevated level of DIMP contamination which has move well beyond the offpost treatment system.

Bottled water has been provided to more than 600 residents with DIMP in their well water. This water has been paid for by the State of Clolrado, with costs shared the first year with the EPA. Due to the widespread nature of DIMP contamination in the offpost area, the state believes that the Army should provide all residents in the study area a permanent, municipal water supply.

CONTAMINATION OF THE DEEPER AQUIFER

Since 1990, testing by the State has revealed the DIMP is present in the deeper Arapahoe aquifer at depths greater than 100 feet. (It is of interest to note that this is the same Arapahoe aquifer that has 14.2 tons of plutonium, 1,800 55-gallon drums and 17,000 1-liter containers of plutonium-laced waste site sitting right on top of it at the Rocky Flats Nuclear Weapons Plant just a few miles away from the Rocky Mountain Arsenal.

One microgram (one –millionth of a gram of plutonium is considered a lethal dose. State health officials have already warned that the waste acids from this product are corroding, and gnawing the tanks. Luckily, Plutonium only has a half-life of 24,000 years, so by the year 26,000 A.D. half of this product will be transmuted into sommething else, possibly into somthing even more deadly, the monster has been created, and continues to grow. All that can be done now is to pen it up in coated steel tanks guarding it to make sure it does not get out as it has so many times in the past, because each one is getting larger, and costing $Billions, and more lives are destroyed.

The levels of DIMP fon, range from a trace to 39.7 ppb. The State has identified approximately 20 domestic wells being used that should be closed because they may be allowing contamination to move down to the deeper aquifer. The Army has not closed any of these wells, and the Proposed Plan does not address this problem. The Army has argued that contamination of the deeper aquifer is a localized occurrence, that it is due to poor private well contstruction and is therefore not its responsibility. The State would like the Army to close all 20 wells.

Again, we must point out if this is not the Army's responsibility, then how did these toxic chemicals get there? Surely most families do not have a recipe for the making of Nerve gas in their family cook book.

The Army cannot deny it, They cannot blame it on someone else. The Nerve agents and pesticides have been fingerprinted to the Rocky Mountain Arsenal. The Army at all times was in control of, and took full responsibility for all the chemical waste, and for all disregard to human life, willful, reckless, and negligent methods of operation by Shell Oil Company.

We have reached a point in our Courtry's history when we must tear loose from the wretched bounds of reckless decision making and reclaim what is our's by human right. We must al all cost, make the Government and Corporate America know that we have had enough of their careless, reckless mistakes that not only endangers the environment but also the health and well being of humans and animals every where.

The problem is, I believe most people can't be bothered, because it hasn't affected them. "YET" or they believe it won't bother them or their loved ones. They are silent killers and will affect everyone, so help us keep it in control for our futures, the future of our children, grand children children's children, every little bit helps to make a better environment. speak up and/or help if you can, and one day you will look back and say I did my part.

What the United States Army, and corporate giant Shell Oil Company done at the Rocky Mountain Arsenal, and the Government done at Rocky Flats, was "Intentional and Deliberate", it was no mistake. Now they have tried for years to cover up and hide the most contaminated piece of land on earth, by cleaning the surface and bringing in Deer, coyotes, Eagles, badgers, prairie dogs, fox, and birds, creating and calling it pristine wildlife refuge on the surface for the rest of it will forever remain the most contaminated piece of Land on earth. As they create this refuge for animals I often wonder why they fenced it all off with a ten foot fence where animals cannot leave even if they had a chance. The animals are put there, and kept their. The false face they created with the scars of a 10 foot impregnable fence.

In the course of the arsenal's history, Army and Shell officials were often warned that their waste disposal policies were endangering the environment. Those warnings went unheeded for years, creating a pattern of destruction that killed thousands of ducks and geese, destroyed farmers and their crops, contaminated hundreds of drinking water wells. Please notice in the picture one of the 8 to 10 foot chain link fences that keep the animals from escaping.

DAVE JENNINGS/Brighton Blade

WILDLIFE AMONG THE RUINS

CHAPTER THIRTEEN

"SMALL OPPORTUNITIES ARE THE BEGINNINGS OF GREAT ENTERPRIZES, AND THE RECORDS OF SOMTHING WELL DONE, IS WHEN YOUR DONE. IT WILL LEAD TO HAPPINESS THAT LIES IN THE JOY OF ACHIEVEMENT, AND THE THRILL OF A CREATIVE EFFORT. THE CHALLENGES WE LOOK FOR OFTEN BRING GREAT EXHILERATION OF VICTORY."

THE TRUE ENTREPRENEUR LARRY D. LAND

"ONE MAN WITH COURAGE MAKES A MAJORITY."
ANDREW JACKSON

LARRY BELIEVES, "NO PROBLEM CAN STAND THE ASSAULT OF SUSTAINED THINKING, ESPECIALLY WHEN YOU ADD TENACITY."

VOLTAIRE

What began as a single audit by the IRS on Larry's business and how he was filing his taxes came mid 1990, requesting he call and set up an appointment with the auditor on the letter, stating he owed an enormous amount of back taxes and fines.

Larry thought this was a little strange as he always had an accountant do his taxes for him, and was never audited in over 30 years of filing taxes always organized with a small business expenses etc. In 1988 Larry had just won by unanimous decision of the House Judiciary SubCommittee on Law in Washington D.C. a Bill and House Resolution to the House of Representatives. Where it would be introduced to the full House of Congress and give them an opportunity, to look through his complaint and the Subcommittee's recommendation to allow Larry to sue the United States of America for felony crimes against the people.

It was Congressman Hank Brown of Colorado who had introduced the bill to the House Judiciary Committee on law. Mr.

Brown knew exactly what had taken place, and what the Rocky Mountain Arsenal was guilty of. He also realized and knew they had created the most comtaminated piece of Land in the World, and Larry sat in his Congressional District. Larry thought He sure hopes that Congressman will be here for him over the years, because he knew full well it was going to be most arduous and difficult, since now he was face to face with the United States. He already knew what they were capable of, and the power they had.

The purpose of goals is to focus our attention. The mind will not reach toward achievment until it has clear objectives. The magic begins when we set goals. It is then that the switch is turned on, the current begins to flow and the power to accomplish becomes a reality. "That reality is our destiny."

Larry D. Land

Everything Larry has done in life he has set his goals, he calls it reaching for the stars. "Writing this book has been a great experience in sharing his knowledge with others. Love is the most important ingredient of success. Without it, your life echoes emptiness. With it, your life vibrates warmth and meaning. Even in hardship, love shines through. Therefore, search for love***because if you don't have it, you're not really living***only breathing." Larry sets his goals, then adds tenacity, and never stops in the middle.

With this book Larry hopes he is able to just touch, and pass on a little love and knowledge to everyone. He hopes neither you nor your children will have to use it, but the knowledge is important and may help you take the precautions necessary if need be to avoid injury to you or your children.

Back to the 1988 audit, it came only a few months after the Department of Justice had done their first deposition on Larry. He could remember how they had really hammered him in his finances and could not figure out how in the world he was able to afford doing all the things that he had accomplished. How he could afford several attornies, and to pay for all the high buck expert witnesses. The best in the World. Now he realized why the IRS, and the fact they were trying to disallow any and all the deductions, business or legal.

It was in October 1988, Larry watching Congress as they are voting on the House Bill presented. nervous, and excited at the same time. They read House Bill 816, and House Resolution 61, for the relief of Larry Land and Marie Land et al. Larry held his breath, for what was about to happen will tie a lot of knots and laws into how chemical waste is handled stored, and controled. It was a giant step forward for Larry and his family, and the world. As the Yea's and Nay's began it was exciting to see early on it was unananimous, Majority vote of the Congress of the United States just past House Bill 816, Now by resolution it will be turned over to the United States Court of Claims where the United States will be facing legal and equitable claims for the devistation, they carry full responsibility for their actions at the Rocky Mountain Arsenal.

At this point the Attornies were in a buzz now trying to draw up the complaint against the United States. News paper and television everywhere for a story.

By this time in 1991, the auditor found out he was dealing with a person that would not just fold up. The next thing Larry knew a letter in the mail again denouncing him for not paying enough tax, and large penalty, amounting again to several thousand dollars he had not even got 1988 set up with appeals yet where he also owed several thousand dollars in back taxes and penaltys. By the time he seen the auditor on 1989, he was also seeing the appeals officer on 1988. both places Larry had given them all the information and reciepts. By time he had got the appeals report back on 1988, he also got the auditors report on 1989. So he called the appeals officer and said well as I told you 1989 is also back from the Auditor. He said drop it in the mail, and he would call when we could get together.

By spring of 1992, the Justice Department wanted another deposition. At the same time frame the attorny's struggling with the Federal Court of Claims, on a Motion in Limine, a Motion for a protection order, and a Motion for Summary Judgment all filed by the defendant.

The Attorney doing the deposition again was hammering away on everything to do with his finaces, financial stability, and financial past. He seemed angry as he was not able to find out what he would like to. At one particular time after he had ask the same question for the sixth time. Larry done as a joke, he ask the attorney what the hell is it you don't understand, I said N-O, NO now for the sixth time. This whole thing was looking, and sounding like an interrogation.

At about the same time the Appeals officer called, and we had to set a time for meeting enough to go over both 1988 and 1989. Larry set up an appointment to work with the Appeals officer, and the time soon rolled around, they met and talked back and forth why the cost factors and deductions. By the time they ended up in an agreement, and they were both happy, 1988 the IRS owed a certain amount back to Larry, and 1989 Larry owed a certain amount to them they brought the two figures together and no money changed hands. Larry kept all his deductions according to all IRS publications.

By the spring of 1992 was here Larry recieved a letter again in the mail from the IRS, claiming he owed many thousands in taxes where again his deductions were not accepted, and because it was his mistake they also wanted nearly $4,000.00 in penalties and interest. this was for 1990, Larry called the number in the letter and made an appointment to see the auditor. The Auditor was very intense, and wanted copies of everything, Larry explained to him he has every right to make the copies he wants, but he cannot have my personal copies not even to borrow. So he sat all day giving them papers to make copies of logging them out and back into the file.

In just a few weeks another letter in the mail this time for 1991 again the same thing there were several mistakes and deductions that were not allowed, and fines and penalties. Larry has his own corporation (Land Systems of North America), and his company NTI, NEW TECH IDEAS. and they had also run into some discrepancies in them and now decided they also wanted to audit them as well.

After the auditor now had 1991, personal Larry brought in his Corporate, and Company records in along now with his personal and we are now in the spring of 1992. Larry spent another compete day with this Auditor, and he wanted copies of everything which he got but under Larry's rules of logging in and out.

It remained awful quiet in 1992 not a word from the Auditor, Larry tried calling a couple of times, and did not get a return call. The 1988 and 1989 tax fiasco was settled. By late spring the auditor had sent a letter in the mail stating what problems he had with both 1990, and 1991 personal taxes. several thousand owed in taxes and penalties. He found no problem whatsoever with the Corporate taxes which was quite surprising since Larry had set up on the depreciation schedule a lot of equipment.

The way Larry worked out the cost factor for the research and development was to make the items workable and marketable, and have all legal documents set in stone. Doing it with an individual company Larry was able to deduct all costs for research and development, and all small equipment under $300.00, Auto mileage, and insurance, he used his home also as his office and shop. Nothing was ever turned over to the Corporation until it was shown the product had made it's first one dollar, and all costs to Market were deductable.

Larry now holds nine patents, and a file full of wanta be's, that are in the research and development stages. He has about 16 copyrights. He is known as the Idea Man.

Patents are, The Land Biometric Systems, using a proprietary, patented process with "Land Film Strip" as the direct input to the "Land Biometric reader". produces digital images in 256 levels of gray scale that can be remotely directed to another location to print out a card or cards. There are two patents on this system, and one copyright, as to how fingerprints are placed on clear acetate cards wheras the fingerprint can be viewed as a forward latent print, or a reverse latent print for comparison, and at the same time.

The Larry holds a patent, and again a copyright on all see thru prints. This is the Land Bio-metric scanner comparator. which allows a finger print image to the blown up for investigation. They can be blown up to over six foot by six foot, images, and then if the investigator wants to show the suspects finger print would match, they then can be overlayed to show the two are a perfect match. Perfection in fingerprinting.

Larry is thrilled about the research and development program he has been able to put together. He takes his ideas, takes them through the Proto-type stage all the way to a working model with a Patent. An Idea to a patented working model, has cost upwards of $30,000.00. Every last penny is tax deductable, the IRS act like they don't understand and even struggle trying to take away ligitimate deductions. Larry is beginning to think it's more like harassment by the IRS. The IRS is nothing more than a tennacle of the Justice Department who are trying their best to belittle, and discredit, and suddenly wants to put him out of business because he is doing everything by the letter of the law, and they are trying to stop him even though he is right. The real thing here is when he filed suit against the United States of America, and they would like to stop him in his tracks. Before the IRS routine there were at least nine attempts made on Mr. Lands Life. explained earlier in this book. Now they are using the tennacles of the IRS, which are merely a tool of the Justice Department, trying belittle and discredit him, and an all out effort to put him out of business He said I was good at playing there game in Colorado, and He can beat them at the tax game also, because what he does is perfect and honest, as per IRS publications, somtimes the law library, up front, he said I cannot tell a lie. The IRS has repeatedly and purposefully even though the legal expenses I have paid are totally deductable every year so far the Auditor has tryed to take that away. When I have to go back down to Denver for meeting with Congressmen, the State Health Officials I get a room out aways so they are like $29.95 very conservative knowing this will be a legal deduction, then meals again deductable I keep them below $6.00. Then when I found out what do the boy's from the Justice Department do. They sure don't stay in a reasonable room. They stay in the $200.00 and $300.00 dollar a night rooms. If there is more than one, do they stay in a double?, "NO" their rooms are all singles. Then they have a per meal limit of $35.00, and what ever else. Larry says he is just speachless when they tell him he cannot take the room or the meals off as a deduction at all for a legal visit.

One could tell when the heat was turned upon the 1990, and 1991 Audit since the Justice Department had just filed three motions with the Court. 1. Motion in Limine (limit investigation by the Plaintiff), 2. Motion of a protection Order (restrict the witnesses the Plaintiff can question), 3. Motion for Summary Judgment. (The Defendant asking the case be dismissed saying the Plaintiff cannot produce any Material Fact, that would allow any reasonable jury to find for the Plaintiff). Larry's attorney's were working very hard to prove Defendants Motions were frivolous and not necessary. This truly explains why this auditor was very difficult to work with and wanted everything, and couldn't agree on nothing. Then in early fall of 1993 he had tried to get ahold of the Appeals Officer, and couldn't so he called the tax court in D.C. and ask them to send the papers so he could file 1990 and 1991 tax Court. They were filled out and sent back with the courts fee. It was set up and docketed with a Court date being spring of 1994. Well Larry had a lot of time to look the tax papers over and what they had demanded. Larry went back into his tax records, reciepts etc. believe it or not Larry had used just what he could in 1991 to break even on the taxes. Leaving several epenses that were not used on the tax document, so he listed everthing, put all receipts in order so he would be ready for the Appeals Officer should he need extra deductions. Finally by late fall the appeals officer wrote a letter sending me what he had got from the IRS, told me to put everything together and give him a call. Well Larry was ready so he gave him a call and set up an all day appointment in little over a week.

Larry went into the Appeals officer all set for a big battle with everything in exact order the way the appeals people like to see it. Larry was met at the front desk by the appeals officer, and taken with his two office file boxes to his office. He had a large desk, the front half was clear, so Larry began laying out his appeal with all receipts in separate little bundles. by the time the day was over they had agreed to the original tax statement, the original deductions that had been taken. then he ok'd all the receipts on the extra expense that were found leaving Larry with nearly $22,000 in deductions, and carried NOL carry forward. This was $22,000 he was not going to use, he said from here on out he will not leave a stone unturned. He told Larry he would take care of it, have it prepared for the court, and will be put mail exactly as discussed, for signatures for the Court.

Larry had gotten an answer in the mail, everything as discussed on appeal had been approved. No more than got that over, and Larry thought well there is four years, but he had this erry feeling because he remembers the last audit he had told the auditor, I had a good time on this one, when is the next one then knowing the defendant (the Justice Dept.) was just dealt a serious blow when they didn't get what they were trying for in Court. They lost the Motion in Limine, the Motion for protective order, (Larry Said imagine that they want a protection order against me,) then the Motion for Summary Judgment, [Judge Robinson Criticized the Defendant for doing what they had done], ("Undisputed facts establish that Defendant has contributed to pollution of the groundwater underlying the lands of the Plaintiffs." "The fact of pollution has been admitted by Defendant while it contests the quantity, and quality of such pollution." "This pollution found in the Land wells has been fingerprinted to the Rocky Mountain Arsenal" Then Just as sure as heck the same auditor sending him the usual nasty old letter from 1992 tax year, telling him how many thousands in back taxes you owed, and tie on that damned penalty of about $2,000. Certain deductions were not allowed even though they had already passed the appeals process as correct deductions.

Larry got hold of the Auditor, and ask him who at the Justice Department had ordered this Audit, of course he denied that, and again had an excuse why they had randomly chose his taxes for the fifth year in a row with only gains for being audited, even though it is frustrating once is expected but five times, and the Justice Department just lost in Federal Court, and now know the United States is going to face Trial on felonious, and wilfull negligence, and reckless wrongdoing, in the operation of the Rocky Mountain Arsenal. Larry says they now will try even harder to belittle or discredit him, by taking away all his rightful deductions by law, and cause as much discomfort in his life as they possibly can.

Larry lives by the old saying, "When things get rough, that's when the tough get going, Larry already had learned to keep everthing in exact order, so he called the Auditor and ask for an appointment. This was the Auditor who liked making copies, Larry told him it's ok I will log in and out each piece. There were three ways to make sure everthing was in order, for every receipt, there was a check receipt, every thing was paid by check. Then them two items will correspond to the deductions taken, on the balance sheet then them totals must match the tax forms used. When he went in for the Audit, the auditor had his desk cleared as if he knew exactly how it would be layed out.

By the end of the day, about two in the afternoon, there were several things he and Larry did not agree on, even though they had been allowed before on appeals. He also then requested the 1993 taxes, and also both 1992 and 1993 corporation tax filings, and Larry said why is that them paper's were not just randomly chosen, He then indicated that there was somthing he had seen in materials audited that he wanted to go ahead and do also 1993 while were right here. with all the NOL carry forward Larry had to work with he was completely satisfied he would have no problems, went ahead and set up the appointment for 1993 tax audit two weeks down the line.

By the time 1993 Audit was over it was already the fall of 1994, and he was about to end still another year, he somehow knew would also be audited, so when he put it together it was put together with skill and organization because it made it much quicker to prepare for the inevitable tax audit. For 1992, and 1993 business and personal taxes he had denied the deductions taken for the battle with the Government, and about half of the expenses taken on the C schedule. However he sent back the statements for the corporation taxes back with no questions. Larry had told him just to bundle it up and send it to him in the mail because he will be going to the tax Court and to appeals. He got it back in the mail and called Washington D.C. for the paper's for the tax Court, and it was filed in tax court. The Court date was set up about eight months out. He knew when the Appeals was ready they would contact him.

Early in the spring of 1995 word came that the Land v. United States of America was scheduled for trial January 15-18, 1996 this trial was to prove there was ample evidence that the United States Army was what caused the pollution problems just north of Denver in South Adams County where Larry had lived and lost his 635 head of livestock to the pollution from the Arsenal both in the water and in the air. Well the attornie's were very busy doing depositions of the Army experts, and witnesses from the Arsenal, and the Justice Department was very busy collecting all the depositions of Land's expert witnesses. This is going to be a very busy year for everyone.

Early spring 1995 Larry got word from the appeals, wanting to set up and appointment to go over 1992, and 93 tax cases giving them the opportunity to settle it before going into tax court. On that day the appeals officer met Larry at the front desk escorted him and his two boxes of records to his office. Larry had went out now an purchased a 2 wheeler to make it easier to carry the load. He spent about 5 hours with the appeals officer.

They thought they had gone through enough and both felt satisfied as to where they stood. The appeals officer said I'll get through this and notify you in the mail as to the results. Larry said thanks I'll be looking forward to hearing from you.

Within just weeks Larry got word back from appeals. He looked it over was pretty satisfied except for a couple of deductions he felt were surely allowable and maybe with a little bit of a different explanation to the Appeals officer he might just accept it. Larry put the explanation together, with him accepting the rest, leaving a few deductions to carry forward. by the end of two weeks the papers were in the mail to sign so they be sent to the Court for approval without having to be there. Then it was about 3 weeks Larry got his letter from the Appeals officer, everything was fine broke even across the board, and even carried a few deductions forward to 1994 taxes. Well this makes six entire years that Larry was able to pick up some NOL carry forward, especially 1991 bringing forward $22,000 extra, and even for 1993 bringing ahead some more deductible expenses. When your in a small business every cent counts.

It's now time to get the 1994 taxes done and filed, this was another good year for allowable deductions, and Larry ended up having a NOL carry forward to 1995 tax year of $17,847.00, and an additional $3,000.00 the appeals officer had given me to carry forward to the 1995 tax year. Larry micro managed his money and was very careful in how he invested in himself, his patents, and his copyrights, for his future and the future of his family. Larry had already written his first book in 1988 copyrighted and published for Washington County, a Field Training Officer's, training guide for jails, and probation. A real valuable tool in dealing with people." An award winner of the year. "It was accepted and used at the Minnesota jail resource center, for the Minnesota Department of Corrections. It was also recognized by AJA. American Jail Association as a valuable tool in training, and dealing with people. It was also recognized by ACA. American Correctional Association for its content as a valuable tool for the field training officer's in learning how to deal with people, and training new staff.

Well here is the fall of 1995, and had already got back my tax refunds, sure enough a letter from the IRS. wanting to audit my records for the year 1994. Lets take a real clear look at what is going on now then tell me the Justice Department isn't controling the IRS when they want, and really this is beginning to clearly appear as harassment on their part. This will be year # 7, and they are saying they are randomly selected. this would have to be about one chance in 300 trillion that my taxes would be selected randomly 7 years in a row. then stop to think where Larry is with his battle with the Untied States Army, and the Justice Department are their attornies. The trial for causation is scheduled for trial January 15, 1996. just around the corner.

Well Larry called the Auditor to set up an appointment to go over the records, as what Larry can see already their trying again to say he cannot deduct what has been approved six times. but appeals, and the tax court. It is plain harassment, almost with a face painted on it. This is really horrible, but at least his taxes have improved in money back, and not being paid out.

Larry had just barely got his refund back, and now they want it all back again. Larry by now was doing his taxes with every thing organized, and clipped together, for an easy layout for the auditor, so he gave the auditor a call to set up a time to visit. He met with this auditor on two occasions, both time trying to take everything Larry used as deductions even money for legal costs for the farm. Right after the first visit he went to the IRS office downtown and got a publication that covered everything that he had deducted, laid this out for the auditor and he still would not budge. So Larry told him just to wrap it up send it me along with what I will need to file it with the tax court, and to appeals. It can easily seen, evidently it's like a contest who can break Mr. Land down instead, letting him win all the audits. This guy was really difficult to talk with, and would not give on any of the deductions even though they had been approved by the Appeals, and the tax Court, and even though they were clearly explained in the publications, he was like a stone wall, not willing to budge.

Arrived in Denver, Colorado for the Trial Larry Land V. United States of America. The United States Army, and Shell Chemical Company had both been closed except for clean up operations. They already accepted responsibility and clean up of the most contaminated place on earth. This clean up first thought to cost about ½ billion dollars, that quickly changed once they got started it went to 1 billion dollars, and in less than two years later they are talking of 3 to 4 billion dollars. It's not really sure at how many billion dollars they now figure but they have projected the clean up effort to be accomplished by the year 2015. However it will never really be cleaned up at all. It may be somewhat contained, and will actually be a hot bed for many hundreds, or even thousands of years to come. What is really pitiful Shell Oil Company made a deal with the United States where as it will not pay more than 35% of the first 1.7 billion cost to them less than 600 million dollars for their part, and they are out of the loop for any addition clean up funds, no matter what it costs. The National Super fund that allows EPA to try cleaning up so many disasters across our nation by the year 2002 will be out of money, and they only got a good start on the clean up of our nation.

Word has just come down August 2, 2001 on the Hudson River disaster, PCB's. Fishing in the Hudson river was banned for 25 years spoken by EPA's Christy Wittman, there is hope that fishing can reopen after the dredging operation. The huge dredging project was approved by EPA, and the Federal Government, with General Electric to foot the entire bill. Although G.E. is unhappy and think it will only make the problem bigger than it is now. That all the PCB's now lying on the bottom of the River. The dredging up of 2.5 billion cubic yards from the rivers bottom will just stir up what is not a very quiet situation. The only thing Larry wasn't able to figure out if the PCB's are laying so quiet why does everything remain so toxic, and the fish are actually so toxic they had to ban all fishing in the river.

Though Larry is just now going to trial, he has already won our battle on a cleaner environment, not only with Rocky Mountain Arsenal, and Shell Oil Company, being closed forever. What Larry began in 1972 has surely opened the eyes of the World, and our Nation about pollution in the environment, and our back yard.

It's now Larry's turn to ask the Court for appropriations from the United Sates of America. It must be decided on a legal claim for the United States Army to be responsible and the Court would decide the reparations. If it is decided on an equitable claim, or as an equitable gratuity, It would be up to the United States Congress by the way of a bill Placed through the House Judiciary Subcommittee, and then to the full House of Congress to appropriate the funds for the Land's reparations.

***** This case has been referred to this Court for determination of facts pursuant to House Resolution 61, adopted August 11, 1988, referring House Bill 816 a Resolution which provides for payment of an unspecified amount to the Plaintiffs in "full and complete satisfaction of all claims against the United States based upon health and other related problems resulting from the operations and activities at the Rocky Mountain Arsenal...." The case is being tried under Plaintiff's "First Amended Complaint" alleging equitable claims against the United States based upon Negligence, Strict Liability, Ultrahazardous Activities, Inherently Dangerous Activities, Trespass, Nuisance, and Deceit Based upon Concealment.

***** The government answered generally denying plaintiffs' allegations with some exceptions. One exceeption is the admission that TCE (trichloroethylene) is one of the contaminants escaping from the Arsenal. The government has also admitted that DIMP has been found on the properties of James Land and Larry Land and that chemical contaminants did flow off the Arsenal in groundwater which passes under the plaintiffs' property. The government asserted affirmative defenses of statutes of limitation, Lack of jurisdiction of the court, Contributory negligence, assumption of the risk, release of a joint tortfeasor and laches.

***** Pursuant to the order September 13, 1990 parties had agreed the trial should be bifurcated on the issues of liability and damages. The defendants moved at that time for a partial summary judgment on the grounds that there was no evidence of negligence on the behalf of the government. The Court denied defendants motion for summary judgment. in Land v. united states, 29 Fed.Cl. 744 (1993) because there were many issues of material fact.

Trial was to the court on January 22, 1996 through January 25, 1996.

Statement of Issues of Law and Fact.

1. Does Colorado law require private entities storing or disposing of toxic chemicals or hazardous and non-hazardous materials upon their property to store and dispose of such chemicals and materials in a manner which will not injure third parties either through the contamination of groundwater or emissions into the air?
2. If the operator of the Rocky Mountain Arsenal had been a private entity, would it have had a duty to investigate the areas where chemicals escaping the boundaries of the Arsenal would contaminate the ground water or the air and warn persons at risk to allow them to minimize their injury or to avoid damage?
3. If the operator of the Rocky Mountain Arsenal had been a private entity, would it have had a duty to disclose information not generally available to the public regarding the source and nature of the chemicals escaping the boundaries of the Arsenal to enable persons injured by such chemicals to obtain proper medical treatment?

B. ISSUES OF FACT:

1. Did the conduct of the United States in the construction, operation, maintenance and use of Basins A, B, C, D, E, or F, on the Rocky Mountain Arsenal violate any duty to the plaintiffs under Colorado Law pertaining to private entities?
2. Was the water in one or more of the wells of plaintiffs contaminated by chemicals or other materials emanating from the Rocky Mountain Arsenal and did such chemicals of other materials cause or contribute to injury to one or more of the plaintiffs resulting in whole or in part from consuming and bathing in such water?
3. Was the air over plaintiff's properties contaminated by toxic chemicals or other materials emanating from the Rocky Mountain Arsenal and did such chemicals or other materials cause or contribute to injury of one or more of the Plaintiffs and their livestock resulting in whole or in part from breathing or otherwise coming into contact with such air?
4. Did the water containing chemicals emanating from the Rocky Mountain Arsenal cause or contribute to the injury, illnesses and death of livesotck owned by the plaintiffs from such livesotck consuming water from plaintiffs' wells?
5. Did the United States knowingly or negligently withhold from the plaintiffs or fail to collect information during the period of 1965 to an including 1975 which information would have enabled the plaintiffs to either avoid the injuries from drinking and bathing in the water from their wells, or breathing or otherwise coming into contact with contaminated air, to obtain medical assistance in the treatment of such injuries.

SUMMARY OF ARGUMENT

***** Plaintiffs' allege injury by contamination of air and water having its source on the Rocky Mountain Arsenal ("Arsenal"). Liability for air contamination is under theories of nuisance and negligence. Air contamination in the present case was both willful and negligent.

***** In Colorado, persons storing liquids upon their property have the obligation, under both statute and common law, to confine such liquids to their property or be liable for all injury resulting from escape of such liquids. The government under the name United States Army breached this duty.

***** Negligence theories are also applicable. Colorado requires the use of best available technology for inherently dangerous, though lawful, enterprises. The government failed to use the best available technology to either confine liquids to its property or to warn innocent neighbors of its escape.

***** Proximate cause is established by expert testimony that the plaintiffs were injured (a) by ingesting, (b) by coming into bodily contact and (c) by breathing toxic substances. The exact chemical or chemicals, ingested or breathed are not required to be identified in the liability phase of the trial. The government was the principal cause of the toxic contamination and under Colorado law would be liable for all damges suffered by the plaintiffs, if it were a private entity, even though other parties may also have had some responsibility.

***** The lease on the South plant was transferred to Shell Chemical Company in 1951, and Shell continued the manufacture of dieldrin, aldrin, and commenced the manufature of nemagon as well as other pesticides and herbicides. Shell dumped untreated pesticides and herbicides into Basin A. The Army manufactured incendiary materials during the period from 1950 to 1954. Wastes from this operation were also placed into Basin A. Wastes from the Chlorine processing plant were placed into Basin C, prior to construction of Basin F.

***** THE EXPANSION FOR NERVE GAS: The North Plant located in section 25, was completed in 1953, and manufactured nerve gas. Liquid wastes from this operation were discahrged into Basin A which was enlarged for this purpose. Basin A proved to be inadequate and its overflow went into the surface drainage into newly constructed

Basins C, D, and E, located on section 26. Each of these basins were unlined and constructed by the placement of dikes on the lower end of natural surface drainage features. No engineering studies were conducted prior to the construction of the new basins. Basin C is the most permeable of the basins. Hydrazine, a Rocket fuel, was manufactured at the South Plant between 1954 and 1964. While the manufacture of nerve gas ceased in 1957. the Army continued to use the North Plant to distill and load munitions with nerve gas until 1969.

***** Following the natural drainage of first creek by diking at the lower ends. The sands and gravel bed for the underground water begin at less than 20 feet, that is what made these basins so permeable the toxic pollution was flowing right through the sands and into the subsurface water aquifers. There for every well to the north and northwest of the Arsenal were heavily contaminated with very toxic pollution. Had the required hydrology or geologic investigation been done by the Army, it would not have allowed Basin A in the first place, or any of the basins after that. The Army had not followed Colorado Law when these basins were constructed. The laws were in place at that particular time period.

***** 1950's CONTAMINATION. The first complaints of contaminated water were made in 1951 by farmers irrigating land northwest of the Arsenal. The Army admits, and acknowledges liability for the contamination and made settlement to the claimants.

***** CONSTRUCTION OF BASIN F. The Arsenal hired an outside consulting firm in 1955 to adivise it on the pollution problem. It contained a recommendation which resulted in the contstruction of what became known as Basin F. Basin F. was only considered temporary, with a life time expectancy of on 10 years. The liner was ¼ to ½ inch thick blown asphalt. It covered 93 acres, and designed to hold 180 million gallons of liquid, with a overflow maximum of 240 million gallons on liquid. relying soley on solar evaporation. Since the inflow exceeded the anticipated rate of solar evaporation, other methods were considered. In 1956 the first groundwater study was to be conducted in the area. The 1956 study predicted that the northeastern boundry of the contaminated area would be approximately two miles from the Home place, by 1957 report by the University of Colorado showed the contaminated area as being 1 and ¾ miles from the Home place.

***** Basin F was filled in 1957 by draining liquids from Basin A. While this was occurring, wave action created a tear 1,520 feet long in the liner. Liquids in Basin F were pumped into unlined Basin C and the tear was mended.

***** BASIN A CONTAMINANTS; Sixty million gallons ultrahazardous toxic liquids had been introduced into Basin A. Among the toxic materials placed into Basic A 300 tons lewisite, 500 tons of mustard, 30 to 50 tons of hydrofluoric acid, 400 tons of sodium fluoride, 183,000 pounds mercuric chloride, and 30,000 pounds of elemental mercury. Nerve gas which could not be brought up to military specifications was dumped into lime pits in Basin A. Pesticides and Herbicides also placed into basin A.

***** RESERVIOR FLUSHING: After basic F tear was mended, the Army was going to pump the millions of gallons of waste they had placed into basic C. and found none. It had leaked millions of gallons of toxic waste into the aquifers. So the Army quickly started what they called reservoir flushing. They pumped clean irrigation water into Basin C. The records state the purpose of such water was to "FLUSH" and "DILUTE" the polluted aquifer. Such waters also had the effect of driving pollutants well beyond the boundry of the Arsenal and increasing the height of the water table. Records did not reflect when or whether this practice ever ceased, but does reflect that water was in evidence in Basic C as late as 1974, a leak at the South plant area provided a mound of fresh water which served the same purpose.

***** DEEP WELL: A deep well was drilled in 1962 to inject waste into fractured pre-cambrian granites at the depth of 12,000 feet. Basin F was modified in November 1964, by constructing a dike to isolate a portion of the south end to create a separate and smaller basin known as Basin F-1 to segregate liquids coming out of the industrial sewer for injection into the deep well. Expenses of operating the deep well were greater than expected. Regular use of the deep well was discontinued by mid 1973 because of the high treatment costs. The deep well was no longer used after 1966. When it was concluded that its operation was causing earthquakes. Alternative means of disposal other than basins were considered to be uneconomic.

***** Spray Raft: A spray raft was placed into Basic F in mid-1963 to increase evaporation by pumping Basin F liquids well over 100 feet into the air. An Army observer stated in 1965 the odor of Basin F was as strong on the north boundary outside the Arsenal as it was on the banks of Basin F. The height of the spray, 100 feet, combined with the liquid loss creates a presumption that aerosol droplets were carried some distance. A 1977 report reflects that the use of the spray raft was abandoned because of odor complaints, after nearly fifteen (15) years.

The Land family had complained that the odors coming from the Arsenal beginning in the 1950's was so strong it would make you vomit. Lois Land called the Arsenal to complain about the odor in 1968 and Arsenal personnel hung up on her. The Land family attributed the aerosol droplets of deadly Pesticids, Herbicides, Mustard, and lewisite agents and then the effects of GB Nerve agents (Sarin Gas) is what caused the death of the family fruit orchard and other trees on the home place, garden, horses, livestock, even the dogs, and eventually Mr. And Mrs. Land. This was a virtual shower in nerve agents.

***** A test presumably done by the Colorado Department of Health on the water in the Land wells, showed 238 ppm Chloride it is considered to be presumptive evidence of contamination of the water, by pollution from the Arsenal at 200 ppm would one would presume 3 to 5% of that water is Arsenal pollution. At the time this was done in 1972, there were no other tests available for detecting what may be in the water. This test showed nearly 5 to 7% of the Land well water was pollution from the Arsenal.

***** It was brought to Judge Robinsons attention, as the Government stoutly defended the action of the hearing officer to require attorneys' appearing before him to be a member of the bar of the Court of Federal Claims. The Clerk of the Court had advised that

Matthew J. Regan and Charles Quinlan, III both Justice Department Attornie's representing the United States, either one were a member of the bar to be practicing before the United States Court of Claims.

***** Mr. Reagan Attorney for the United States, states with regard to a fact that is not in dispute here. The United States has stipulated in the past and do not back away from that stipulation, that the groundwater flows North from the Arsenal and in fact the Contaminant DIMP came from the Arsenal, and the Contamination DIMP was found in Plaintiff's Larry and James Land's well water's.

***** Mr McClaren Attorney for the Plaintiff's, The Plaintiff's do not accept the limitation upon our case that the government attempts to put on it, that it is DIMP, and only DIMP.

***** Judge Robinson stated he understands fully, as if he would not let that happen.

SUMMARY OF THE ARGUMENT: A Congressional Reference confers jurisdiction on this Court to determine whether the Plaintiffs' claims are based upon the existence, (a) of a legal claim, (b) of an equitable claim or (c) as a gratuity. This Court has ruled that a claim based solely upon some moral principle without a showing that the government committed "wrong" would constitute a gratuity. A legal claim is one where the government has waived its sovereign immunity and the claim falls within the scope of such waiver. An equitable claim is one where the plaintiff would prevail if the litigation were among private parties. Equitable claims are based upon state law unless preempted by federal law.

***** Colorado law requires the owner of the property on which a manufacturing process is being operated or a hazardous activiity is being conducted to restrict polluting materials to the property. Failure to restrict such materials constitutes a breach of affirmative duties. Liability results where the escaping materials cause injury with or without a showing of negligence. A polluter of waters within the State of Colorado is liable for damages resulting from the polluted waters even though there were others contributing to such pollultion.

***** JURISDICTION: Legal remedies are usually based upon either of two statutes in which the United States has waived its sovereign immunity. The more common remedy is 28 U.S.C. *2509 commonly known as the Federal Tort Claims act. ("FTCA"). The FTCA confers jurisdiction to the district court to determine negligent acts of employees of the government. The second remedy is 28 U.S.C. *1491 commonly known as the Tucker Act. The Tucker Act confers jurisdiction in this Court for claims based upon the constitution and other matters not sounding in Tort.

***** A Congressional Reference is a totally independent source of jurisdiction for claims against the United States. Much of the reasoning in Defendant's brief utilized in reaching the conclusion that plaintiffs should be required to show negligence on behalf of the government in the present case is based upon reasoning behind the FTCA. Jurisdictional limits under FTCA cases have no application to a Congressional Reference. No specific response to arguments in Defendant's Brief relating to the FTCA cases will be made except to note that the Defendant's underlying premise is in error.

***** Applicable Law: "Put otherwise, it has been said that the test of 'equity' is whether the claim asserted would be recoverable against a private party."

***** TRANSPORTATION OF CHEMICALS FROM THE ARSENAL: Dr. Waddell expressed the opinion that all of the known compounds, excepting elemental mercury, in Basin A could be transported to the shallow aquifer groundwater under the property of the "Land's home place. This opinion is confirmed by the 1975 testing which found DIMP and Dieldrin on the Home place. The government stipulated that other chemicals could be transported to the home place. With the presense of DIMP you also have the associated presence of sodium fluoride, and extremely toxic substance, and the other materials associated with the manufacture of nerve gas. With DIMP comes the associated presence of the compounds identified by Dr. Teitelbaum in his work for the plaintiffs in Daigle. Dr. Waddell testified that water which passed under the southern portion of the Home Place would pass under the northern portion where James Land had his well.

***** Not only does the evidence show that it is probable that the toxic compounds were transported from Basin A on the Arsenal onto and under the Home Place, it also reflects that there were no sources of contaminants in the groundwater excepting those located on the Arsenal. This was an important element of the circumstantial proof. The Arsenal had a test program called the 360 degree program. The program tested water as it came onto the Arsenal as well as when it left the Arsenal. The contamination to the North were directly linked to basin A, and basin C which were the major source of contamination of the groundwater under the Home Place. There was no other means of contamination on the northern boundry of the Arsenal before it reached the Home Place. Let's remember the Arsenal was using basin C over 10 years to flush contaminants from beneath the Arsenal. This was stamped top secret and classified by the Army and the Pentagon.

***** ILLNESS AND DEATH: The final element to the circumstantial case is whether or not there was illness and death associated with the contamination. Prior to the construction of the Arsenal the Land family lived on the Home Place, raised animals and conducted dry land farming without problems while raising a family of ten (10) The first indications of illness were in association with the air contamination in the latter part of the 1950's and during the 1960's. The worst of the problems of buring eyes and nausea were at the same time that the Arsenal was using the spray raft to help increase evaporation of Basin F. The liquid of Basin F. were shooting sprays of the deadly toxins one hundred to one hundred fifty feet into the air. At times you would feel a light mist, and no clouds in the sky, and when the mist began stinging and irritating your skin, burning your nose and you would have trouble breathing also nausea and shaking, trembling.

***** During the 1960's caused a great deal of illness for Mr. And Mrs. A. C. Land. We often wondered what was happening, deep down we probably knew what was happening. The Land's had a beutiful Orchard, Apple's Plumb's Blackberries, Cherry trees, Peach trees, some of these trees had been there since the 30's, but during the 60's and the Arsenals spray raft, it killed all the trees in the area. Every one living in the area north of the Arsenal, during the 1960's became quite ill. The death rate of cancers in that area to the north, northwest, and northeast during the 70's and 80's skyrocketed, per-cap-i-ta (per unit of population) they had the highest death rate in the World.

***** Contrary to the findings in the report, Plaintiff's evidence of air contamination was that the worst of the air contamination occurred prior to 1972. Basin F remained a source of air contamination after removal of the spray raft whenever the wind was from the south. Monitoring devises installed for project Eagle were not designed to monitor the compounds identified as escaping from Basin F. in 1988

***** then there is the evidence of the cattle, they became ill and died without regard to their age or their source. The cattle from the box-elder Farm were moved to the new place prior to most of the others. Even before drilling the new wells they were showing signs of becoming ill. Larry lost probably three of the smallest calves before having his new well drilled by April 12, 1972. At that point in time new animals were bought and began arriving, Most of the cattle were at least 12 to 14 weeks old and had been weened at 7 weeks. With in two days of having the new well drilled the pregnant heifers began to abort all their calves, most of them were 8 months along, none of the calves lived. Then the larger animals showed the common signs, red blistered eyes, sores in mouth and nose, ears drooping and heads hanging low, severe diarreha, Vomiting. staggering, difficulty getting up, huge abscesses, teeth turning black and falling out, inhability to eat, lap at the water, show signs of difficulty breathing, then would fall over unable to get up, go into convulsions, with the legs kicking in rythem to the eyes that were pulsating, then death soon followed. Trying hard not to believe what happened James Land Larry's brother had bought a group of cows carrying calves. His well was located at a different place than Larry's, but within days of moving the cows to the place they started aborting all calves and they themselves were starting to show the signs of being poisoned. Jim took them completely off the water and began hauling in water in huge tanks from a well far from the Arsenal. he managed to save about 5 calves to full term, only losing one cow. They immediately improved and were able to stop lapping at the water.

***** After Larry's cattle were moved to a different property the deaths ceased because the water was changed, the cattle immediately improved healthwise, the sores in the nose and mouth disappeared, their eyes cleared up. the breathing was not labored, they stopped staggering, the heads and ears came up to normal. Everyone had worried somewhat that they had shipping fever or pneumonia, or Hemmoragic Septicemia, which are very contagious diseases in cattle. The gentleman who trusted Larry and allowed him to use his corrals, were next to his nearly 300 head Dairy, the animals had only a barbed wire fence between them. So if it were contagious he could lose his etire heard also. Larrys. animals all got better and none of his cattle ever got sick. The Government had been trying to call it pneumonia, but it was not. Larry had lost over half of the 635 head of livestock to the poisons, and was rather happy that at least he saved some of them. However while they cease to die, and appeared to behave as young animals often behave, they ceased to grow. what ever had caused the illnesses and deaths had stopped their growth.

***** Larry and his family, his brother and his family, and his parents, the man working for Larry and his family, were able to get away from drinking that water. Their health began the long road to improvment for some, but Larry's Mom and Dad were both dead before their time, they didn't make it through the 1970's, Bill Phinney the man working for Larry, died before he was 42, Larry's wife suffered a miscarriage, and it would take years for them to recover with disabilities.

***** The only expert witness who testified that he had witnessed the illnesses and deaths of these cattle was Dr. Scott, a veterinarian who performed necropsies on many of the dead animals. It was DR. Scott who tried various treatments and ruled out all illnesses except those associated with drinking the water. It was this expert, who testified that it was toxic substances in the water that caused the illnesses, deaths and the resulting stunting of the growth patterns of the survivors. His opinion is no less valid even though he was never able to identify the toxicant in the water. Such identification, while critical to a single coumpound toxic case, is not necessary to a circumstantial case. The report of the necropsies that were done by the Colorado State Uniyersity, the pathologists diagnosis being pneumonia, and was most probably secondary to other unknown causes leading to a general disability. This finding supports the opinion of Dr. Scott that it was the toxic substances that was the primary cause of illness and subsequent deaths.

***** The circumstantial evidence is buttressed by the absence of algae and fish dying in the stock tanks, dead insects and oily film on the water, the dead and dying birds atound the stock tanks, and the fact that water would consume solder joints but not the galvanized iron of the sotck tanks and the fact that water would destroy concrete that was in contact with it. The livestock were seen lapping at the water as if to say they did not want to drink it. Each of this facts are sufficient of themselves to establish contaminated water.

***** In 1975, the CDH confirmed that pollutants were leaving the north boundary of the Arsenal and issued a Cease and Desist Order against the Arsenal. Since that time, Arsenal officials have spent millions of dollars building a containment system to treat pollutants escaping from the very portion of the North boundary of the Arsenal through which the contamination flowed from Basin A to the Home Place. The containment system built after 1975 was very similar to the one designed by the Corps of Engineers in 1961.

***** Dr. Teitelbaum, recognized as an expert in toxicology, testified that the illnesses of each of the members of the four families was the result of exposure to toxic substances both in the watter and in the air. In the damage portion of the bifurcated trial, Dr. Teitelbaum will suggest identification of some of the toxic substances and will testify as to the effect of each such substances upon the health of the plaintiffs.

***** Mr. Regan [attorney for the Defendant]: Stated yes, your Honor. "I just wanted to state with regard to Mr. McClaren's comments, it is not in dispute, and we have stipulated in the past and do not back off that stipulation, that groundwater flows north from the Arsenal and in fact the contaminant DIMP came from the Arsenal, and the Contaminant DIMP was found in plaintiffs' well and well water."

***** Larry says “Mr. Regan just identified part of the pollution in the Land wells is fingerprinted directly to the Arsenal.”

***** “So to the extent of the relevance of this testimony as to developing where it flowed, I see no point for that, and maybe it’s duplicative.” “The government admits the polluted water flows to the North of the Arsenal, under the Land’s property, and in fact DIMP found in the Land well water proves beyond a shadow of doubt that pollution from the Arsenal is in fact in the Land well’s.

***** “The Court recieved an exhibit from the Plaintiff, and placed into evidence, that the finding of DIMP in the Land wells is prima facie evidence that other pollutants from the Arsenal are also in the Land Wells.

***** The plaintiff’s Attorney Sam McClaren sternly reminded the Court, that we, do not accept any limitation upon our case that the government attempts to put on it, that it is DIMP only and only DIMP.

***** The Court Judge Robinson stated twice from the Bench “OH I UNDERSTAND, I UNDERSTAND.”

***** GOVERNMENT’S DEFENSE. The governments proof on DIMP is irrelevant for any one of three reasons. THESE REASONS ARE; (1) What the defense presented applies to a single compound contaminate and has no application to a multiple compound contamination proven by circumstantial evidence; (1) Improper proof was made on the level of contamination; and (4) The incorrect standard of the level of DIMP pollution was utilized. The facts and correct authorities reflect that DIMP may well have been one of the compounds which contributed to the illnesses and deaths.

***** SINGLE COMPOUND DEFENSE; A single compound defense such as DIMP ignores that the Colorado Law uses Circumstantial evidence that will establish multiple contaminants.

***** In 1972 the Colorado Department of Health found in the Land wells the product they had used for 30 years to determine the range of contamination from the Arsenal in any well, and whether or not it was contaminated by the Arsenal. Chloride was the material used, at the 100 to 200 PPM chloride in the water would suggest Arsenal pollution. Land's well tested at 250 PPM. Chloride in 1972 when the livestock and people got sick. This information was secreted from the Lands by the Health Department. Their water contained a minimum of 6 to 7% Arsenal Pollution. This means 6 to 7 mililiters of toxic contamination in every liter of water, a person working in the hot sun may consume easily 3 to 4 liters of water, meaning they were consuming 24 to 28 mililiters of Ultra Hazardous Toxic pollution every day. The live stock consuming double to triple that amount on a daily basis. Where ever one would find Chloride DIMP would surely be present along with small amounts of 100's of other chemicals from the Arsenal.

***** The government's own reports reflect that cattle were killed by the contamination in the 1950's before DIMP existed. The identity of the contaminate that killed the cattle in the 1950's remains unknown to date. The Multiple contaminates which have been identified by the Arsenal are but a few of the thousands of the compounds present which entered the groundwater by seeping out of unlined Basin A in the 1940's and 50's. Dimp is but one of the contaminates that was unknown in 1972 which has been subsequently identified. Whether DIMP was prsent in sufficient quantities to cause injury is irrelevant unless the government can exclude all other contamination in the shallow aquifer under the Home Place. The additional contamination as well as contamination by DIMP has been admitted by the Defendant.

***** The government's defense was based upon the assumption that plaintiffs must prove a specific compound to prove causation of illness and death in a toxic tort case. However the Judge explained to the government attornies more than once, "There is no room for the FTCA in the Court of Claims" Cases previously cited reflect error of this assumption. The government then assumed that DIMP is the sole contaminate. The government's own records reflect the error to this assumption. Assuming without agreeing that the government did prove that DIMP was not the cause of the deaths and illnesses, such proof does not prove that DIMP could not have contributed to death and illness in the multiple coumpound toxic "Stew" found in the groundwater under the Arsenal, in the groundwater under Land's, and in their wells.

***** Compounds, which can and do cause deaths, illnesses and other injuries to people and animals, were present on and under the property of the Land's and in thier ground water in addition to DIMP. The identity of these compounds will be commented upon more fully in the damage portion of the bifurcated trial.

***** Even if DIMP had been the sole contaminant, the use of testing results in 1975 under conditions very different from those in 1972 was not the proper means of determining the level of contamination. The importance of whether the gravels and cobbles representng the paleostream were saturated with shallow aquifer groundwater has alreaady been discussed, and understood by this Court. Even if such gravels and cobbles had been saturated, the 1975 water samples were diluted by water from an uncontaminated deeper formation and waters because of the difference in the rates of pumping in 1972 and 1975. The pumping rates in 1975 were not all that typical of the pumping rates of 1972 when there were hundreds of cattle present, and Larry was pumping the wells as told in an attempt to clean them up. The third difference is the fact that by 1975 the shallow aquifer contamination had moved three thousand feet since 1972. In addition, the use of Basin F would have reduced the degree of contamination as well as the level of the aquifer.

***** The Armie's use then also of basin C. where they had pumped into it non contaminated irrigation water with no other reason than to flush and dilute the contaminants disbursing them further out. This made it impossible to replicate the 1972 conditions by 1975 testing.

***** In this case subsequent testing is insufficient to overcome a circumstantial case. Where there is evidence that Pollution ran north from the Arsenal under the Land property, certain toxic chemicals were found to be in the Land water, by the Army, Shell Oil Company, and the Colorado State Health Department. They then abandoned all testing of the Land wells, while they stated there is no way in Gods green earth could chemicals get from the Arsenal to your well, Statement by Mr. William Dunn of the Colorado State Health Department. Larry says "he often wondered how the army, and Shell had got by with polluting 35,000 acres of Land to the North of the Arsenal, and in fact all the way to Brighton Colorado, for so many years, He believes Bill Dunn did secret the truth of what was happening." You see he was the cheif chemist for the Health Department and was in charge of the monitoring program. "It is believed the truth was purposefully covered up by this man." He had got by for so many years With the Army, and Shell Oil Company, saying essentially the same, When they knew first hand Arsenal pollution was found in the Land wells.

***** Colorado Law provides two alternative methods to prove causation in a toxic tort case. The first method involves showing that a single toxic compound was present in sufficient concentration to cause injury. The second method involves circumstantial proof of the probability of the presence of one or more contaminants, which may or may not be identified, which are shown to have probably caused or contributed to the cause of injury.

***** The government defended their entire case admitting that chemicals from the Arsenal were found in the Land wells. (AGAINST the LAW of the STATE OF COLORADO, they used a quantitative analysis set up by EPA at 600 ppb, for DIMP was allowable, and clearly was not the safe standard 8 ppb used by Colorado which the Land's were using. EPA did not have any right in Colorado to set a standard for anything, they had to go through the State Health Department for any Standard. EPA had testified to a standard that did not exist in Colorado. In a Court of Law that was suppose to be hearing the case under Colorado Law, not the Armie's Law or EPA's Law. Larry says "how can they do that in a Federal Court of Law, yet the Colorado Health Department, nor the Attorney General's office stood up to say you cannot do that in this State.

***** Undisputed facts establish that defendant has contributed too pollution of the groundwater underlying the land of the Plaintiffs.

The fact that pollution ahs been admitted by the Defendant while it contests the quantity and quality of such pollution as such, Plaintiffs have under Colorado Law established an equitable claim in proving causation as a fact in this trial, and for the all damages resulting from the polluted groundwater.

***** Judge Robinson stated undisputed facts that "Jurisdictional limits under FTCA cases have no application to a congressional Reference. No specific response to arguments in Defendant's brief relating to the FTCA cases will be made except to note the the Defendant's underlying premise is in "Error."

***** Judge Robinson also made it very clear in his Court the plaintiff has established an equitable claim in this reference, The defendants have an obligation to confine contaminants, it is absolute, and no negligence is required for liability, and this contamination was not confined by the defendants.

***** One accurate means to determine the level of contamination DIMP in 1972 would be what is characterized as chemical fate and transport testimony. This analysis involves a determination of the level of contamination at its source and a computation of the amount transportd In this case, the transport calculation is to the Home Place. Dr. Waddell confirmed the flow of contaminants from Basin A and C to the shallow aquifer groundwater adjacent to wells located on the Home Place. Defendant admitted that chemical contaminants entered the groundwater on the Arsenal and that such groundwater flowed under the Home Place. The admission together with Dr. Waddell's testimony provided the transport portion of the equation.

***** The chemical fate portion of the equation is supplied by Dr. Walton in his 1960 report. The criteria for testing established in 1960, utilized by the HEW report in 1965 and still in effect in 1972 was that the presence of water with chloride level in excess of 200 ppm was evidence that 3 to 5% of the water had its source on the Arsenal. Plaintiffs' Exhibit 43 AR is a November of 1972 water test on Land well #1 which reflected a chloride level of 235 ppm. This number is conservative since the November, 1972 rate of pumping was the reduced rate resulting in dilution of contaminated water.

***** Application of the Walton criteria on the level of contamination results in the determination that DIMP concentration in the shallow aquifer on the Home Place in November of 1972 was in the range of 1,440, to 2,400 ppb. The reliability of this calculation is proven by the fact that on May 20, 1974, a well up gradient to the Home Place, located at 9760 Peroria, one half mile south of the Home Place and one-half mile north of the Arsenal, tested 2,200 ppb of DIMP. Further support for this calculation is found in testing in the wheat field where tests of water from three shallow wells within feet of each other showed test results 310 ppb, 4,960 ppb, and 6,000 ppb. These test results are all from the shallow aquifer within the same paleo valley adjoining the Home Place. Application of the chemical fate-transportation calculations results in determining Dimp concentration on the Home Place in 1972 of two to three times the EPA's safe equation, of 600 ppb, and 2,250 times greater than the maximum safe standard of 8 ppb allowed by the Department of Colorado Health Department. Should be enough to compel this Court to find the Plaintiff's satisfied their burden of proof of proximate cause even if the issue were to be decided soley on DIMP.

***** As a portion of the Courts decision, The Judge stated, for the record, "The need for oversight activities to enforce removal and remedial actions is particularly acute in situations in where the United States is a CERCLA defendant. If only the Enironmental Protection Agency were granted oversight authority, the State would be forced to leave the interests of its people in the hands of a law enforcement agency with an inherent conflict of interest. There is no reason to believe that the United States is immune from the conflicts that arise when a liable party is responsible for enforcing its own cleanup activities."

***** The comment has equal application to the EPA's conflict of interest in adopting a standard for Dimp. In this instance the federal agency [EPA] adopted a standard for cleanup of DIMP by another federal agnecy [the Arsenal] using DIMP studies prepared by the same federal agency [the Army]. For further background on the effect of statutes giving CDH jurisdiction over the Arsenal See United States v. State of Colorado, 990F.2d1565(10th Cir. 1993).

***** The Colorado standard for DIMP is 8 ppb. See regulations adopting such standards. The record reflects that at least one of Defendant's expert witnesses was aware of the Colorado standard, that EP could set no standard for Colorado. Defendant chose to ignore the existence of the Colorado standard. Application of the Colorado standard creates a presumption of causation.

***** THE DEFENDANTS DUTY TO CONFINE; The common law of Colorado requires that polluted water be impounded upon one's property or the polluter is liable for damages resulting from discharges into the waters of the State.

***** Liability of owners for damage. The owners of the reservoirs shall be liable for all damages arising from leakage or overflow of the water therefrom or by floods caused by breaking of the embankments of such reservoirs.

***** The supreme Court expanded upon this concept and stated that it was the injury which is really prohibited. The disposal is required to be in a manner which does not cause injury. The proof of causation even in a toxic tort case is proof of a wrongful act. Plaintiffs have gone further and have proven numerous wrongful acts including the failure to study the hydrology in the construction of Basin A, the failure in 1961 to construct the containmnent facility; The failure to comply with requirements of the Walton Report; The failure to comply with requirements of the HEW Report in 1965 and the failure to warn adjacent owners so that they could have protected themselves.

***** Abuse of Discretion. On or about August 8, 1995 the Hearing Officer requested Mr. McClaren to participate in a telephone conference with himself and the counsel of record. Since due to internal problems of the Land's Attornie's Harding and Ogborn suddenly moved to withdraw as counsel for the Plaintiffs. Mr. McClaren had represented the plaintiffs in 1974 through 1988 in obtaining a Congressional Reference. Once the reference was obtained Mr. McClaren recused himself from acting as trial counsel because of the necessity that he testify as a witness. Mr. McClaren agreed during the telephone conference to represent the Plaintiffs once again until they were able to obtain new counsel. The Hearing Officer made it quite clear that he considered it important that nothing be done beyond to delay the trial scheduled for January 22, 1996 in Denver Colorado. Mr. McClaren noted at that time that it might be necessary for him to act as trial counsel and to testify, as an expert in Hydrology, Geology, and an expert working with chemicals. The Hearing officer suggested that an effort be made to find additional witnesses to minimize such testimony. The Hearing Officer also suggested that a statement could be taken in lieu of cross examination.

***** Pursuant to the instructions of the Hearing Officer, Mr. McClaren signed a joint status report dated September 5, 1995 although he had neither previously entered an appearance nor had he been admitted to practice before the Court of Federal Claims. The Hearing Officer was advised that Mr. McClaren was seeking the consent of the Plaintiffs for him to enter an appearance. Subsequently, Plaintiffs determined that it was impossible for them to obtain competent counsel within the time constraints of being prepared for trial on January 22, 1996. Under these circumstances, Mr. McClaren agreed to act as trial counsel since his failure to do so would work an extreme hardship upon the Plaintiffs' even though it remained necessary for him to testfy.

***** With this in mind Mr. McClaren immediately registered with the Bar to allow him to practice before the Court of Claims, so there would be no incident at trial. Under date of January 17, 1996, only four days prior to trial, the Hearing Officer announced his ruling on all pending motions. The motions filed by the Plaintiff under date of November 22, 1995, their Appendix G filings together with a motion to designate Dale Wingeleth as an expert on chemistry. Under date of December 20, 1995 Plaintiffs' filed a motion seeking to clarify that Mr. McClaren could also testify as an expert. The Plaintiffs had previously designated Mr. McClaren as a witness but had not specified that he would be offering opinions as an expert. All of Plaintiffs motions were vigorously opposed by the Defendant.

***** Plaintiffs' motion to add Dr. Wingeleth as an expert was denied. Plaintiffs' motion to allow Mr. McClaren to testify as anything but a fact witness was denied. No mention was made of any particular requirements in allowing Mr. McClaren to testify.

***** On the third day of the trial, Plaintiffs announced their intention to call Mr McClaren as a witness. Plaintiffs announced that Mr. McClaren's testimony was to be obtained under a question and answer format with Mr. McClaren asking the questions as well as answering them. The Hearing Officer stated that Mr. McClaren could be a witness only if questioned by another attorney and that attorney had to be admitted to practice before the Court of Federal Claims. Finding an attorney in Denver, Colorado in a brief period who is admitted to the Court of Federal Claims is nearly, if not impossible. For this reason, Plaintiffs had no other choice but were forced by the Court to proceed to make the best of a bad situation.

***** Not only were the requirements of the Hearing Officer a complete surprise to the Plaintiffs, and their Attorney Sam McClaren, they were unduly burdensome in that an attorney admitted to practice before the Court of Federal Claims was required. This requirement of the Hearing Officer was contrary to Rule 13 of Appendix D of this Court's rules which allows attorneys admitted before the highest Court of any state to practice before the Court of Federal Claims in Congressional Reference Cases. All Motions were filed in ample time to allow the Defendant to prepare for trial.

***** At page 29 of the Government's Brief, the Government stoutly defends the action of the Hearing Officerrto require attorneys appearing before him to be a member of the bar of the Court of Federal Claims. Plaintiffs have been advised by the Clerk of this Court that Justice Department Attornies representing the United States of America, Mr. Matthew J. Regan, and Charles Quinlan, III, each of whom has participated extensively from the beginning and in the January 22, 1995 Trial, as well as all pretrial matters, the Summary Judgment Trial, were never and have not been admitted to practice before this court. Sam says, "Apparently, there is a different standard for Plaintiff lawyers, than for Government Counsel.

***** This Court should in the exercise of its discretion, and in the interests of Justice and the peculiar circumstances of the present situation, grant Plaintiffs' motions to add Mr. McClaren and Dr. Wingelith to their list of experts. *****The Extreme sanction of exclusion of expert testimony is not only a hardship but is not warranted. The government has no basis to claim surprise or prejudice in the sense required for sanctions. The Government should be allowed to take more depositions if it should desire. The week prior to trial would have been an appropriate time to do so.

***** The Attornie's for the Defendant Mr. Matthew Regan, and Mr. Charles Quinlan III, stated before the Court. "Surely it is no stretch to assume that a licensed attorney is aware of the general position that in order to appear as an attorney in a proceeding before a particular court, one must be admitted to practice before that Court as a member of the Court's bar or pro hac vice. Thus plaintiffs had adequate notice that they would be required either to locate a local members of the Court's bar or to have a local attorney admitted to the bar pursuant to 'Appendix D, Par. 13; there is no record of representation that they attmepted to do either. Plaintiffs' Brief ignores the fact that Plaitiffs did not even have any attroney present to do direct examination of Mr. McClaren. Therefore portraying the necessity for (and the decision to exclude) Mr McClaren's dual persona testimony as turning upon the requirement that the questioning be conducted by a member of this Court's bar is disingenuous. However Defendants Brief totally ignores their responsibility, as Federal Attornie's they should be aware one must be admitted to practice before the Court of Claims as a member of the Courts Bar.

***** While it is answerable that the Plaintiff's had already stretched their financial strength, and had decided to have Mr. McClaren ask the questions and answer the questions and the Judge had been agreeable to that if necessary but then denied that at the last minute as any possibility, causing Plaintiffs' a severe loss of expert testimony. Truly this is extreme sanction was not warranted, and beyond common sense.

***** Plaintiff then notified the Judge was that neither Mr. Matthew J. Regan, Mr. Quinlan III, Mr. Frank Hunger, Mr. J. Patrick Glynn, Mr. J. Charles Kruse, nor the Attorneys of record for the Army Major Jonathan Potter, or Captain Thomas Cook, were "not" members of the bar as ordered by this Judge to practice before the Court of Claims, and could not possibly act pro hac vice.

***** PLAINTIFFS' "NEED" FOR EXPERTS; Toxic tort cases may "need" to be proven by experts. While the present case is similar to a toxic tort case it is a water pollution case. Moreover, it is a water pollution case with a unique twist. The source of the water pollution was on a government installation where access was denied to the public. Not only was a cloak of secrecy placed upon the government's activities, such mantle was extended to the activities of its lessee, Shell Oil Company.

***** Plaintiffss' principal, and in most instances—sole source of information, is either from government reports or business records maintained by the Department of the Army in connection with the Arsenal. Most of these records were not available to the Plaintiffs in the 1972 to 1975 time period. Although most of these records were prepared by the government's own experts, they lack connection to the Plaintiffs' property. The principal use of Plaintiffs' experts will be either to direct the court's attention to some scientific fact about which it should take judicial notice or to utilize the information in the report to direct the court's attention to a conclusion which the government's experts have failed to include.

***** Further checking with the Clerk of the Courts, after the Trials have concluded we find that Mr. Matthew J. Regan, and Charles Quinlan III were in fact Members, They joined the Bar of the Court of Claims, However this was not done until they were well into the trial, through all discovery, summary Judgment, and other pre-trial matters. In other words knowing everything they know must do before becoming members of the Court of Claims. As for J. Charles Kruse it has been understood from the Clerk of Courts that he is not a Member of the bar to practice before the Court of Claims, He is the special litigation counsel, for the Torts branch and the civil division making it mandatory that he be a member of the bar in order to practice before the Court of Claims. He signs all litigation, in and out of the Court, and takes full responsibility of all legal documents that go before the Court of Claims, and for his trial counsel, his stand by counsel of the United States Army. As for the remaining two counsels, Frank W. Hunger, Assistant Attorney General, and J Patrick Glynn Director of the torts Branch, who we understood were not members of the bar for the Court of Claims. However in asking the Clerk of the Court for information on these five attorney's status, it was then denied and Larry was told if he wanted the information he may write to see if he could have it but it would not be released over the phone. Mr. J Charles Kruse, Mr. Frank Hungar, Mr. J. Patrick Glynn, nor the Attornie's of record United States Army, Major Jonathan Potter, and Captain Thomas Cook. It is their Title and it was up to them to display that title to the Court and the Plaintiff, however they did not one way or another so it will be assumed they were not members of the bar of the Court of Claims, Apparently there is a different standard for Plainntiffs' lawyers than for Government counsel.

***** This is exactly what the government does with all information right or wrong, public information or not, they have a tendancy to conceal everything." Larry Says, "he hopes this helps to explain the Dilema he was in when his cattle were dying, The family inured and dying, to this day the Army has secreted and classified Data that may have helped the Doctors, in learning. the names of chemicals that were in the Land wells besides DIMP. which was finally discovered in 1975 long after the fact and the governments trying to blame the entire problem on DIMP. By using a standard Colorado law does not and will not accept. Yet the Court used the Army's, and EPA's equation of 600 ppb of DIMP, The Plaintiff was not allowed by the Court to use the Standard that is the Law of the Land, State of Colorado, .008 ppb. The Court knew, but ignored the fact, and the Judge would not allow the Colorado Standard to be placed as evidence.

***** In addition to the contamination of the Plaintiff's ground water, the air was contaminated by nerve gas emissions, during the 1950's, 60's and 70's, and to include the Pesticide emissions from

Shell Chemical Company, and the emissions from all the surface storage many hundreds of millions of gallons.

HAZARDOUS SUBSTANCES RELEASED BY SHELL OIL COMPANY AND THE UNITED STATES ARMY AT THE ROCKY MOUNTAIN ARSENAL: DENVER COLORADO

***** It was very clear stated George Donnelly, the man who was in charge of the RMA's Engineering Department from 1954 to 1967, then too RMA's Facilities Division from 1967 to 1976. "THE DEFENDANT, UNITED STATES ARMY, AND SHELL OIL COMPANY KNEW BASIN F, WAS LEAKING CONTAMINANTS OFF-POST BY 1969. ["TRANSCRIPT SWORN TRIAL TESTIMONY BY GEORGE DONNELLY"].

***** Available records show that sixty-four domestic, stock and irrigation wells in areas northwest of the RMA were contaminated by wastes from the RMA by December 1965, nevertheless, Defendant's off post monitoring conducted from 1960 through 1964 did not test for the presence of these contaminants. Chlorinated pesticides, Aldrin & Dieldrin. [Trial testimony by B. Anderson].

***** The groundwater monitoring program conducted in response to the 1950's contamination was stopped in November 1964, despite the fact the U. S. Department of Health and Welfare Ground-water reports states that the water well monitoring programs should be continued. No ground-water testing, either on-post or off-post was conducted between 1964 until approximately 1973-1974. [Sworn trial testimony by B. Anderson.]

***** Larry says, The Army was asked by the U. S. Health and Welfare Department to stop all surface storage, and especially Basin F. as it was leaking toxic chemicals off the Arsenal. They again warned them in 1965 that Basin F was leaking, But then by 1969 the Army Hygiene Department, The U. S. Health Department both warned them they were using Basin F. on the premise it was in fact polluting the Countryside both to the North and North West of the

Arsenal. Then two Generals from the Pentagon came out to inspect the problem, and soon discovered at least part of the protective liner in Basin F. was missing disolved. "The two Generals then had all material regarding Basin F. marked Classified, A statement was made by one we sure as hell don't want the public to find this out. Then they continued using the Basin, for an additional 15 years. The information on Basin F remained classified until 1989, when parts of it was released.

***** Basin F was originally constructed with a life expectancy of 10 years, So when the Army after 10 years discovered it was leaking and was in fact polluting the Countryside, they continued using it another 15 years under the premise it was leaking. The Basin was originally built to hold 180-million gallons of liquid waste safely. However after the Army had found out Basin F was in fact leaking, they increased the capacity to 250-million gallons putting the Basin at a dangerous over flow point, then called on the Pentagon to classify everything, this was including all their mistakes and negligence for the last sixty (60) years.

***** This sounds like a huge bunch of unruley people who cannnot agree on anything or do anything right. I've heard of Imp-pro-vise, but not leave a trail of devistation in your wake. I'm still trying to find out what they were thinking when they added an additional 70-Million gallons to the Basin that was already leaking, and past its life expectancy. [Do ya think they thought maybe it would make it stop leaking]. How does one call that type behavior the United States Army, rattling the doors at the Pentagon in Washington D.C. I'm truly ashamed of their behavior, I hope you are also.

***** PLAINTIFF'S ARGUMENT AT TRIAL; I. IN ORDER TO ESTABLISH THEIR EQUITABLE CLAIMS IN THIS CONGRESSIONAL REFERENCE ACTION, PLAINTIFFS NEED ONLY SHOW SOME UNJUSTIFIED GOVERNMENTAL ACT, EITHER NEGLIGENCE OR SOME OTHER TYPE OF TORTIOUS CONDUCT ON THE GOVERNMENT'S PART, WHICH CAUSED DAMAGE TO PLAINTIFFS.

Defendant acknowledges "Plaintiffs' claims in this Congressional Reference action are based on alleged Tort liability." Defendant's Brief at p. 1 (emphasis added). Nevertheless, Defendant stubbornly clings to its myopic view that the liability element of an equitable claim can only be established by a showing of negligence.

Defendant has failed to recognize that negligence is but one of a myriad of types of tortuous conduct which can result in tort liability. Defendant has also failed to recognize that Plaintiffs' Amended Complaint seeks recovery on the basis of five different Types of tortuous conduct by Defendant, stating claims for: (1) Negligence; (2) Trespass; (3) Nuisance; (4) UltraHazardous Activities; (5) Fraud/Deceit.

Defendant's meritless assertion that negligence is the only type of wrongful conduct which can give rise to liability in a Congressional Reference action is directly refuted by numerous claims Court decisions on the showing sufficient to establish governmental liability. In a very recent case the Court held that the Plaintiff was entitled to recover from the government, even though the plaintiff made no showing of negligence. Rather, the Court found the government to be liable on the basis of misrepresentation and overreaching by the Bureau of Land Managment concerning the parties' options when a fire obliterated some of the trees which were to be taken by the Plaintiff's Sawmill pursuant to a partial contract. The Spalding court provided an excellent analysis of the showing sufficient to establish an equitable claim. Equity contemplates a remedy to a claim in a variety of circumstances where relief would be otherwise unavailable in a traditional juridical setting. Thus while equitable claims, like legal claims, may be founded on the Constitution, a statute, a regulation, or a common Law principle, Such bases are not absolutely necessary to warrant relief in equity. It is true, nevertheless, that most equitable claims are based on a traditional legal principle, and in such cases, the standards to be applied are those which govern the subject area of the law generally bearing on the claim....In Judging whether there is an equitable claim, however, these legal guidelines need not be rigidly applied, but rather may be based on an ad hoc examination of the facts to determine whether the defendant assumed some sort of obligation requiring recognition, even though the obligation is not enforceable, for whatever reason, in a court of law.

What you see above is the guidelines used in setting up a case in the Court of Claims, based on Tort claims only, remembering there is no room in the Court of Claims for the FTCA. Federal Tort Claims Act., and need not be rigidly applied for a Equitable Claim, nor is it enforceable, for any reason in a Court of Claims. Nor will any decission be made using the Tort Claims Act in the Court of Claims The Claimant must, under the prevailing view, demonstrate that the federal government engaged in some wrongdoing that caused the alleged injury....

***** An equitable claim in a Congressional reference must rest on some unjustified governmental act that caused damage to the claimants. Thus in order to recover under an equitable claim theory the claimant must show that; (1) The Government committed a Wrongful act, and (2) this act caused damage to the claimant. It may arise when the government aquires benefits through overreaching by its agents or by misleading misrepresentations by government agents that are beyond the scope of delegated authority.

***** Contrary to Defendant's assertions, this Court has never limited Plaintiffs to a "Negligence Only" theory of recovery. Any doubt about Plaintiffs' ability to recover for Defendant's torts, other than and in addition to negligence, is removed by this Court's Order on March 26, 1992. As when the Defendant's Motion for a protective order, was denied When the defendant was not able to show good cause. This Court recognized that trespass is a type of wrongful conduct which liability attaches itself, and may be imposed upon the government.

II. PLAINTIFFS POSSESS MORE THAN SUFFICIENT EVIDENCE TO RAISE A GENUINE ISSUE OF MATERIAL FACT ON EACH ESSENTIAL ELEMENT OF EACH CLAIM RAISED IN PLAINTIFFS' FIRST AMENDED COMPLAINT.

***** In Colorado liability for trespass requires only an intent to do the act that itself constitutes, or inevitably causes, the intrusion. Specifically, Trespass is the physical intrusion upon property of another without the permission of the person lawfully entitled to the possession of the real estate...'One is subject to liability to antoher for trespass, irrespective of whether he thereby causes harm to any legally protected interest of the other, if he intentionally… enters land in the possession of the other, or causes a thing… to do so…

***** Compensation of injury resulting from trespass can include (1) the diminution of market value or the costs of restoration, (2) the loss of use of the property, and (3) discomfort and annoyance to the occupant.

***** A landowner who sets in motion a force which, in the usual course of events, will sets in motion a force which, in the usual course of events, will damage property of another is guilty of trespass.

***** A. THE PLAINTIFF'S HAVE SUBSTANTIVE, COMPETENT EVIDENCE TO SUPPORT EVERY ESSENTIAL ELEMENT OF THER TRESPASS CLAIM.

During the taking of Gerald Barbieri's deposition by Mr. Kruse Defendants counsel stated and did stipulate as follows. "If it's of any value to you in this litigation, The United Sates is not denying that DIMP was formed at the Arsenal at some time between the time it was built and the current date." It was also stipulated to by Mr. Regan Defendants Attorney that "DIMP was formed at the Arsenal and is the same DIMP found under the Plaintiffs property, and in the Land Wells for animal and human consumption, and also Hygiene use of the water". This in intself is sufficient to establish Defendannt's RMA activities inevitably cuased "the Physical intrusion [of DIMP, and other toxic chemicals] upon property of the [Plaintiff's] without the permission of the Plaintiffs., and under the direction and control of the RMA. This constitutes substantive evidence more than sufficient, and giving rise to question the facts, on the issue of whether Defendant's trespass caused plaintiffs, personal injuries, and property damage. As Doctor Teitelbaum will explain in testimony later that it would not be explained by any other Toxicological mechanism than the consumption of water, and the inhalation of the air contaminated with chemicals which eminated from the Rocky Mountain Arsenal. Likewise veterinarian toxicological expert testimony constitutes sufficient evidence to give rise to the Defendants trespass caused the death or eventual destruction of all but three of Plaintiff Larry Land's cattle. His Expert opinion the presenence of DIMP, and other toxic chemicals in the "water supplies available to man and livestock" on Plaintiff's property "Contributed to the diseased conditions and tissue damage observed by Dr. Scott, and the Colorado University.

***** There is no question the Defendant intended to manufacture chemical warfare weapons and dispose of the toxic waste from the Unltrahazardous Activities at RMA. This is the only intent required to establish trespass in Colorado. "The intent itself constitutes, or inevitably caused, the intrusion."

***** B. PLAINTIFFS HAVE SUTSTANTIVE, COMPETENT EVIDENCE WHICH IS MORE THAN SUFFICIENT TO ESTABLISH EACH ESSENTIAL ELEMENT OF PLAINTIFFS' CLAIM FOR NUISANCE.

***** The very same evidence discussed above, which establishes each essential element of Plaintiffs' trespass claim against the Defendant, also establishes the essential elements of Plaintiffs' related claim for Nuisance. The essential question is whether the Defendant's has unreasonably interfered with the Claimant's use and enjoyment of his property. The Defendants RMA had on that property they had open pit storage of liquid waste in unlined ponds, and leaking basins. they knowingly allowed to escape by air, but by polluting the underground water as well, killing vegetation, killing animals, and devastating, and killing humans as well. """When they were made aware of the devistation they had caused at the RMA; the Pentagon immediately classified everything about the Arsenal. Totally depriving anyone from getting medical assistance, as they lay dying. Congresswoman Pat Schroeder wrote many letters to the Army trying to get us help, letters never answered and if they were answered they contained only denials. She once said about, "The Army, they're about as slow as molasses". No help or assistance was ever given the Land's in their dilema only denials by the United States, and the United States Army. It wasn't until 1989 only part's of the classified information was once again made public, and only after the Army had done a thourogh job of covering up a devistation as they have all over the World. The Army was still denying in letters from the Brass at the RMA. in 1978 after the Army had in fact discovered Arsenal pollution in the Land wells in 1974, and 1975. It was then they refused to set up an scientiffic program for testing the polluted water wells.

***** C. PLAINTIFFS' HAVE SUBSTANTIVE, COMPETENT EVIDENCE WHICH IS MORE THAT SUFFICIENT TO ESTABLISH EACH ESSENTIAL ELEMENT OF PLAINTIFFS' CLAIM FOR STRICT LIABILITY OF ULTRAHAZARDOUS ACTIVITIES.

***** The same facts discussed above with regard to Plaintiff's Claims for Trespass, and Nuisance, also support Plaintiffs' claim that Defendant is strictly liable for its untrahazardous activities which caused Plaintiffs' personal injuries and Plaintiff Larry Land's Cattle Losses. Under Colorado Law, Strict liability is imposed upon parties who engage in ultrahazardous activities.

***** It is stated in Law, One who carries on an abnormally dangerous activity is subject to strict liability for harm to the person, land or chattels of another resulting from the activity, even though the party carrying on such activity has exercised the utnost care to prevent the harm.

***** The manufacturer and/or demilitarization of chemical warfare weapons and the disposal of toxic chemical wastes by Defendant at the RMA most certainly constitutes Ultra-Hazardous Activities. Plaintiffs have presented to this Court more than sufficient evidence to give rise to material questions of fact on the issues of whether Defendant's engaging in these Ultra-Hazardous activities resulted in the contamination of Plaintiffs' ground water, and the air caused Plaintiffs' ground water, and the air caused Plaintiffs' personal injuries and Plaintiff Larry Land's cattle losses, thus, Defendant may "be held liable for harm resulting to persons and property from [toxins] stored at or leaking from" the RMA.

***** D. PLAINTIFFS HAVE SUBSTANTIVE, COMPENTENT EVIDENCE WHICH IS MORE THAN SUFFICIENT TO ESTABLISH EACH ESSENTIAL ELMENT OF THEIR FRAUD, AND DECIT CLAIM.

***** Plaintiiffs' fraud/deceit claim is based on the facts that; (1) Defendant knew that its chemical warfare weapons manufacturing and/or demilitarizing and toxic chemicl waste disposal activities were contaminating the groundwater below and the air above Plaintiffs' land; and (2) Defendants failed to disclose those material facts to owners of the land adjoining the RMA, including Plaintiff. The court described the essential elements to establish a prima facie case of deceit by nondisclosure:

***** The elements of deceit by nondisclosure are: (1) concealment of a material existing fact that in equity and good conscience should be disclosed; (2) Knowledge on the part of the party against whom the claim is asserted that such a fact is being concealed; (3) Ignorance of that fact on the part of one from whom the fact is concealed; (4) The intension that the concealment be acted upon; and (5) action on the concealment resulting in damages…The elements of fraudulent concealment may be inferred from circumstantial evidence.

***** In a fraudulent concealment action, a person generally has a duty to disclose to another with whom he deals, facts which "in equity or good conscience" should be disclosed; A Court recognizing a false representation may be established by a defendant's failure to disclose a material fact which in good consience should have been disclosed;

***** "In November 1959, Defendant was told that [t]he serious nature of this contamination should be recognized as one requiring early action".

***** The effort of the United States Army with Basin F for surface storage of toxic chemicals, to preclude the off-post migration of toxic chemicals being manufactured, demilitarized and/or disposed of at the RMA, this effort was unsuccessful. Defendant knew in 1965, and again in 1969, that Basin F. (Lake F) was leaking. When they realized it was leaking large amounts of toxic chemicals, the pentagon gave orders to classify it and not allow the public even to be aware it was leaking. The defendant failed to disclose this material fact. "As I understand it as classified, secreting the information from everyone, This was ordered by General Olinchuck, from the Pentagon, after he had discovered the lining in Basin 'F' had in fact been liquidized, and was non existant. His order secreted that information for 20 agonizing years for all those who suffered and died.

***** George Donnelly, who was in charge of the RMA's Engineering Department from 1954 to 1967, and the RMA's Facilities Division from 1967 to 1976 testified as follows:

***** Mr. Donnelly testified the Army Environmental Hygiene Agency recommended in 1965 that steps should be taken"...To eliminate Lake 'F', as soon as possible, and thereby remove much of the present environmental hazard of exposed surface storage ot toxic wastes."

***** Mr Donnelly stated the Army Environmental Hygiene Agency was concerned about the toxic effects on the environment of surface storage of this industrial waste in Lake 'F-1' and Lake 'F'

***** Mr. Donnelly also testified and did understand that it was recommended that the current policy of disposing of still and tank bottom wastes in Lake 'F' should be discontinued.

***** The mucom report [an independent contractor], that is also in evidence, stated the 'principal persons involved in evaluation work on-site.' Mr. Donnelly admitted he was the third person listed in the report who was involved in evaluation work on-site.

***** Mr. Donnelly admits, "The membrane was absent in one test hole in Lake F-1, and in one hole in Lake 'F' ... Hydrocarbons in the waste are believed to have dissolved the membrane."

***** Mr. Donnelly understood at the end of 1969 it was Mucom's position that basin 'F' was leaking.

***** Mr. Donnelly testified, the investigation team that was sent by General Olinchuck to evaluate the waste disposal basins concluded that it was prudent to assume Basin 'F' was leaking.

***** Mr. Donnelly testified and understood from his discussions with General Olinchuck on the day he was at the Basin, and reading from the Mucom reports which we have just looked at, that the Army was operating on the premises that Basin 'F' was leaking.

***** George Donnelly, testified after 1970, he knew that generally speaking the things being discarged into Lake 'F' were causing problems.

***** George Donnelly agreed and testified that [page 5 trial exhibit 145]. stated "A physical inspection of the liner was conducted in early December, 1969 by draining the F-1 portion of the Basin and digging by hand to the membrane location". "THE INSPECTION REVEALED THAT PORTIONS OF THE LINER HAD DETERIORATED AND, IN FACT, WAS NON-EXISTENT IN SOME AREAS".

***** Mr. George Donnelly testified "A portion of the major basin area an area where industrial petroleum still bottom wastes were dumped, was inspected". "THIS AREA ALSO SHOWED LINER DETERIORATION. THE BASE SOIL MATERIAL UNDER THE ORIGINAL LINER LOCATION SHOWED HIGH COLORING IN THE AREAS OF COMPLETE DETERIORATION. The extent and severity of the deterioration has not yet been determined. Mr. Donnelly admittedly stated he was aware of this.

***** Mr. Donnelly testified Paragraph 3.(C)B. The Basin has a characteristic odor. This odor has been detected in certain monitoring wells and presently exists in a well just downstream of the Basin.

***** George Donnelly agreed that because the wells had the same odor as Basin 'F', what was in Basin 'F' was in the wells.

***** George Donnelly's sworn testimony "Defendant clearly knew by 1969, at the latest, that Basin 'F' was leaking toxic contaminants off-post, but Defendant did not re-initiate off-post well monitoring. Nor did Defendant disclose this material fact to persons living in the area north of the RMA. On the contrary, Defendant has continued to deny any leakage of toxic wastes from Basin 'F' in this case.

***** G. Watson testified, "Defendant has also failed to disclose the material fact that its demilitarization of GB nerve gas program, 'Project Eagle phase II', resulted in the emission of nerve gas from the RMA, which polluted Plaintiffs' air. When Plaitiffs' and other land owners located to the north of the RMA. filed a petition dated December 20, 1972 telling the RMA that the air pollution coming from the RMA. were causing health problems, Defendant blamed this pollution on Shell".

***** Testimony of Richard A. Jacobs, Jr., "GB Nerve Agent Disposal Procedure, Defendant has continuously adhered to this position, even though Defendant's own experts were telling Defendant by July 1973 That the GB and M-34 GB cluster demilitarization process was emitting nerve agent in the stack exhaust in quantities exceeding the emission standard".

***** Carlson testimony, "As a landowner engaged in the ultrahazardous activities of manufacturing and/or demilitarizing chemical warfare weapons and disposing of toxic wastes, Defendant clearly had a duty, 'In equity and good consciencce' to disclose the very material facts regarding the contamination of Plaintiffs' environment. In addition, Defendant surely knew that these material facts were being concealed by the nondisclosure of Defendants responsible agents and that plaintiffs were ignorant of the material facts of the contamination of their air and groundwater with toxic chemicals created by Defendant at the RMA. Defendant's failure to disclose these material facts, when Defendant knew or should have known that Plaintiffs, farmers and ranchers, did not have the technology to discover the contamination of their environment and had no reason to suspect such contamination by the almost invisible chemical intruders, provided sufficient circumstantial evidence from which one can infer Defendant intended that its concealment be acted upon by Plaintiffs' fraud/deceit claim.

***** E. PLAINTIFFS HAVE SUBSTANTIVE, COMPETENT EVIDENCE WHICH IS SUFFICIENT TO ESTABLISH EVERY ESSENTIAL ELEMENT OF PLAINTIFFS' NEGLIGENCE CLAIM.

***** Plaintiffs' negligence claim is the only claim contained in Plaintifffs' First Amended Complaint which was directly challenged by Defendant on Defendant's erroneous assertions there was no specific evidence of negligence on Defendant's part. "Defendant could only reach this conclusion by utterly disregarding reality".

***** To recover under a negligence theory, a plaintiff must show that the defendant breached a duty of care owed to the plaintiff and thereby caused the plaintiff's damages....A legal duty to use reasonalbe care arises in response to a foreseeable risk of injury to others. When the standard of care involves questions beyond the competance of ordinary persons, expert testimony may be required to establish that standard.

***** Defendant clearly owed a duty of care to Plaintiffs to conduct its ultrahazardous activities of manufacturing and/or demilitarizing chemical warfare weapons and disposing of toxic chemical wastes in such a manner so as not to poison Plaintiffs' environment and persons. Plaintiffs' expert evidence is more than sufficient to create genuine issues of fact concerning whether they have suffered substantial personal injuries and cattle losses as a result of Defendant's contamination of Plaintiffs' environment.

***** Even if we were to ignore the common law and statutory applications of strict liability, the government would remain liable under negligence theories. The common law in Colorado requires that businesses which are dangerous to the public be conducted with the highest degree of care, skill and diligence in the constsruction and maintenance of it's equipment.

***** The fact that the highest degree of care was required of an electric utility leaves little doubt that a similar degree of skill would be applied to the manufacture of chemicals designed specifically to kill things...A summary of the facts leaves no doubt but that the government failed to exercise the required skills in the following areas.

(1). The initial design of Basin A together with the subsequent expansion of Basin A and construction of Basins C, D, and E, without conducting hydrological surveys or percolation tests. Evidence of this liability is further provided by the payment of the claims for the 1950's pollution and the construction of the boundary containment system.

(2). The introduction of irrigation water in Basin C strictly for the purpose of "flushing" contaminants from the Arsenal, "Out into the country side to the North of the RMA", Through out the sixty's and the seventy's, was not only negligent, it was willful, wanton, and reckless. It was also deceitful, to cause them to do somthing like this they by fraud hid the fact chemicals were known to be leaking from the arsenal, and another one of the ideas was to dilute what had already escaped and was escaping.

(3). The failure to locate and repair the "Mount of water" from under the South Plants area had the same effect as the use of water in Basin C. Both were resulting in the flushing action. This wrongful action is now believed to have been purposefully constructed by the Army knowing full well it would have a flushing effect of toxic chemicals from beneath the Arsenal. Was at best—Negligent, and at the worst—willful and purposeful.

(4). The failure in years subsequent to 1959 to follow the recommendations in the Walton report.

(5). The failure in the years subsequent to 1965 to follow the recommendations in the U.S. Department of Health, Education and Welfare reports.

(6). The failure to either repair or discontinue the use of Basin "F" after discovery of the Fact that it was "IN FACT" leaking toxic waste out into the Country side at that point in time had already polluted and contaminated over 35,000 acres of land off the Arsenal property. This was confirmed by George Donnelly the Head of the RMA's Engineering Department, along with the man in charge from the Pentagon General Olinchuck when they realized the general public were about to actually discover they were being poisoned both by air and water. It is said that, "General Olinchuck" stated we can't have the Public find this out and ordered everything about Basin "F" "CLASSIFIED" along with many other toxic and polluting facts. This information has only been peace meal released by the Army Pentagon only since 1988-2001.

***** AIRBORNE TRESPASS. The escape of volatile elements from Arsenal operations in general and Basin "F" were well known and held classified by the Government, at the "PENTAGON". The evidence will show other willful conduct in addition to the willful spraying of liquids from Basin "F" and Basin "A" when authorities knew or should have known that vapors, and a spray mist invaded adjoining properties. Knowledge or willfulness is not material to a cause of action. However the Army was fully aware of this problem they had created, it was wilfull, and yet their best way to hide the truth was to ignore it. The Army had many well trained PR. men to hold the public at bay for as long as they needed, and are still today year 2001 holding the public at bay Trying to tell them the untruth of what they have done and what they are now doing at the RMA.

***** DUTY TO IMPOUND; The common law of Colorado requires that polluted water be impounded upon one's property or the polluter is liable for any discharge into the waters of the State. Hence the discharge of any pollutant is a breach of the duty to impound. The Colorado Supreme Court ruled that it was no defense that the polluter was acting as agent and that there might be others, who were not parties, which might be liable. The rules relating to surface water are applied to underground aquifers as well.

***** ULTRAHAZARDOUS ACTIVITY; The state of Colorado has long recognized strict liability for ultrahazardous activities. Such rule was adopted with the following comment. "The true rule of law is, that the person who, for his own purposes, brings on his own land and collects or keeps there anything likely to do mischief if it escapes, must keep it of his own peril; and if he does not do so, is prima facie answerable for all damage which is the natural consequence of its escape."

***** FRAUD; Colorado recognizes fraud by concealment or deceit as an actionalbe offense. The essential elements for revovery are (a) that there was a failure to disclose a material existing fact that in equity and good conscience should be disclosed. (b) Knowledge on the part of the party against whom the claim is asserted. (c) Ignorance of the facts concealed on the part of the one from whom the fact was concealed. (d) An intent that the concealment be acted upon. (e) Action upon the concealment. (f) damage....Fraud may be committed by the suppression of the truth as well as by suggestion of falsehood, privity is not a necessary element. The difference in the law for Plaintiff's claim based upon fraud and their claim based upon negligence depends upon the knowledge of the government. The Walton report is evidence of the duty of the government to take action to minimize damages from its illegal activities. The Walton report establishes the obligation to investigate the facts regarding the pollution and inform adjoining owners. The discussion in the negligence portion of this trial establishes that the government did not fulfill such duty.

***** In addition, there is evidence already introduced to this Court that such action was willful. The evidence is clear that the government suppressed evidence of the leaking of Basin "F". In addition, the HEW 1965 report reflects that the conclusions under the Petri report on the boundaries of pollution were incorrect. Despite the availability of such information, the government continued to assert the validity of the Petri report.

***** The konikow report even further discredited the Petri report. Despite the discrediting of such report, the government continued to assert such results as conclusive up to and after the granting of the Congressional Reference.

***** SUCH "STONEWALLLING" was particularly damaging to the Plaintiffs in the present case. Because of the ability of the government to conceal the activities on the Arsenal and to conceal the nature of the contaminants which it released, it was impossible for Plaintiffs to receive the medical treatment which might well have mitigated their damages.

***** Defendant admitted plaintiff's charges which provides in part that a groundwater survey…established the probability, and the certainty, of contamination of the wells on the property where Plaintiffs resided. "FACT" diisopropylmethylphosphonate (DIMP) does not occur in nature and can be obtained only as the hydrolysis of GB Nerve Gas, (SARIN NERVE AGENT)

***** Undisputed facts have been established before this Court, the defendant has contributed to polllution of the groundwater underlying the lands of the Plaintiffs. The fact of pollution has been admitted to before this Court by Defendant under the Land property, and in the Land wells. While it contests the quantity and quality of such pollution. As such, Plaintiffs have under Colorado law established and equitable claim for the amount of damages resulting from the polluted groundwater. Plaintiffs are not required to prove negligence to establish such equitable claim. Defendant has committed a "WRONG" by its failure to confine the pollutants to its property.

CHAPTER FOURTEEN

THE CONCLUSION OF THE FOREGOING BY EXPERTS, THAT WILL SHOW THE MOTION FOR A CAUSUAL CONNECTION. TO REACH AS A LOGICAL, AND NECESSARY END BY REASONING. BEING DECISIVE PUTTING AN END TO DEBATE OR QUESTION, BY REASON OF IRREFUTABILITY.

The government's contamination of the water underlying plaintiff's properties and the contamination of the air over such properties, and the property itself, was in violation of numerous duties which would have been imposed upon the government had it been a privaate party. The combination of both such contamination have resulted in damage for which the government in equity should be responsible. such damages represent equitable claim and are not a gratuity.

Plaintiff's Attorney Sam McClaren placed a motion before the Court for permission to designate, and call to testify Samual L. McClaren as an expert in the fields of Civil engineering, geology chemistry, hydrology & law. Also Plaintiff's motion to designate Dr. Dale Clifford Wingeleth, Ph.D. to testify on chemicals, and unintentionally manufactured chemicals at the Arsenal, to include deficencies in the disposal of chemicals by both the Army and Shell Chemical Company. Also pre 1972 chemical testing and the current technology in testing water. The Court refused to allow Mr. McClaren, or Mr. Wingeleth to testify at this trial.

***** Judge Robinson; "Defendant has failed to convince the Court that operation of an Ultrahazardous or abnormally dangerous activity in the instance and under the attendant circumstances, i.e., a few miles, from working famland, one mile from the Land property, with knowledge that basins were leaking, and basin "F" was leaking is not a "wrongful" act and that compensation for injuries resulting from such activity", are surely warranted.

***** The Court finds that such activity will be considered "WRONGFUL", conduct while defendant falls under the scope of a Congressional Reference".

***** Judge Robinson," As a matter of law, Courts will hold every reasonably prudent and careful [person] to the exercise of the utmost care and diligence in protecting the public from the dangers necessarily incident to the carrying on of a hazardous business."

***** Judge Robinson; "This Court is satisfied, after resolving all doubts as to the facts, presumptions, and inferences in favor of the Plaintiff's, that genuine issues of material facts exist as to plaintiffs' claims including those sounding in negligence. Defendant has failed to convince the Court that no issues of material fact exist as to Defendant's negligence or wrongdoing".

***** Judge Robinson; "Plaintiff's have placed into evidence that the Army's conduct on the RMA. manufacturing and demilitarizing chemical weapons and disposing of chemical wastes, created a substantial risk". "THIS COURT AGREES".

***** Judge Robinson; "In the U.S. Court of Appeals 10th Circuit. Stated that Colorado could consider such activity ultrahazardous or abnormally dangerous. 'This Court accepts that the disposal and manufacturing of toxic munitions is ultrahazardous or an abnormally dangerous activity.

***** Ruling by Judge Robinson; "For a determination of negligence, or any other substantive claim, the law of Colorado governs".

***** The Court reminded the Defendant, "Jurisdictional limits under FTCA cases have no application to a Congressional Reference. No specific response to arguments in Defendant's Brief relatng to the FTCA cases will be made except to note that the Defendant's underlying premise is in error".

***** "The remedy set forth by this Court is what's known as the Tucker Act. The Tucker Act confers jurisdiction in this court for claims based upon the constitution and other matters not sounding in tort". This is where the circumstantial evidence was allowed and approved by this Court, it follows hand in hand with all the laws of the State of Colorado".

***** Judge Robinson stated, "The activities of the Rocky Mountain Arsenal have resulted in one of the worst hazardous waste pollution sites in the Country, ultimately the Army identified fourteen specific sites which needed Intrim Reponse Actions, (IRA'S) **IMMEDIATE AND RAPID RESPONSE** To stop the spread of toxic, and Ultrahazardous contaminants so as to protect human health and the environment, Basin "F" was at the top of the list".

***** This statement, "is a report by Dr. Daniel T. Tietelbaum M.D. / Tolicologist, agreed to by Judge Robinson, signed and dated, made a part of the Courts records. "Review of the records which were supplied to me establishes unequivocally that the Land family's groundwater was contaminated with materials from the Rocky Mountain Arsenal. In paticular, the finding of DIMP…establishes beyond any question that the source of the material in the groundwater under the Land's property and in their drinking water and hygiene water wells was the Rocky Mountain Arsenal. No other source of this material is conceivable in that area. This finger-printed chemical establishes without any question that the toxic material which reached the Land family had, as its source, the Rocky Mountain Arsenal GB manufacturing facility. In addition, the finding of IMPA, a metabolite of DIMP, further identifies the material as having a source in the Rocky Mountain Arsenal".

***** Planintiff's," rely on a report by Dr. Frederick W. Oehme, D.V.M., ph.D., which opines that the cause of the death or eventual destruction of all but three of the calves was the presence in the well water of the man-made compound DIMP and it's breakdown product IMPA".

***** The Plaintiff's "also rely on a recent report by Dr. Daniel T. Teitelbaum, M.D., concluding that the cause of the health problems suffered by plaintiffs and Larry Land's cattle was DIMP and IMPA contamination of the Land well water and the air".

***** HYDROLOGY; Dr. Waddell, “the Plaitiff's expert on hydrology testified that contaminants from the Arsenal would migrate in the ground water from sources on the Arsenl to the Plaintiff's wells. He further testified that contaminated water would be present more of the time in the James Land well than in the Hollenbaugh well because of Occurrence of gravels in his well at lower elevations than the gravels in the Holenbaugh well. The government admitted that water contaminated with DIMP could and did migrate from the Arsenal to the Home Place. Dr. Waddell testified compounds which would migrate along with the DIMP were Sodium Fluoride, Nemagon, Dieldrin and Aldrin. These are all substances identified by Dr. Teitelbaum as being extremely toxic and hazardous to consume”.

***** Wrongful act; “This Court has ruled to recover on an equitable claim, Plaintiff's must show that the government committed a wrongful act, it need not be negligent”. “A trespass was ruled to be a wrongful act, it need not be negligent”. “A trespass was ruled to be a wronguful act, Citing of this Court”.

***** Duty to confine; THE COURT RULED, “The lower owners were entitled to have the waters preserved in their purity, that fish might swim, that their stock might drink, and that the water might be applied to all domestic uses.”

***** Duty to Confine; THE COURT RULED, “The common law of Colorado requires that polluted water be impounded upon one's property or the polluter is liable for any discharge into the waters of the state.” “It is the polluter's duty to determine the boundaries of the pollution, and to warn those persons in the areas of concern.

***** Ultrahazardous Activities; THE COURT RULED, “It is unnecessary to determine if the manufacture of war gases is an ultrahzazrdous activity since the result would be identical to that under the reservoir cases. In fact, the foundation for ultrahazardous activities is based upon the reservoir cases.”

***** Breach of Duty; The Court rules, “by previous law, Inherently dangerous activity in Colorado, “Such classification leaves no doubt that the manufacture and storage of toxic war gases and pesticides is also on inherently dangerous activity." "One carrying on an inherently dangerous activity...must exercise the highest possible degree of skill, care, caution, diligence and forsight with regard to that activity, according to the best technical, mechanical, and scientific knowledge and methods which are practical and avaiable at the time of the claimed conduct which caused the claimed injury. The failure to do so is Willful Neglingence in every sense".

***** Failure to warn; "In 1965 when the Arsenal failed to notify the Lands of the findings of the HEW report that contamination was adjacent to their property, and they were living in an area with elevated chloride levels. Such a warning would clearly have allowed the plaintiffs to avoid property damage and their injuries.

***** False Statements, "were particularly damaging to the plaintiffs. Because of the ability of the government to conceal the activities on the Arsenal and to conceal the nature of the contaminants which it released. Making it totally impossible for plaintiffs to receive the medical treatment the situation demanded. Largely because of such misstatements the Plaintiffs suffered additional injury which might well have been mitigated had they been told the truth."

***** Expert Witness Doctor Scott DVM., Dr. Scotts testimony about his knowledge of the Land Livestock program. Dr. Scott testifed he knew Larry from back in the 1950's, Larry grew up with cattle. I took a liking to him when he was just barely a highschooler. He was always out grooming the horses and the cattle, the little ones and the big ones he had a real expertise in Animal-husbandry. I could remember he was always asking questiions about animals, how would he be able to do certain things, He was an expert on his way to the top.

***** Larry called me out to his Box elder Ranch where he was raising cattle in the early 1970's. He had a lot of baby calves, and was doing real well. He wanted to keep growing, and had ask for some adivise in just about everthing one needed to know about raising calves, to breeding techniques. He done all his own de-horning, castrations, branding, and giving all the necessary vaccinations, to treatment as necessary, there isn't much he didn't know.

Mr. Land, "had developed his own special breeding program, he shared the information and ask that it not be discussed, so I will try to tell you where he was headed. It was one of widest of wide outcrosses for dairy cattle ever known. The program was to have it's beginning in mid August on 1971, he had right at sixty heifers, being groomed for the program, The ideal weight for them at the time of breeding was 850, to 900 pounds. It was an artificial insemination program aiming for much smaller calves especially for the first and second calf heifers. It was also designed through his program to help build the immune system to be much stronger than anything we know today. By the spring of 1971 Larry already had the first six of the new breed of livestock. Just as true as could be and exactly what Larry had told me the calves were much smaller, just about as strong and full of Pep as one could be. They seemed to do very well and he had no problems with them. By August 1971 Larry began the breeding program for up to 60 heifers. At that time he had over 900 head of livestock, he had most of them running on pastures where ever he could find it. Over 300 on the flat top mountains outside Golden Colorado, another 300 at the Devel's Thumb ranch on the other side of Berthoud pass. He had room for the rest at the Box Elder Creek Ranch, and that is where the smaller ones were taken care of. Everything was going so well for him, his death and loss rate was very low running below ½ %."

***** Larry had purchased from his Father the Home Place where he was born and raised, he was seeing his hopes and dreams coming true. At the Home place he had to build all new corrals, pens, and refence the property so it could be pastured. He constructed several lien twos, buildings with a roof and back facing the wind to protect the livestock in the winter time. His aim was to remain at about 1000 head, hopefully expanding as time and finances would allow. Larry began moving a few of the larger animals to the new home place as he was working on it, he was using the water from his parents well. The well driller was not gonna make it out until about the 1st of April, in the mean time he got all his plumbing in, corrals and buildings built, and ready for the new well. He had got a small truck load of calves in on March 30, 1972, he had moved over more livestock so the water consumption was going up rapidly. The first thing that was noticed the livestock were lapping at the water at first, and some of them seemed sluggish, two of the new ones were real sluggish, and ears began drooping. This was not a good sign, and by April 1, a calf had perished, weak, eyes red, sores in nose, seemed to stagger and become lame then could not get up, and went into convulsions, and died. We noticed the water seemed to be tinted red in color.

***** As a veterinarian, "I had no problem when Larry had asked for advice on raising young calves, I had practiced 20 years here as a veterinarian and one of the things I pride myself in doing was to encourage, and coach farmer's in accordance with what I perceived their talent to be. I was sure Larry could do an excellent job, he had already showed at the Box-elder farm what he could do, and do with excellence. What happened between the 30th of March 1972, and April 12, 1972 was only a precursor of things to come. When the well was complete Larry ask the driller why is the water so red in color, he said it's only iron just pump it steady for a 24 hours, and we can test the static level tomorrow. Well the next day came and the well level was checked, and was found to be steady at about 22 gallons per minute, however the water was still very red. Larry called the State, and the Tri County Health Department to test the water and tell him if it was good to use, and why was it so red. He was given the same answer it was iron and the more you pump it the sooner it will clear up. The Tri County Health came back the next day with a slip calling it good potable drinking water."

***** Dr. Scott testified "The calves began going down hill, and dealing with every conceivable type disease. It appeared to be at first pneumonia, Bronchitis, Shipping fever which is a highly contagious secondary infectious diseases. The fact they died with what appeared to be shipping fever is not unusual. The difference here was the fact that they also had many other symptoms and even the cattle which were bought local, and the 135 head brought in from the Boxelder farm many of the over two, and three years old. They all showed the same symptoms, and had similar disease patterns, as those which had been shipped any distance at all."

***** Dr. Scott Testified, “I had enough experiece to suspect poisoning. I told Larry that soon after we began to treat the sick animals. It wasn’t until about 1, April 1972 some began to look slow and droopy. On March 25, 1972 we have vaccinated 30 calves with nazalgen. This was to prevent respiratory infections, such as pneumonia, bronchitis, and shipping fever in the calves. There were a few cattle seemed down an out somewhat droopy ears, sores in nose, staggering by 2nd or 3rd of April. Larry had started making the new water available to the cattle on April 12, 1972. Larry also had brought over the large Gold fish he kept in all his stock tanks at the Box-elder farm, and placed them in the tanks here, the tanks were all automatic self filling, tanks were kept full as it was used. The fish were good sized at least 6 to 8 inches in length. The water was red, but everyone including the State Health Department had stated it was ok. The cattle all lapped at the water as if they didn’t really want to drink it.”

***** From about April 12, 1972 We medicated and treated calves daily. The death losses were terrible. I did not attempt to keep score. cattle would usually die of what appeared to be pneumonia. The cattle exhibited unusual incoordination. Many had persistent diarrhea, vomited, eyeballs were red and ulcerated with what appeared to be keratitis or “pink eye”. The mucous membranes were often ulcerated the mouths and noses of all the cattle. Teeth seemed to be sore and discolored, and there were some with what appeared to be arthritis, they would stagger, eventualy fall down and go into convulsions that were unstoppable. The death rate was high. The pregnant heifer’s all within days of feeding them the water from the new well, aborted all calves. The cattle would just before falling down going into convulsions would appear to be losing total control head down, tongue out, staggering, unable to stand, beller as if in a lot of pain, and then fall down in full convulions. It was the younger 6 to 12 week old calves that seemed to be affected faster than the older ones. There was no doubt in my mind at that time that the cattle were being poisoned. The gold fish Mr. Land had in all the stock tanks, bright orange gold fish would lose all scales, and turn half white before going belly up. There would be dead insects in and around the water tanks, as well as flopping, or dead birds, not only in the water but nearby the water tanks.

***** “The final cause of death first appeared to be pneumonia, bronchitis, shipping fever type deaths and that is because they are secondary infections that take over immediately after the animals risistance, and immune system has been weakened by somthing else such as a toxic pollution contaminant. It was the other signs and symptoms that were not associated with the secondary infections that tipped me off to the fact it was poisoned water. We began putting two and two together when the humans began showing the same signs and symptoms as the livestock, then remembering the dead birds, the dead fish, the dead insects, the water would not grow algae, as would have been normal.”

***** Dr Scott then testified, “he had enough experience to suspect poisoning. I told Larry that soon after we began treating the sick animals. I talked with Larry several times about moving the cattle to a different location, but where. I had talked with a friend of mine Jerry Sitzman a local dairyman, to allow Larry to move his sick and poisoned animals over to his farm and use the spare corrals. He agreed, and I was that sure they had been terribly poisoned, and also knew I was taking an extreme chance because if it were in fact pneumonia, or shipping fever we would have a very serious and contagious disease, with Sitzman’s dairy cattle having only a barbed wire fence separating the two heards. Larry had only around 270 head of cattle left out of over 635 animals. He move them all within in two days”.

***** “The proof that the water was the problem soon became very apparent, the calves got better immediately. In a few days the calves began to play and run, their ears pointed upwards, the eyes began clearing up, the sores in the noses and mouths soon cleared up. It was a very dramatic and happy circumstance, at least for a while. There was never a bit of trouble with the Sitzman cattle, and I was certainly gratified for that. If there had been a contagious problem with the Land cattle the Sitzman cattle would have been vulnerable, because the only separation was a barbed wire fence. I spent a very anxious week for fear that I might have spread a bad problem to the Sitzman dairy who was one of my clients. None of the sitzman cattle got sick. There was no symptoms of any respiratory disease in the cattle which had certainly been exposed to whatever the Land cattle suffering from. The Land cattle were looking real good. unfortunately our optimism was short lived. We soon realized that the cattle were pemanently changed and would never develop into profitable animals, or even break even animals. They would never grow more that half the size of their peers. I know for certain the Land cattle were poisoned. I know the well water on Larry Lands' Farm was polluted, and contained toxic contaminants from the Arsenal. I stop and look back for a moment and am sure the air was contaminated as well."

***** Dr. Scott testified, "At the Land farm before the animals were moved, they seemed to respond to medication, but would relapse when the treatment stopped. The lymph nodes in the head and neck of some of the calves swelled and abscessed. They did not respond to routine treatments. In face the use of sodimum iodide seemed to make the lumps worse. Many of the calves had what seemed to be the mouth form of calf diphtheria. On 06-07-72 we sent calves to the diagnostic laboratory as CSU in Fort Collins Colorado. The results of this study were inconclusive, because in order for them to detect the toxic chemical, they needed the name, and technology, and the United States Army had that classified as secret, and were unwilling to share that information. The cattle were poisoned by toxic chemicals which did weaken the immune system allowing the secondary infections to take over. We had no way to test the water to detect what the toxic chemicals would be, the Army had classified all that information where only they would know. No commercial laboratory had the technical information to detect many of the most toxic chemicals that was present in the water. DIMP was not detectable in the water in 1972, it was 1975 the Army admitted it was there, and did not release the technology until 1979 before the local laboratories received any information".

***** Dr. Scott, "In spite of the inconclusive test results on the water, I was not surprised, since the Army had all technology for detecting their pollution classified. I finally convinced Larry to move the sick animals to another farm around the 20th of July, the cattle were very sick. Many of them were so sick they had to use hydraulic loaders to get them on and off the trucks. They had what appeared to be "Shipping Fever" or some other form of the Bovine mucosal disease complex. Clinically this is a very contagious problem".

***** Dr. Scott, states we moved the cattle to Jerry Sitzman's dairy farm. He had many young cattle of his own, which were only separated from the Land cattle by a barbed wire fence. I spent a very anxious week for fear that I might hhave been wrong and spread a bad problem to Jerry who is also my client. None of the Sitzman cattle got sick. There were no clinical symptoms of any respiratory disease in the cattle which had certainly been exposed to whatever the Land cattle were suffering from. We know that most pathogens are only able to attack an animal whose resistance is lowered by some sort of stressor, [such as a toxic chemical]. Something in the Land Cattle diet was causing a stress. We fed the Land Cattle the same ration at the new place, the same diet with the exception of the water".

***** Dr. Scott testified, "The change in the cattle when they ere placed on different water was dramatic. Many of the calves seemed to recover. In a few days some of the calves began to run and play which they had never done on the Land farm. The number of new sick cases dropped within two weeks to near zero".

***** We began to relax and anticipate normal development, there were a few which became chronic lung afflictions. We sacrificed one of those that was chronic to an autopsy. With Dr. Baum, a Vetrinarian from the State Health Department, and Mr. Dunn the chief chemist from the Department of Health had done nothing to find the problem. The autopsy dislosed Pneumontitis, abscessed lungs and other evidence of septicemia. Mr Dunn promised to sample the water and mark it [ETKTM] everything known to man. If there was any testing done we never seen the results of anything.

***** The cattle did not grow and develop normally. The chronic would eventually die. Autopsies were performed on some and the findings usually included lung abscesses, damaged lungs, heart muscles in some were enlarged, flabby, and diseased, the liver's were notably destroyed abscesses appeared on other organs. The gland that helps stabelize immunity were totally destroyed.

***** On August 5, 1972 we returned 2 animals to the Land farm, observing them, taking moving pictures to record their actions and growth. We took pictures on several occassions, as they became sick showing every sign and symptom that had been described. They were near death when they were move off the place and again we took pictures as the began the long recovery of being poisoned. With in days it was again noted the sores in the mouths and nose were healing well, the red irritated eyes were returning to normal, the ears began to perk up straight, and they were able to move with much less pain, the staggering, jerky motions disappeared.

***** Dr. Scott testified, “I’m not sure how much detail to put on this preliminary note. The final outcome was very bad. Larry was forced to sell most of the damaged cattle at a severe financial loss. The cattle were stunted and unfit for breeding purposes. All his hope of maybe at least paying his bills was dashed, all his hopes and dreams with the livestock were destroyed. John Thimmig, and myself then declared that the veterinarian bill for Larry Land would be deferred, until he recieved repairations from the United States of American [United States Army]”.

***** Dr. Scott looking back, “One of the most dramitic findings in my opinion was the skeletons of the dead cattle. There was evidence of arthritis in every joint, the bones were not white as would normally be, they were brown in color, the teeth were poorly formed, and loose in the sockets.

***** Dr. Scott testified, “Larry bought more calves, and raised them on the Land farm since 1972. Some of them were purchased from Minnesota, and Wisconsin, some purchased locally. In order to do that we must not use the water. The water had to be imported in large tanks, It was really an impractical situation. I guess Larry wanted to say see I could have done it had the water not been polluted with Ultrahazardous and toxic chemicals from the arsenal”.

***** Dr. Scott stated, “I know Larry was buying and selling livestock, he would buy them one day and take them to the market the next day or two and really had discovered a whole new method of making a living with livestock. He was a licensed buyer, both State of

Colorado, and Nationally. He could buy and sell livestock in one State and sell them in another. Larry said he was doing real well and hope to pay off all his bills left from the loss of his cattle. He went to the mail box one day and out came a letter from the U.S. Marshall's [Department of Justice], Washington D. C. It stated clearly that his bond for buying and selling livestock was revoked on order of the Justice Department. That each animal purchased and sold in under thirty days would be prohibited, and for each animal not held for a minimum of 30 days he would be fined up to $5,000.00 and six months in jail. Because of his financial loss of livestock in 1972".

***** Dr Scott stated, "You know where that came from, and why. I told Larry to keep his chin up and you'll get thru this. Larry had begun a small trucking business, and had just bought his first semi tractor trailer rig, along with three bobtails. He also began working for the Denver sheriff Department, and was still doing his Auction work also. [A man of iron]. By 1974 he had put together a herd upwards of 400 head of Herford, Angus, and Black Baldies, half steers half heifers he had all them at another farm where the water was good".

***** Dr. Scott Continues, "He had them all trim and buffed ready for pasture. He rented pasture wherever he could find it, some were on the flat top mountain overlooking Golden Colorado the home of "Coors" and some more over Berthod pass to Devil's Thumb, and a few pastures local. He said I know this is gonna be hard, but I have to prove who I am, and who I was before the Arsenal almost destroyed me and my family. I guess I have to say, and show the World [it's not over til it's over], [and it's not over yet] I can do it.

***** Dr. Scott stated, "to make a long story short Larry's cattle had done real well on the pasture, they were what he called long, lean and lanky, and he had it timed out perfectly with them coming out of the pastures at 750 pounds give or take a few pounds they were ready to be put straight into a feeding program feed lot style where they are figured to gain four pounds per day. At the end of 120 days in the feedlot, they will come out weighing, right at 1235 pounds going to slaughter. Larry had raised this group off cattle, with a ¼ percent death loss, that is one out of four hundred. They dressed ninety percent prime, and ten percent choice. This man is an expert in animal husbandry no matter where he chooses to go from here.

***** Dr. Frederick W. Oehme DVM, PhD., “The presence in drinking water of the man-made compound diisopropylmethyl-sulfonate (DIMP) and its break down product (IMPA), as well as some chlorohydrocarbon insecticides, confirm contamination of the water supplies available to man and livestock. Although the sick calves showed signs and lesions associated with pneumonia and environmental stresses, the degree of involvement and the longterm duration of such affects, when taken in combination with the continuing exposure to the contaminated water supplies, support a detrimental effect of the contaminated water upon the calves’ health. The lack of an expected response to the extensive use of antibiotics further supports the harmful effect that the man-made chemicals had upon the livestock.

***** Dr. Oehme continued with his expert testimony, “It is my opinion as a trained, and experienced veterinarian with Board Certification in Veterinary Toxicology, and based upon my experience and knowledge of livestock diseases and the effect of chemicals on animal health, that the presence of these man-made chemicals in the water consumed by the calves adversely affected their health, contributed to the disease conditions and tissue damage observed by Dr. Scott and The Colorado State University, and led to loss of productivity and livestock death on the Land farm”.

***** Dr. Daniel Teitelbaum M.D. P.C. Medical Toxicologist, and also Veterinarian Toxicologist, Testimony to the Court. “As you know, I previously had some contact with Mr. Larry Land and the Rocky Mountain Arsenal emissions, and groundwater contamination problems in the 1970’s. It was therefore, interesting for me to have an opportunity to look at the body of material again in 1991. I have tried to reconstruct the setting in which the problems for Larry Land and his family arose and to assess the medical consequences of the pollution problems associated with the Rocky Mountain Arsenal emissions”.

***** D. Teitelbaum continues, “You were kind enough to provide me with an extensive collection of documents which are relevant to this case. I have reviewed all those documents along with all the medical records, the depositions, and groundwater assessment documents and the various government documents which have been supplied to me in connection with this case. In addition, I have reviewed the minimal amount of scientific literature which is available concerning the materials emitted from the Rocky Mountain Arsenal, and their potential impact on the Land family and their cattle. I have also had an opportunity to discuss some of the aspects of this case with Dr. Robert Scott, Who was the veterinarian who treated the Land family’s animals during the acute phase of this problem. With all of this information in mind, I am now prepared to write a preliminary report to you regarding my testimony, that will be sufficient to establish a causual connection, concerning my findings thus far in this case”.

***** Dr. Teitelbaum continues his report and testimony, “Please let me emphasize that there are many more items of information which have not yet been provided. There are numerous reports and documents prepared by various government agencies, including the Department of Defense, which are not readily available in the open scientific literature. Indeed, it is shocking how little peer reviewed information is available on the materials emitted from the Rocky Mountain Arsenal. Most of the technical information is to be found in reports generated by the Dept. of Defense in various obscure settings. I believe it will be necessary for me to review all of these reports and to see the original data upon which the studies are based before a final opinion will be possible. I have not yet had an opportunity to examine the plaintiffs in order to establish their current medical status. Nonetheless, in the context of the information which is available to me at the present moment, I have formed certain conclusions and I wish to present them to you at this time. This expert testimony will be more than enough to establish the Causual Connection in the Land family’s case”.

***** Dr. Tietelbaum continues with this report. “The historical background of the Land family’s problems associated with the Rocky Mountain Arsenal is well known to you, and has been presented to the court. I will not take the time now to detail all of the information which is available in various medical charts and depositions. Suffice it to say that various members of the family lived on the land adjacent to the Rocky Mountain Arsenal for periods of time which ranged from a few years up to and over fifty, (50) years. During that period of time, the Army and Shell Chemical Company conducted both research and manufacturing activities on the Rocky Mountain Arsenal grounds. They disposed of materials in such fashion that the groundwater to the north and west of the Arsenal became contaminatied with both the by-products of the manufacture of pesticides and Chemical weapons, "CHEMICALS OF DEATH" with the materials, themselves. Since most of this information is, and remains classified by the Dept. of Defense, and is not readily available to me or anyone else in the medical or toxicological field. Therefore my ability to evaluate the cause and effect issue in this case is hampered".

***** Dr Teitelbaum's report continues, "In addition to the groundwater problems which afflict the Rocky Mountain Arsenal region, there has been a substantial problem with the air emissions from the Arsenal, both during the time of the Arsenal's active operation Ultra hazardous and super toxic weapons factory and Ultra Hazardous and super toxic pesticide manufacturing plant, and later during the period of time that various disposal activities took place. As recently as 1989-1990 the Arsenal waste had proved irritating and detrimental to the health and welfare of citizens who live in areas adjacent to the Arsenal. Recent studies have identified by-products of the cyclodiene insecticides, various war gases, Ultra toxic nerve agents, and other materials both in the air emissions and in the sludges at the Rocky Mountain Arsenal. While these current problems cannot be directly related to the Land family's distress, it is clear that the same products were being manufactured at the Arsenal at the time the Land family cattle and the individual plaintiffs lived on the land adjacent to the Arsenal. The respiratory, gastrointestinal and other problems suffered by patients in recent times due to the air and dust coming from the Rocky Mountain Arsenal represent useful information in judging the origin of the problems noted by the Lands in 1972 and 1973".

***** Dr. Daniel Tietelbaum's Testimony continues, "Review of the records which were supplied to me establishes unequivocally that the Land family's groundwater was contaminated with materials from the Rocky Mountain Arsenal. In particular, the finding of DIMP, (diisopropyl-methyl-sulphonate). also known as Phosphonic acid-methyl-bis-trans-1-methyl, ethyl-ester, Chemical Absract No. 1445-5-6, registry of Toxic Effect of Chemical Substances, No. 59196 and IRIS Database NO. 303 establishes beyond any question that the source of the material in the groundwater under the Land's property and in their drinking water and hygiene water wells was the Rocky Mountain Arsenal. No other source of this material is conceivable in that area. This finger-printed chemical establishes without any question that the Ultra-toxic material which reached the Land family had, as its source, the Rocky Mountain Arsenal GB manufacturing facility. In addition, the finding of IMPA, a metabolite of DIMP, further identifies the material as having a source at/in the Rocky Mountain Arsenal".

***** Dr. Teitelbaum's report and testimony continued, "Various quantitative studies of water contaminants have been done. These studies demonstrate that DIMP reached the groundwater under the Land property, and also reached their drinking water wells. On at least one occasion IMPA, a breakdown product of Dimp, was also identified. Because the technology for detection of this material was in its infancy at the time the studies were done in the early 1970's, their reliability must be questioned. Many of the water samples which were submitted by the Land family were not analyzed for these materials both because there was no reason to suspect that it was present, and because once there was some suspicion that it might be present, technology to carry out the analyses was not readily available. Nonetheless, concentrations of Dimp from a few parts per billion to a few parts per million were detected in the drinking water wells of the Land family. This chemical undoubtedly reached the Land family not only in the water vehicle, but by air pollution as well. By the same token, it is likely that this material reached the cattle who were brought to the Land family's property and may be presumed to have been in the free drinking water and other sources of water used for the animals as well as the air on the farm. I will refer here for a moment to Dr. Scott's materials which he repeatedly expressed his conviction that the source of the animal's distress was the water they consumed, and the air they breathed, once they arrived at the Land feed lot".

***** Dr. Teitelbaum's report in testimony continues, "A petition which was filed on December 20, 1972, by a significant number of residents of the region to the north and west of the Rocky Mountain Arsenal, including a number of members of the Land family, records the presence of noxious and obnoxious odors which were emitted from the Rocky Mountain Arsenal. These odors are evidently quite similar to those which were noted by the residents of the Henderson area to the northwest of the Rocky Mountain Arsenal during 1988 when the Basin "F" was reopened and the chemicals in the sludge were allowed to evaporate into the air. Studies demonstrated that among the materials present in those emissions were breakdown products of sulfur mustard, Cyclodienes, and other irritating, toxic and carcinogenic substances.

***** Dr. Teitelbaum's report continues, "In order to make some sense of the multiple medical records and the depositions which were provided to me, my staff did a preliminary review. They abstracted the medical documentation which is available. In addition, I read and reviewed the medical records and depositions of the various plaintiffs. A number of striking associations come from this review. These can be summarized in just a few sentences.

In the time frame, 1970 through approximately 1973, with the majority of the emphasis on 1972, the Land family had problems associated with the cattle and wheat on their farm, and a series of medical problems which can best be characterized as headache, airway irritation, eye irritation, bronchial irritation, nausea, diarrhea, and mental and emotional changes. These noxious signs and symptoms were intense during the period of time which is mentioned in the petition filed with the governor in December 1972. These complaints are similar to those which were present in the residents in the Henderson area following the opening of Basin "F" and the release of the same materials which were probably present in the air during the early 1970's, in 1988 and 1989. I am convinced that the headache, airway problems, bronchitis, and coughing and the aesthetic disturbances which were noted both in 1972 and 1988-1989 are traceable to these air emissions.

I have reviewed the extremely limited toxicologic information which is available on DIMP. This material has been tested only in small studies. Diarrhea and hypermotility have been noted in dogs dosed with high concentrations of Dimp. While it is impossible to draw any conclusions from the study to which I refer which was performed by the Dept. of Defense in 1980, It does appear that when the overall group of studies are reviewed, there are concentrations of DIMP in the water which can cause diarrhea and disturbance of gastrointestinal function such as those experienced by the Land family members.

The irritant nature of the materials has not been adequately tested. I am unable to draw a conclusion concerning DIMP, but it does appear that the sulfur mustard breakdown products may well have been responsible for many of the airway and eye problems".

***** Dr. Teitelbaum's report in testimony continues, "The Land family has a panolply of complaints. several of these are most probably related to some chemicals that we are not aware of at this time, and are highly likely to be associated with the exposures which occurred as a result of contamination of their drinking water, their hygiene water, and the air they breathed, which chemical pollution eminated at the Rocky Mountain Arsenal, during production, and disposal of same. Most of these complaints are consistent, reasonable, and are temporally related to the groundwater, as well as the air contamination problem. It is my intention to examine some or all of the members of the Land family in order to determine what their current clinical condition is. I will then be better able to focus on other problems which were associated with the Rocky Mountain Arsenal groundwater, and air contamination problem. This will allow for the preparation of testimony that will be required for the damages part of this Trial, which may be some months away".

"Studies of the groundwater for substances other than DIMP, and IMPA, and some of the sulfur compounds, the Lewisite compounds, as well as the halogenated hydrocarbons, were generally not remarkable. They did not provide any other explanation for the family's difficulties and so the probable cause remains the Arsenal's Chemical Pollution of both water and air".

"A recent abstract which appeared in Government Reports, and Announcements Index Issue 4, 1989 over the signatures of Cataldo et al. from the Fort McClelan, Alabama, International Worshop on Chemical simulants indicated that DIMP was being used as "Nerve agent stimulant in the open air." This constitutes an extremely interesting observation. It suggests that at the time the Land family was living adjacent to the Arsenal, the production of DIMP was not as a by-product from GB. It rather may have been a primary chemical product being manufactured at the Arsenal. This paper indicates that 2 choloro-ethyl, -ethyl sulfide was also being used as an open air simulant for blister gases. This material has also been found in the air and in the Basin F Sludge. Both of these findings suggest that we do not know yet everything there is to know about the contamination of the environment from the Rocky Mountain Arsenal. I suspect that these materials were being produced. They may even have been released at the Arsenal for testing purposes and they may have reached the Land family in the air as well as in the groundwater during these processes".

***** Dr. Teitelbaum's Testimony, "It is my opinion that the records substantiate the contamination of the groundwater under the Land's property with materials which emanated from the Rocky Mountain Arsenal. I believe that the records further support the fact that the various members of the family would have consumed quantities of these materials in drinking water, cooking water, the hygiene water, and inhaled air which was contaminated by the Rocky Mountain Arsenal Chemical Pollution in the period of the 1960's into 1970-1972, into 1973, through 74".

"I believe that the Land's suffered a consistent syndrome of headache, airway and digestive distress, and mental disturbance which cannot be explained by any other toxicologic mechanism than consumption of water, using it for hygiene purposes, and inhalation of the air contaminated with chemicals which emanated from the Rocky Mountain Arsenal. I believe that the sentinel events which occurred with the Land's cattle support not only the water contamination, but also the air contamination with emissions from the Arsenal quite like those which we observed in 1988 and 1989 during the period of time the Basin "F" work was being done".

"The paucicity of information about DIMP and IMPA, the absolute and complete absence of any properly done chronic toxicology studies and unknown mutangenicity and carcinogenicity status of these chemicals make it clear that, at the very least, these people who have suffered a significant exposure to chemicals emitted from the Rocky Mountain Arsenal and who became ill as a result of those emissions and who are now in a position of complete medical uncertainty, need thorough medical examinations and long-term medical surveillance. This unfortunate family may prove too be the Governments first set of laboratory animals exposed on a long-term basis to these materials. For their benefit, and for the benefit of society in geneal, they need to have ongoing medical surveillance and a much more thorough medical work-up than has been done to this point".

***** Judge Robinson asserted at this point regarding nuisance, "Lowder v. Tina Marie Homes, Inc., 601 P.2d 657 (Colo. App. 1979) stating general rule that liability for nuisance may rest upon any one of three types of conduct: (1) an intentional invasion of a person's interest's. (2) a negligent invasion of a person's interest's. (3) conduct so dangerous to life or property and so abnormal or out-of-place in its surroundings as to fall within principles of strict liability); and Thompson v. Andover Oil Co., 691 P.2d 77 (Okla. App. 1984) (Court stating that negligence is not an essential element of a cause of action for nuisance and need not be proved. In this case, Defendant's liability for nuisance "rests on a duty" There has been sufficient evidence presented to this Court to establish Defendant's breached that duty, which caused the contamination of Plaintiffs' groundwater and air". "Established in sworn testimony of Defendants own expert witnesses, G. Barbieri, and G. Watson".

***** MOTION FOR DESIGNATION OF ADDITIONAL EXPERT WITNESS; "At least four years prior to the trial Sam McClaren the Plaintiffs' Attorney tried and asked the Court for the designation of additional expert witnesses. The Court ignored those motions, and made no order at all until the Plaintiff's were ready to call Dr. Wingeleth, and Sam McClaren, third day in trial. Plaintiff's called on their expert witnesses, the Defendant immediately objected, and the Judge denied Plaintiff's Motion to call their expert witness

Dr. Dale Clifford Wingeleth, Ph.D., Chematox Laboratory, Inc., Boulder Colorado. His testimony would include the following (a) an explanation of the deficiencies in the governments' processes used in manufacture or destruction of nerve gas and its associated chemicals and by-products; (b) an explanation of reports of the government in finding certain chemicals in an acquifer although such chemicals had not been intentionally manufactured on the Arsenal; (c) a description of the techniques for water testing in and prior to 1972 and the effect of the absence of knowledge of chemicals either used or manufactured on the Arsenal in testing Contaminatied waters; (d) a description of current technology in the testing of water; (e) an explanation of the deficiencies in the disposal of chemicals by Shell Chemical Company; (f) an explanation of the manufacture of organic chemicals to include the use of resin beads. Dr. Wingeleth's direct testimony will be approximately 3 hours. This denial dates back to a time in 1992 when the government had filed a motion for a protection order. I didn't understand at the time and thought the Judge had denied that order. As you look back and then look at the expert testimony Dr. Wingeleth would have been able to give, I truly think he could have turned this whole case upside down against the government, and they knew it, stopped at nothing to stop it".

***** Then lets talk of the Plaintiff's Attorney who is also an expert in geology, hydrology, and chemistry and Mr. McClaren was also our expert witness. He became our Attorney when the Law firm who was handling this for us, and had already spent hundred's of thousand's of dollars building this case. Upon returning to Denver after "WINNING" the Summary Judgment trial in Washington D.C. They called a meeting with everyone present including Sam McClaren, and without explanation, suddenly and mysteriously withdrew from the case completely. That is now Mr. McClaren became the expert witness and the Attorney. As I had said he had tried to clear the groundwork with the Courts since 1992 in order to testify at the trial. For him the Judge had agreed that he would ask the questions as an attorney and answer them as an expert. The Judge had agreed with this process for him at trial earlier last year, However when it became time for him to testify the Judge stated to Mr. McClaren he could be a witness and testify as an expert only if questioned by another attorney, and that attorney had to be admitted to the bar for the Court of Claims. He gave the Plaintiff's one hour to find someone to act as the attorney during his testimony, finding an attorney in Denver, Colorado in a brief period who is admitted to the Court of Federal Claims is nearly impossible. For this reason, Plaintiffs' elected to make the best of a bad situation.

***** During that hour Mr McClaren also was told the Attornies representing the United States weren't in 1992, and probably are not now members of the Bar to practice before the Court of Claims. This needs to be checked. It was impossible for the Plaintiffs' to find an attorney that is a member of the Bar of the Court of Claims, and was unable for another expert witness to testify. Then thinking back at what the Defendants had said that any licensed attorney would know in order to appear as an attorney in a proceeding before a particular court, one must be admitted to practice before that court. Perhaps we have two Attorney's Mr. Regan, and Mr. Quinlan III practicing before this court as legal counsel for the defendants, were not licensed to practice before the Court of Claims until several years after the trial had begun and where many of the decissions such as the one were looking at here. The defendants attorney's when the motion for a protection order was filed, did not belong to the bar, or were licensed before the Court of Claims, and the court rules here against the Plaintiff's when the motions for Dr. Wingeleth, and Mr. McClaren had been filed before the Court multiple times, then the Court waits until the last four days before the Trial to deny motions naming Sam McClaren, and Dr. Dale Wingeleth as expert witnesses for the Plaintiff. Several times in hearings before the Trial it was discussed that Sam McClaren would ask the questions before the Court then give sworn answers to those questions. Yet at trial the Trial Judge Robinson refused and then denied Sam McClaren the right to give expert testimony, then gave the Plaintiff's one hour at lunch time to find an attorney who could ask the questions of Mr. McClaren, and belong to the bar of the Court of Claims. Since this would be very rare that any attorney in Denver would even be licensed to practice before the Court of Claims, There was a decission to follow another rule of the same Law that would give Mr. McClaren the opportunity for his expert testimony, and bring in another attorney, called (PRO-HAC-VICE.) to ask the questions, He would be licensed to practice before the Federal Courts, for this one time basis, However Judge Robinson disallowed that also, specifically stating he would have to be licenced before the Court of Claims, and this would be impossible.

***** ABUSE OF DISCRETION. It was by a pleading dated July 2, 1995, Harding and Ogborn moved to be allowed to withdraw as counsel for the Plaintiffs. This was after a supposed settlement conference. The attornies of record came back with an offer of up to maybe $10 thousand dollars that would not begin to pay the bills, The Law Firm had already peaked $300, thousand dollars into the case. There behavior was very awkward, was a take it or leave it situation. They warned at that time if we didn't take it they were pulling out. Plaintiff's answer was to reject the offer and at that moment Plaintiffs ask Sam McClaren to assist. Then On or about August 6, 1995, just 5 months before trial Judge Robinson called for a telephone conference. Mr McClaren noted at that time that it might be necessary for him to act as trial counsel, and to testify. The Hearing Officer suggested that an effort be made to find additional witnesses to minimize such testimony. Mr. McClaren found that witness Dr. Dale Wingeleth, and the Judge denied the motion to bring him on board to testify. The Hearing Officer had been advised that Mr. McClaren was seeking the consent of the Plaintiffs for him to enter an appearance. Plaintiff's had determined that it was impossible for them to obtain competent counsel within the time constraints of being prepared for trial on January 22 1996. Under these circumstances, Mr. McClaren agreed to act as trial counsel since his failure too do so would work an extreme hardship upon the Plaintiffs' even though it remained necessary for him to testify. The Hearing Officer used the very highest Abuse of Discretion, to The Hearing Officer used the very highest abuse of discretion to, deny the two most important witnesses for the Plaintiffs'.

***** Not only were the requirements of the Hearing Officer a complete surprise to the Plaintiffs, they were unduly burdensome in that an attorney admitted to practice before the Court of Federal Claims was required. This requirement of the Hearing Officer was contrary to Rule 13 of Appendix D of this court's rules which allows attorneys admitted before the highest court of any state to practice before the Court of Federal Claims in Congressional Reference Cases. All motions had been filed in ample time to allow the Defendant to prepare for trial. Much of the confusion evidenced by the Report of what was or was not in the record could well have been avoided. Appropriate measures to lift the restrictions of the January 22, 1992 order should have been made prior to further hearings or this trial.

***** Judge Robinson had returned to the Court that afternoon, as if he were a changed person. Any and all motions before the Court for the Plaintiff's were denied. The Judge became very demanding of the Plaintiff's attorney telling the plaintiff's they did not prove that DIMP had caused the injuries. There was evidence in the records that the Water, was found to contain Nemagon, IMPA, DIMP, Deildrin, Endrin, and these are chemicals they specifically tested for and found. However the Judge would not listen to anything at this point. The Plaintiff's attorney ask to have Colorado Law placed in record that Dimp over 8 ppb is dangerous. The Judge would not allow the safe standard of Colorado put in the record and stated were using the standard set by EPA which is 600ppb, and then demanded the Plaintiff prove there were other chemicals in the water besides DIMP, and why didn't these people have blood tests done in 1972, He was told there were no tests available at that time to detect DIMP, IMPA, or for any of the other Chemicals that were in the water and in the air. The Judge was told when the Army had done the Testing in 1975, DIMP well above the legal limits of Colorado were found in the water along with IMPA, Deildrin, and Endren, two insecticides that are so deadly and sytemic they had been banned in the United States.

***** The Judge leaning very heavy toward answers that would be needed for a Tort Claim, and remember he is the one who put on the record there is no room in the Court of Claims of Tort Claims. He seemed to be throwing all the case the Plaintiff's had presented under Colorado, Circumstantial theory where if a person is able to show one chemical from the polluter he would be liable for anything elso on his property.

***** It was easy to see he was pushing for the Federal Tort Claim Act, FTCA. At that moment Judge Robinson Invoked the Federal Tort Claims Act in theory ending the case being heard under Colorado Law, the ending Plaintiff's hopes using Circumstantial that had been approved by the Court, and doing exactly what he said would not happen in the Court of Claims, "There is no room in the Court of Claims for the Tort Claims Act.

***** DISCRETIONARY FUNCTION; Judge Robinson involked the FTCA. Federal Tort Claims Act, FTCA Claims based on the exercise of a performance, or failure to exercise or perform a discretionary function or duty on the part of a Federal Agency or Government Employee are not cognizable under the FTCA. The purpose is to protect the Government from liability that would seriously handicap efficient Governmental operations. Whether or not the discretion is abused is immaterial. United States V. 331 f.2d 498 (10th Cir. 1964)

***** Like it says whether or not discretion is abused is immaterial. The reason for this was to protect the United States from admitting any liability through the Court. having an unknown thousands that lived anywhere near the Arsenal, that were or even may have been injured by the pollution. from getting a lawyer who would file a claim for each one that may have had a share, and the Government would not have allowed that to happen. At this point they knew already they had lost under Colorado law, and with the use of Circumstantial law. Even for the Court to have made a decission it would be of record, and open the doors to everyone. They took the power of the liability issue in an equitable claim away from the Plaintiff's, by the little known law yet such an "enormous power" to be used only when the United States feels vulnerable. see above, that "states even if they abused discretion is immaterial". That means a ruling of this nature can excuse them from anything.

***** Judge Robinson could not have handed down from the Court a Lawful Claim, or an Equitable because that would have been a Court decission that would have had an amount for damage attached, leaving the government liable in a Court of law for anything that was similar. However Judge Robinson did hand down an award that would need a bill before Congress, and a decission as to the amount of reparations in satisfaction of the wrongful injuries suffered, this had to be enacted by the Congress of the United States.

***** Judge Robinson did not think this had to go before any Court, Federal, or the Court of Claims. He thought it should have been handled discretely by Congress in the first place. By making his decission this way he took equitable claims away, and he took equitable claims away, and he took lawful claims away from the Plaintiff's sending the Plaintiff's back to Congress. This means the Land's would have to depend on their Congressmen, and Senators from Colorado in order to put a bill before Congresss. He gave it back to Congress with an order for a Gratuity. A Gratuity is the act of making amends, or giving satisfaction for wrongdoing, and wrongful injury. This is done by a bill for relief to those injured that is approved by Congress. That act was given by the Court of Claims to the Plaintiff's with all facts present in the record before the review panel i.e. Court of Claims. This was returned to Congress for action in February 1997. The Plaintiff's can understand what has taken place and did not want any harm to come of this to the United States. The Clean up process at the Arsenal is being followed closely under the Scrutiny of the Colorado Health Department. This hopefully will be a clear lesson to the Army in their clean up of other sites throughout the United States, and how chemicals can be handled without such massive destruction, death, prolonged illness, and huge costs to the Taxpayer's of this nation. The tab on the Rocky Mountain Arsenal clean-up last heard was 2.6 billion dollars going up every day since, and to the year 2050 and beyond, to even begin to honor their "priority one" clean up obligations. More money has already been spent in 20 years of clean up than the entire 45 year cost of the Arsenal operations. The clean up will take almost double, and beyond the operations time frame, for the Army just to clean the surface, the deadly chemicals will forever be there with life time expectancy of many thousands of years. Clean up operations have now been over twenty years and 2.6 billion plus, and it is already projected to the year 2050, and beyond just for clean up. With millions of tons of the worlds deadliest chemicals, and possibly bio-warfare, along with experiments on bio-chem warfare. Very much like the Anti Wheat Spores that were made back in the 1950's, at the Rocky Mountain Arsenal. This I suppose along with all the rest of the deadly chemicals, some of which we may never know or even be made aware of, will be entombed at the Arsenal. the Army didn't know then or do they know now how this can be destroyed. they had already spread this wheat spores made specifically for the Russian wheat belt during the Cold War Era. over the entire wheat belt in the United States just to see what effect it was going to have. Little do the people in Colorado, Nebraska, Kansas, Iowa, and Minnesota, possibly Oklahoma, and Texas. Yes this happened I have no fear of telling the truth, this is on file with the United States. What remained of the wheat spores was spread over hundreds of acres at the Arsenal and covered up discretely, that was in 1950's and it is still there today 2001.

***** What Mr. Kruse, Defendants Counsel stipulated to is more than sufficient to establish the essential elements of trespass and Nuisance in this case. It is a factual basis establishing the liability element of Plaintiff's trespass and Nuisance claims.

"The United States is not denying that DIMP was formed at the Arsenal at some time between the time it was built and the current date. United States is not contesting that point".

"Mr Kruse stipulated that, DIMP found off-post originated or had its source at the Arsenal".

"Mr. Kruse stipulated the United States is not contesting that there is no other source of DIMP found off-post than that of the Rocky Mountain Arsenal being the only source of any DIMP found off post." "DIMP was found off post in Plaitiff's soil and in the plaitiff's well water in the early 70's. This alone is sufficient to establish Defendant's RMA activities inevitably caused" "the physical intrusion of DIMP and other toxic chemicals upon the property of Plaintiffs without the permission of Plaintiffs."

***** STATEMENTS ATTESTED TO BY JUDGE WILKES ROBINSON, IN THE CASES LAND et al V. UNITED STATES of AMERICA.

*** "This finger-printed chemical, to which there is no other source of this material is conceivable, establishes without any question that the toxic material which reached the Land family had, as its source, the Rocky Mountain Arsenal GB manufacturing facility"

Judge Wilkes C. Robinson

*** “The RMA halted monitoring in 1960, and refused to resume testing.”

Judge Wilkes C. Robinson

*** “The RMA was aware Basin F was leaking in 1965, and 1969, the protective membrane was missing. The Army understood that it was operating Basin F on the premise that it was leaking.”

Judge Wilkes C. Robinson

*** “Activities on the Arsenal have resulted in one of the worst hazardous waste pollution sites in the Country.”

Judge Wilkes C. Robinson

*** “The Department of Health supposedly informed Larry Land in correspondence dated June 11, 1975 of the presence of chemicals in the house well. Tri County Health Department as well have said on July 7, 1975 it had indicated the RMA was the source of contamination.”

Judge Wilkes C. Robinson

*** “Larry disagrees and states; “Neither the State or Tri County Health Department notified the Land’s of anything. They still had the Chief Chemist coming out to the farm to deny the death of illness of the cattle were due to any toxic chemicals in the water. He seen the water running red, he seen the test results that Larry had run in private Laboratories, yet the State tested for none of those chemicals or not that he would have had knowledge was leaking from the Arsenal. The Chief Chemist at that time was the only person allowed to come out on any calls by the Land’s, always stating there is no way in Gods green earth could chemicals from the RMA get into your water. I believe he knew all and making sure the mail was never sent to the Lands’. keeping it low key as he had been able to for the Army for Years. If he was may he live with guilt, he never done anything that would have showed any difference. Never was the water tested for anything that may have come from the Arsenal. He seen first hand the 300 to 400 mesh size resin beads in the water, knowing they were not natural, and were believed to be a tracer for the RMA to see the boundary of pollution, Larry had microscopic photos made of the tiny tracer’s so small they could move anywhere water could move. yet he done nothing. Even after showing him the chemicals found in the water by out side labs he tested the water for none of these things, Larry to this day has the lists that he gave directly to Mr. Bill Dunn for testing and heard nothing except there is no way in Gods Green Earth your water could be affected by the RMA."

*** "In regard to whether the activity at the RMA constitutes ultrahazardous or abnormally dangerous activity, Defendant has failed to convince the Court, that Colorado would not consider it as such. Defendant was aware as far back as the 1950's that the disposing techniques posed a risk too all those neighboring the RMA. Colorado stated it could consider such activity ultra hazardous or abnormally dangerous. This Court agrees and accepts that the manufacture and disposal of toxic munitions is ultrahazardous or abnormally dangerous activity."

Judge Wilkes C. Robinson

*** "The Defendant Attorney's tried using the legislature to relieve the government of liability, and showed no special circumstances evidencing any intent by the legislature., Memorandum does not evidence an intent to relieve liability for either the production or destruction of toxic munitions. More over the defendant did not show that it's manner of storing and disposing of toxic muntions was authorized by the legislature."

Judge Wilkes C. Robinson

***** "The referencing statute in the present case contains no concession of liability. However, Congress would not have needed to refer this case to the Court for "claims against the United States based upon helath problems and other related injuries resulting from the operation and activities at the Rocky Mountain Arsenal merely to affirm that the FTCA or RMA enabling legislation barred any recovery for liability arising out of the operation of the RMA."

Judge Wilkes C. Robinson

*** "Plaintiff's have been damaged by the polluted groundwater as well as polluted air, and soil. An obligation to confine contaminants in Colorado is absolute, no negligence is required in order to prove liability. Pollution from the Rocky Mountain Arsenal was fingerprinted to the Land wells, as well as the property, and the air above, Plaintiffs are not required to prove negligence, since Defendant has already done a "Wrong" by its failure to confine the pollutants to its property."

Judge Wilkes C. Robinson

*** "Trespassing would make the Defendant automatically liable for Nuisance, Ultrahazardous Activities, Strict Liability, Fraud by Concealment or deceit as an actionable offense.

Judge Wilkes C. Robinson

*** "Upon general principles of law it is so entirely clear that defendant is liable in damages for this pollution of a stream which has injured plaintiff, that we do not cite authorities or deem it necessary to argue such a self-evident propostion."

Judge Wilkes C. Robinson

*** "Under the Laws of the State of Colorado, any private party, Corporation, or Business would be liable to the Plaintiffs if they had as the RMA did, (1) failure to confine contaminants is absolute. The probability of and certainty was established by the Plaintiff, as well as the fact wastes from the RMA did migrate into the Land wells, and was proven a fingerprinted source being the RMA."

Judge Wilkes C. Robinson

*** "Judge Robinson stated several times the Defendants conduct does fall within the scope of a Congressional Reference, and fact this Court finds that such activity might well be considered "Wrongful or even Negligent".

Judge Wilkes C. Robinson

*** "The Plaintiff proved to the Court the Common Law negligence required in Colorado to prove a Prima Facie case. (1) The existence of a duty owed by the defendant to the plaintiff; (2) A breach of that duty; (3) The breach actually, and proximately caused injurie's;."

Judge Wilkes C. Robinson

*** "In Colorado it is axiomatic that while engaging in a particular activity every person is "bound to exercise that reasonable care and caution which would be exercised by a reasonably prudent and cautious person under the same or similar circumstances. As a matter of Law, Courts will hold every reasonably prudent and careful [person] to the exercise of the utmost care and diligence in protecting the public from the dangers necessarily incident to the carrying on of a hazardous business."

Judge Wilkes C. Robinson

*** "Plaintiffs argued that the Army's conduct on the RMA,, manufacturing and demilitarizing chemical weapons and disposing of chemical wastes, created a substantial risk. "THIS COURT AGREES". "The point is well made that the Army has, from the 1940's been engaged in extremely risky activities that demand the utmost care and diligence to protect the public." "They gave the Public no due care or warning of the hazards, as was attached automatically by Colorado Law".

Judge Wilkes C. Robinson

*** "This Court is satisfied, after resolving all doubts as to the facts, presumptions, and inferences in favor of plaintiffs, that genuine issues were presented held firmly by material facts as to Plaintiff's claims including those sounding in negligence. The Defendants failed to to convince the Court that material fact did not show the Defendants was negligent or wrong doing." "IT WAS HELD IN THE AFFIRMATIVE IN FAVOR OF THE PLAITIFFS."

Judge Wilkes C. Robinson

***** "The Judge in summing up the fact he would not allow any further witnesses for the Plaintiff, He stated he had heard all he needs to hear, and is aware of what he must do. He almost insidiously repeated to the Plaintiff you did not prove that DIMP done the damage. Yet he wouldn't even allow the Plaintiff to call two of the most important witnesses that also could easily have linked the causual connection between DIMP, what happened and a multitude of other toxins. No matter how many times the Judge was told that to link the damage to only one chemical DIMP would be Involking the

Federal Tort Claim Act, FTCA, and he knew exactly what we meant as he involked the FTCA."

***** "IT was noted by Dr. Leonard Konikou, of the United States Geological Agency in 1973 when he had a meeting with the United States Army, and Shell Chemical Officials had openly admitted that the claims by the Land's of Nerve agents and pesticides in their well water were true". It was the Dr. Konikou reports of Geology, and Hydrology in that area clearly showed the plumes of toxins migrating from the Arsenal at that time polluting over 35 thousand acres off the RMA.

Chapter Fifteen

THE FINAL SOLUTION?

The Judges final decission had already been made, it was just a matter of giving him time to reduce it to a report.

LOOKING BACK

To Judge Robinson decission on the defendant's Protection order that he had used against the Plaintiff in naming any further witnesses, as if they had won their motion to stop the Plaintiff's, and in fact they had lost and the Judge had denied the Defendant's any right to use that protection order. He stated "In denying defendants motion for a protective order, this Court recognized that trespass is a type of wrongful conduct for which liability may be imposed upon the government under Colorado Law. For the Judge to make such a profound abuse of discretion when he denied Plaintiff's their right to other expert witnesses on the grounds the Defendant's had won their protective order March 26, 1992 to disallow the Plaintiffs any further witnesses and discovery. Yet it was the Plaintiff's who had won, and the protection order for the Defendant's was denied, with the Judge stating the Defendant under this Courts rules has failed to show 'good cause' for the issuance of a protective order.

***** "The contamination of the air and the underground waters to the north of the Arsenal is well established. The government is liable for damages resuting from such contamination under theories of strict liability and for breach of the duty to confine such contaminates to Arsenal property. Failures to confine were both negligent and willful. The government had the duty to determine the extent of pollution and to inform adjoining owners of the dangers from such pollution. The government not only failed to fulfill its obligatory duty of determining the extent and dangers, but suppressed such information denying the adjoining landowners the abillity to mitigate their damages."

***** "The government's contamination of the water underlying Plaintiffs' properties and the contamination of the air over such properties was in violation of numerous duties which would have been imposed upon the government had it been a private party. The combination of both such contamination have resulted in damage for which the government in equity should be responsible. Such damages represent equitable claims and are not a gratuity.

***** It seems the Plaintiff's were forced to go one step further in the Court of Claims. A motion of exceptions from the report of the Hearing Officer The Honorable Wilkes C. Robinson, Judge. to be heard in Washington D.C. by a three panel Judgeship., much like an Appeals Court.

***** We have all heard the story about the man with a pet two thousand pound gorilla who responded to the question of where does the gorilla sleep? with the answer of "Any where it wants to!" Neighbors who lived in the vicinity of the Rocky Mountain Arsenal ("Arsenal") have compared their situation to that man. The Arsenal did pretty much as it wanted to do prior to the passage of the comprehensive Environmental Response, Compensation and Liability Act of 1980 ("CERCLAL"), Pub. L. No. 96-510, 94 Stat. 2767. The result of the Arsenal's unsupervised activities were summarized by Judge Baldock in Daigle v. Shell Oil Co., 972 F.2d 1529, 1531, (1992) as follows: The combined acitivities of the Army and Shell Oil Company, have resulted in one of the worst hazardous waste pollution sites in the Country.

***** "This controversy stems from the year 1956, when the Army constructed and began using Basin F, a ninety-three acre hazardous waste surface impoundment on the RMA. The Army, as operator of the Arsenal, used Basin F to impound hazardous waste generated from it's chemical warfare agent, chemical product and incendiary munition manufacturing activities. In addition, Shell used Basin F under lease from the Army to impound hazardous waste generated in it's herbicide, and pesticide manufacturing activities on the Arsenal. The combined activities of the Army and Shell on the Arsenal resulted in one of the World's worst hazardous waste pollution sites in the Country, and Basin F is only a small portion of the problem. Army officials have estimated that the twenty-seven square mile Arsenal has 120 contamination sites which contain huge quantities of liquid and solid wastes, some of which are unique because of the mixture of private herbicide and pesticide manufacturing activities along with the Army muntitions manufacturing activities."

***** The contamination at the RMA. was so extensive that the average person's first reaction would be to deny that the government could have created a situation so harmful to it's citizens now and for years to come."

***** "This will breifly summarize the experiences of one extended family's experiences with the Arsenal, and Colorado agencies denying that anything wrong had occurred. As time passes and more information is collected and agencies became better informed, most agencies change their position. Doctor's who examined the Plaintiffs also tended to deny that anything wrong had occurred until they were better informed. The Arsenal continues to deny any wrong doing. The purpose here is to set forth the information which was ignored by Judge Robinson, the Hearing Officer, in this action, even having been given this information first hand as evidence, hopefully the panel may correct the errors in the report and findings of Judge Robinson."

***** "This case had been referred for determination of facts pursuant to House Resolution 61, adopted August 11, 1988, referring House Bill H.R. 816 which would provide for payment of an unspecified amount to the Plaintiffs in "full and complete satisfaction of all claims against the United States based upon helath and other related problems resulting from the operations and activities at the Rocky Mountain Arsenal…"

***** First amended complaint, alleging equitable claims relating to personal health and injury to livestock, against the United States based upon Negligence, (wrong doing),, Strict Liability, Ultra Hazardous activities, Inherently Dangerous Activities, Trespass, Nuisance and Deceit and Fraud based upon concealment.

"The government answered generally denying Plaintiffs' allegations but admitted that TCE (trichloroethylene) is one of the contaminants escaping from the Arsenal; That DIMP had been found on the properties of James Land, and Larry Land. "Therefore admitting to the air contamination"; and "That there were chemical contaminants that did flow off the Arsenal in the groundwater which then passed under the Plaintiffs' property". """Then admitted that DIMP "Diisopropylmethylphosphonate" and IMPA "Isopropylmethylphosphonatealpha", and along with these tests that the Army had done on the Land wells, also discovered toxic levels of Deildrin, and Endrin". This was filed and placed into evidence with this Court. There was evidence using the Colorado safe standard of DIMP 8 ppb, that the levels found in the Land's well water, far exceeded that standard. There were other tests that were done and placed directly into the evidence before this Court, Chloride was the only indicator found at 200ppm that meant contamination and above that they used a percentage. Thru 1975 this was the Army or State only method to identify the possibility of Arsenal Chemicals in the water without naming out a single chemical except chloride. Land wells showed a reading of 238 ppm and 250 ppm chloride meaning the Land wells contained at least 5% to 8% toxic and Ultrahazardous pollution in their well water". "Tests were also ran showing the Land wells contained toxic levels of Nemagon". "Mr Kruse Counsel for the United States confirmed the fact the United States was not denying that DIMP was formed at the Arsenal at some time between the time it was built, and the current date". "Mr. Kruse also stated affirmatively that the United States did find DIMP off-post that it originated and had its source at the Arsenal". "The government asserted affirmative defenses of statutes of limitation, lack of jurisdiction of this Court, contributory negligence, assumption of the risk, release of a joint tortfeasor and laches" "Mr Kruse stated the United States is not contesting the """fact there is no other source of DIMP except the Arsenal".

***** Defendant's admission that the RMA was the only source of DIMP off post or on the Plaintiffs property, and in their well water. Knowing there were no tests available until 1975 to even detect DIMP. The DIMP found on the Land, and in the well water on the Land's property in July 1974, and July 1975, is sufficient to establish

Defendant's RMA activities inevitably caused "The physical intrusion of [DIMP and other toxic chemicals] upon the property of the Plaintiff's, and in their well water without the permission of the Plaintiffs.

***** "The bifurcated first trial was solely on the issue of whether the Defendant had committed a wrong which caused injury. CAUSATIVE CONNECTION, What caused the effect. Toxic chemicals traced directly to the RMA, and it was the Army that discovered those chemicals in the Land Well Water and on their Land. This was held before the Hearing Officer January 22, through January 25, 1996 in Denver Colorado. April 15, 1996 recommending a finding that any relief would be a gratuity there fore taking away any legal and equitable right for the Plaintiff in this case to any charge of punishment, where the Court would have the power to set any amount for 'retribution', with a Gratuity, the Court is giving it back to Congress, who would place a Bill before the House of Congress to determine and appropriate the funds for the Lands'.

***** "Gratuity, Rests, on an equitable or honorary obligation. The Nation, speaking broadly, owes a "debt" to a [party when its] claim grows out of general principles of right and justice; when in other words, it is based upon considerations of a moral or merely honorary nature, such as are binding on the honor of an individual, or a nation. Although this debt obtained recognition in this Court of Law. It was grounded in the prcepts of the conscience, and not in any sanction of positive law. as on legal, or equitable claim, the Court referred it back as an equitable, or honorary obligation, to the Congress of the United States, for recognition as a gratuity, and for a determination on appropriations".

"This equitable entitlement has been established in the White Sand Ranchers, much like the Land case, "where the plaintiffs suffered an invasion of rights", for which compensation was due them. An award and payment for which no legal or eqitable entitlement was established.

STATEMENT OF ISSUES OF LAW AND FACT.

A. Issues of Law:

1. Did the Hearing officer properly apply Colorado law in requiring proof of a specific compound in order to prove causation?
2. Did the Hearing Officer fail to apply Colorado law by ignoring circumstantial proof of causation?.

B. Issues of Fact:

1. Were the findings of the Hearing Officer that the proximate cause of the illness and death of cattle was from causes other than the contamination of well water supported by evidence?
2. Did the Plaintiffs prove a circumstantial case that Defendant's contaminants were responsible for or contributed to illnesses of Plaintiffs and to the illness and death of cattle?

"Colorado law provides for two alternative methods to prove causation in a toxic tort case. the first method involves showing that a single toxic compound was present in sufficient concentration to cause injury. The second method involves circumstantial proof of the probability of the presence of one or more contaminants, which may or may not be identified, which are shown to have probably caused or contributed to the cause of injury.

***** "Evidence of a multiple contaminants placed into unlined Basin A where they entered the ground water and by natural forces were transported to Plaintiffs lands where they contaminated the well water which plaintffs were using for domestic purposes constitutes unrefuted circumstantial evidence of causation. Such well water had not been contaminated prior to the Defendant's contamination reaching the shallow aquifer adjoining plaintiffs' wells. The evidence proved the probability that there were no sources of contmination other than sources located on defendant's property. The pollution found in the Plaintiff's wells was fingerprinted to the RMA. Each such source was a point where Defendant had failed to properly dispose of toxic substances. Upon such contaminants entering the well water, people and animals consuming such water became ill and died. The Animals and people which survived, had been relocated to use other water then ceased to become ill and die. The animals which survived ceased to mature and each carried their disabilities but lived, the humans also each carried on with a multitude of disabilities which will severely change their lives and shorten their life spans, and the lives of the children and their children's children, which ended up with a multitude of learning disabilities, heart problems, and others. The source of chloride found in 1972 1973 plain and simple showed 5 to 8% toxic arsenal contaminants in the Land Wells at that exact time frame which chloride at 238 to 250 ppm clearly proving the well water contained from 1800 ppb to 2800ppb of DIMP, (SARIN Nerve Agent) in the Land wells. This was placed as evidence before the Court and stands in the records, however Judge Robinson did not even consider it as evidence. 1974, and before, Chloride was the only source of testing available that would show what percentage of Arsenal pollution was in the well water. The amount of DIMP. in the well water was enough too kill and injure what ever may drink it or bathe in it.

***** "In a test for fluoride there are two kinds, Calcium fluoride and Sodium fluoride, the test available for either one at the time in1972 when everything was injured or killed the Fluoride content of the water was 1.7ppm and 2.3 ppm the fluoride standard had been set at 2.4 ppm because the only fluoride that existed in normal water supplies was calcium fluoride. No tests again were available to detect man made sodium fluoride separately, which by itself is a deadly toxic nerve agent it has a more electropositive element "O[r is much more radical]" Sodium fluoride is felt to have been in small quantities in the well water that checked out simply as fluoride. If we were to take the 1.7 ppm found of fluoride divide by 10 and just use 1/10 of that and call it sodium fluoride we would have .170 ppm or 170 ppb. would have been very toxic and soak right through the skin. Even a small quantity of sodium fluoride because of it being R[adical] in the water is now believed what caused very disableing teeth and bone disease that we live with today as disabilities, it is called in-part Chondro-Calcineosis of the bone. very painful disability of the ligaments, bones, and the teeth. Baby teeth were very deformed, everyone's teeth and gums turned black and teeth became very brittle. The animals teeth turned black and fell right out of the gums. Larry had animals that could not eat regular hay, and grain, everything had to be ground up for them to even survive".

***** "The defendant contended in Elam, that the injurious compound must be identified for the plaintiffs to prevail. Judge Robinson rejected this with a comment that "[t]he identiity of the toxic substance to which harm is attributed, however, may be shown by circumstantial evidence". A cattle case specifically holding that circumstantial proof, and need not identify specific chemical contaminant in Rusch v. Philips Petroleum Co".

***** Application of the Walton report on the level of contamination results in the determination that DIMP concentration in the shallow aquifer of the Home Place in November 1972 was in the range of 1,440 to 2,400 ppb. The reliability of this calculation is proven by the fact that on May 20, 1974, a well up gradient to the Home Place, located at 9760 Peoria. one-half south of the Home Place and one-half mile north of the RMA, tested 2,200ppb of DIMP. Further calculations is found in testing in the wheat field where tests of water from three wells within just feet of each other showed testing results 310ppb, to 4,960ppb, to 6,000ppb. These test results are all from the shallow aquifer within the same paleo valley adjoining the Home Place. Application of the chemical fate-transportation calculations results in determining DIMP concentration the the Home Place in 1972 at 1,440ppb, to 2,400pp which would be two the four times the standard of 600ppb set by EPA. Plaintiff's have satisfied their burden of proof of proximate cause even if the issue were to be decided solely upon DIMP.

***** Jeff Edson Colorado State Heath Department, "states that EPA has no business setting a standard in colorado for the RMA, that Colorado sets the Standards on the water. If the Court decided the Land case using Colorado Law, they would not, and could not use what EPA, and the Court called it's Standard at 600ppb instead by law would be obligated to use Colorado's Standard of 8ppb". But They Did, "Judge Robinson used EPA's standard to make his decission, and ignored the plaintiff trying to remind him he needed to use Colorado Law, However he denied the use of the Colorado Standard and used EPA's standard, clearly a violation of Colorado

Law”. While supposedly making a decission under Colorado Law. Jeff Edson said, “Even a Federal Judge cannot hear a case under Colorado Law, in Colorado, using Colorado Law, then insert what he called EPA’s standard, as the standard to be used in Colorado. EPA has no right to set any standard in Colorado”.

***** Plaintiff’s circumstantial proof consisted of the following.

(1). Hydrological evidence showing
  a. Movement of contaminated water in the shallow aquifer
  b. The timing of such movement to the home place
  c. The existence and importance of a high water table and high withdrawal rates from plaintiff’s wells

(2). The Identity of contaminants introduced into the groundwater by the defendants

(3). Movement of contamination into plaintiff’s wells. and

(4). Sickness, illness, Injury, and Death of people and animals, and other evidence of contamination of the well water and the air, and the soil”.

***** Hearing Officer’s approach, “totally ingnored, what was stated by himself “[caused]” by the presence of DIMP. His statement is unsupported by the records placed before this Court, When Judge Robinson became mind set DIMP only, and not caused by the presense of DIMP in his report, and not seeing it as the presense of other toxic chemicals, and the use of Colorado Law which has the only Standard for DIMP set at 8ppb. While he used the Federal standard that was set much higher at 600 ppb by EPA who does not have a legal right to set any standard for Colorado. Please see Jeff Edson statement below, Yet they did in a Federal Court of Claims hearing the Case of Land v. RMA, United States of America using Colorado Law The presence of Dimp was important as a known finger printed chemical, and would tie the unknown chemicals in the water to the Arsenal. This Trial was based on Colorado Law not Federal Law, yet Judge Robinson used the Federal standard for DIMP, and would not consider using or allowing the Plaintiff to place Colorado standard into evidence”. Judge Robinson would strip it out as fast as plaintiff tried to introduce it into evidence.

***** Federal statute provides that State standards, as described by "Jeff Edson, clearly outlines Colorado standards prevail over any Federal standard. In the present instance, it has been judicially determined that the Colorado Department of Health is the appropriate body to oversee the Arsenal cleanup". See State of Colorado v. United States, 867 F.Supp.948(D. Colo. 1994). As a portion of the decission the Court made the following comment:

"The need for oversight activities to enforce removal and remedial actions is particularly acute in situations in where the United States is a CERCLA defendant. If only the Environment Protection Agency were granted oversight authority, the State would be forced to leave the interests of its people in the hands of a law enforcement agency with an inherent conflict of interest. There is no reason to believe that the United States is immune from the conflicts that arise when a liable party is responosible for enforcing its own cleanup activities".

The Land's have notified their Congressman, and Senator's, and past Congressman and Senator, of how Judge Robinson brought in Federal Law to make his decission, and that he personally involked the Federal Tort Claims Act in order to end the trial.

1. Made his decission under the Federal Tort Claims Act. FTCA. When he ruled at the beginning of trial that there is no room in the Federal Court of Claims for the FTCA.
2. The Decission would be made using Colorado Law, but when it came time to apply Colorado Law on a safe standard for the water on DIMP. He used EPA's standard which was an illegal standard, and supported illegal by Colorado Law.
3. "Then we had a Trial Mr. Kruse, stand up before the Court and admit directly to Judge Robinson", "it's on the transcript, his admission for the United states that he has more than once admitted to the Court and the Plaintiff in this case, The United States is not denying that

DIMP was formed at the Arsenal between the time it was built and the current date. The United States is not denying that DIMP has been found off post". "Defendant's concession that the RMA was the source of DIMP found off-post, coupled with the substantial evidence that DIMP was found in Plaintiff's Soil, in the air, and Plaintiff's well water, and is sufficient to establish Defendant's RMA activities inevitably caused "the physical intrusion [of DIMP and other toxic chemicals] upon the property, and in the water, of [Plaintiffs] without the Plaintiff's permission".

4. "Undisputed facts, and admissions, establish that the defendant has contributed to the pollution of the groundwater underlying the land of the Plaintiffs". Court records.
5. "Plaintiff's under Colorado Law established an equitable claim for the amount of damages resulting from the polluted ground water below, and the air above".
6. Defendant's have comitted a "WRONG" by its failure to confine the pollutants to its property".
7. Jurisdictional limits under FTCA. cases have no application to a congressional reference. There will be no specific response to arguments in Defendant's breif relating to the FTCA. cases will be made except to note that the Defendant's underlying premise is in error. By Judge Wilkes C. Robinson

***** This is only just a few of the violations made of Colorado Law, and out and out violations of the Court on the Judges part. a lot of the rest can be seen throughout the book. Since Congress sent this bill to the Court of Federal Claims having them be the "FINDERS OF FACT" and to report it all back to the Congress of the United States. So a decission could be made on the basis of Facts Issues and findings. None of this was reported back to Congress, yet it makes up the transcripts of the case.

***** CONCLUSION; of the Court reported back to Congress. and put on file in the House offices Building. The Findings and 6 page report of Judge Wilkes C. Robinson. Review panel about the same.

***** [The Hearing Officer] shall append to his findings of facts, and conclusions sufficient to inform Congress whether the demand is legal claim, or equitable claim, or a gratuity.

CONCLUSION AND ORDER

"THE COURT CONCLUDES THAT AN AWARD FOR DAMAGES IN THE PRESENT CASE WOULD BE A GRATUITY".

Larry said he would admit this decission made him very angry, but he knew at the end of the trial what that decission was going to be, but he did look forward to an equitable claim. Every thing that was evidenced before this Court should have at least been a decission of Equitable Claim. He along with every one else didn't really even know what a Gratuity is.

"Larry, his family and Attorney scurried around like beaver's in a pond getting everything ready for the Appeals. which is a Three panel Judgeship that is appointed to hear the case. You've already seen the facts, findings, records, and the many error's that were made by Judge Wilkes C. Robinson. You have seen the methods used for the United States to invoke the Federal Tort Claims when they know they are in trouble, and in a place of losing the trial and being found liable in a legal or equitable claims. In the Appeal my Attorney Sam McClaren had listed all the error's even the fact the Judge would not allow plaintiffs' to use the Colorado Standard for water standard, and forced the use of the Federal Standard used by EPA. As stated by Jeff Edson, The "Colorado Health Department has the only standard in Colorado or under the Law. EPA has no standard on water in Colorado". Then I found Federal Law that provides Colorado standards on water will prevail over the Federal standards. It states EPA is not the proper standard, Federal Statute provides that of Colorado Health Department is appropriate to oversee the Arsenal's clean up".

***** Jurisdiction: Legal remedies are usually based upon either of two statutes in which the United states has waived its sovereign immunity. The more common remedy is 28 U.S.C. *2509 commonly known as the Federal Tort Claims Act, (FTCA). The FTCA confers jurisdiction to the district courts to determine negligent acts of employees of the government. The second remedy is 28 U.S.C. *1491 commonly known as the Tucker act. The Tucker Act confers jurisdiction in this Court for claims based upon the constitution and other matters not sounding in tort.

***** A Congressional reference as we have been through in this trial is a totally independent source of jurisdiction for claims against the United States. Much of the reasoning in Defendant's Brief utilized in reaching the conclusion that plaintiffs should be required to show negligence on behalf of the government in the present case if it were based upon reasoning behind the FTCA. Jurisdictional limits under FTCA cases have no application to a congressional reference, or in this case. No specific response to arguments in Defendant's motion relating to the FTCA cases will be made except to note that the Defendant's total underlying premise is in error.

While it may be difficult for some to see, what the difference between Legal Claim, Equitable Claim, and an Equitable Gratuity.

(1). Legal Claim; Is a claim that would be heard under the Tort claims act, A legal claim must show complete negligence on the part of a government employee to have caused the event. The Defendant in this case could also be held liable for Criminal charges both under the law and in a civil action.

(2). Equitable Claim: a claim against a governmental agency, that would show wrong-doing on the part of someone. within the agency. In a Congressional Reference it must rest on some unjustified governmental act that caused damage to the claimants. This could lead to Criminal Charges, & other actions against the government.

(3). "Equitable Gratuity" In a Congressional Reference, that is handled by the Court of Federal Claims. An award of monies based upon moral considerations without finding a degree of Government fault, but with all facts, records and the conclusions sufficient to inform Congress an award of a "GRATUITY" in this instance is in place. With all facts and findings in the record before this Court the Finders of Facts in this case. In the sense used in a Congressioanl reference. The decission to pay an award of a "gratuity" involves matters of social policy and social conscience which, all trial commissioners possess, but which only Congress may exercise, from the findings of facts in this case an award of monies based upon moral considerations

> by the Court and by reference resolution H. R. 61 to Congressional, House Bill, H.R. 816. Judge Robinson stated review of the records which were supplied to me establishes unequivocally that the Land Family's groundwater was contaminated with materials from the Rocky Mountain Arsenal, RMA.

***** Resolution H.R.61 was returned to the full House of Congress by the United States Court of Federal Claims February 1997 stating an award to Plaintiffs on all facts present in the record before this Court would be a Gratuity. Which only Congress after considering the facts and findings, and considerations. In a Congressional reference, an award of monies based upon moral and honorary considerations to include Social Policy and Social Conscience placed upon Congress is an award of a "Gratuity". For which Congress takes that responsibility upon themselves. to appropriate the necessary funds that would fill in the blank space on the Bill H. R. 816 that was passed by the full house of Congress in October 1988. Where it has been sitting in the House Office files since 1988, awaiting action by the Congress. A gratuity can only be handled by Congress and must be handled by the Congressman of your District, where this took place. That would go back to the District held by Congressman Hank Brown, and is now held by Congressman Bob Schaffer. and the Office once held by Congresswoman Patricia Schroeder who came out and visited the farm and could see first hand the devastation. She tried hard to assist in any way possible, but out of Congressional respect she with any assistance needed or need be, She turned it over to Congressman Jim Johnson, who tried his heart out to get a bill passed before Congress during the early 70's and 80's up until he turned the office over to Congressman Hank Brown, who's office conducted large scale investigation, [There is one man who stands tall from Mr. Browns office and that is Mac McGraw to whom I hold the highest respect and regard, just for who he was, and what he stood for, he was always there to hold me up when I felt like giving up, and he would encourage me to keep going]. Mac and others of his staff would visit the farm, and various sites North edge of the Arsenal. It was Congressman Hank Brown and his staff who fought long and hard and finally got a Bill passed by the House Judiciary Committee, which sent it on the the Full House of Congress. It was somtime around September, October 1988 that the bill was heard before the Full House of Congress, As Larry was watching it on closed circuit TV he said my heart is in my mouth. He said Finally! its for real the final vote. It passed the Full House of representatives by a full and complete unanimous vote by the Congress of the United States of America. THE BILL H. R. 816; Be it enacted by the Senate and House of Representatives of the United States of America is authourized and directed to pay out of any money in the Treasury not otherwise appropriated, the sum of $___________, dollars to Larry D. and Marie A. Land et al.

***** "The "Bill" passed unanimous by the Full House, same as when it passed the House Judiciary Subcommittee, it was referred to the Full-House. Then the House of Representatives filed the bill in the House Administration, and they referred it to the Court of Claims by House Resolution 61 H.R.61 where it became Congressional Reference No. Cong. 1-88 where Judge Robinson was to report at the Facts and Findings involved in this case. If it had been a legal Claim, the Court would arrange any settlement decissions. If it had been an Equitable Claim, the Courts would be making any settlement decissions. The Decission of the Court was an award of an "Equitable Gratuity" which then is referred back to the Full House of Congress for the appropriations to HOUSE BILL H. R. 816. This bill was already passed by the House Judiciary Committee and is not necessary to repeat the steps before.

***** The Congressional Reference 1-88 has been sitting in the files at the House Administration Offices now for years. The problem Larry is running into the Congressman, Hank Brown, Congressman Patricia Schroeder who would always listen, or their staff would listen and do everything they could do to help.

***** The Bill 816 like he said has been sitting in the House files now for five (5) years with no action by the Representatives in Colorado. No one will take the responsibility of the first step in resolving this situation. Larry has sent everything to each of them, giving the exact instructions that need to be followed in order for the United States to resolve this issue, and resolve it as was awarded by the Court of Claims based on all facts, findings, and reports in the file. The Court cannot figure the amount of a Gratuity, this is done by Congress only, and it is done by the Congressman, or Senator where it happened and where it began. We look at the point in trial when the United States Attorney admitted in the Court to the Judge, that DIMP. a nerve agent along with other toxic chemicals were found in the Land's well water, in the air, and on the ground. Like Larry says just give me back part of my life before it's too late. Larry was just 30 when this happened, it changed his life and his family's life forever. He has never got to build the house he had promised his wife 30 years ago, and now he suffers total disability, and it all started there. He lost both his parents long before their time, Larry and Marie lost one of their children.

U. S. Senator Ben Nighthorse Campbell, Colorado
U. S. Senator Wayne Allard, Colorado
U. S. Congresswoman Diane Degette District 1 Colorado
U. S. Congressman Scott McGinnis Colorado
U. S. Congressman Bob Schaffer District 4 Colorado

***** Larry said, "It's not what we expected and yet it is, since the United States is enforcing the Clean up, and stopping the pollution to our environment. Not only the United States Army, but the Armed forces, and little and big Corporations alike, they got by with polluting the environment for so many years. "CERCLA" is a good tool, it also helps to Keep tabs on the Environmental Protection Agency. who so many times have cuddled up with the polluters by making rules to fit their individual needs".

***** "The Individual State Health Departments have taken a more active roll for their states, and its environment. "CERCLA" gives them some of the muscle they need in cleaning up the pollution that has gone so wrong and for so long. Living in our environment today can be hazardous to your health, be wary, and take precautions of whats out there, and whats around you, if you feel threatened tell your Health Departments they are there to work with everyone, they will help protect your health, and the health of your family, with a cleaner and brighter environment".

***** Larry has sent everything necessary to each Representative, and each Senator listed above in Colorado, and they all have available their computers that they can bring everything in front of them. He also asked each of them to assist him on the serious problem he has had with the IRS. They have tax years 1988 through 1995, the bill cleared Congress in 1988, they began the depositions in 1990, and they couldn't figure out how he could get so much and have so many top notch people working with him and nothing shows up in his taxes so they started auditing 1988, and havn't stopped yet. They have refused to allow him to take off expenses for anything legal, or for travel expenses, but then when he began winning all the audits, and even found expenses he forgot to take one year came to over $22,000.00 dollars in deductions. Then in 1995 they began to disallow everything that he was able to deduct, and that included everything he had won in appeals since 1990. Before 1988 Larry had been filing taxes for over 30 years was never questioned or audited. It seems real strange when he had a Bill pass Congress giving him the right to sue the United States, that very year is the one they begin auditing, of course they say they are being randomly chosen. It was right after the first deposition the Audits began, they were grabbing at straws trying to disallow everything he was allowed by IRS regulation to deduct until he would win out in appeals. Then in 1995 is the very one that began shortly after the trial in Denver, and before the 2nd trial in Washington D.C. They came on taking all deductions away, and trying not to give any back according to the law. The last time he met with an IRS Attorney who agreed to allow him to amend from the standard deductions to the Schedule A Itemized deductions, he said don't forget to take your state tax is also deductable, and when Larry presented the Itemized deduction schedule to the Attorney Blaine Holiday, the deductions allowed were greater than that of the standard schedule, he became very upset pushing the papers aside bringing the standard back up and pointing and tapping his pen on the Gross Income showing, and stated that is the only figure I'm interested in and flatly would not accept the amended form. He would not even then acknowledge that the itemized deduction schedule even existed. Larry wrote Blaine Holiday a letter 2 ½ weeks before the case was to go to trial, sent him a copy and the original to be filed with the Court, of the amended Itemized deduction schedule A. He personally stated that He would file lit with the Court as a stipulation for everything being amended.

***** However the day of the court hearing, Blaine Holiday Attorney for the IRS had only brought with him a stipulation agreement containing what we had agreed to, the taking out of the standard deduction all the Trial expenses, and $29,000 dollars worth of expenses that were taken as research and Development regarding work done on Larry's Patents and Copyrights, which by law are deductable. Then took away all mileage used and taken for research, to include the rights to use his office in home deductions, and all utility expenses, clearly tax deductable items found in IRS publications.

***** I found it shameful that the Hotel rooms for those involved in defending the case for the United States cost $200.00 per night for doubles, no they did not share a room, they were allowed a seperate room for each of them, they didn't double up, and there were several enjoying their stay in Denver. While Larry had to find somthing reasonable and stayed in a motel for $29.00 per night, actually ate very little and lost weight during the trial, his meal ticket was $12.00 per day, and of course he had to rent a car to get around on a special for $70.00 for the week. Not only did the Government spend $200.00 for a room each, They also had a meal ticket of $100.00 per day, Yet Larry was not allowed to deduct any of the expenses he incurred for the trial of his case, and he was exceptionally conservative really because that is all he could afford. He wasn't getting a free ride on the taxpayers.

***** Larry found out at trial that Blaine Holiday was not filing with the Court the amended Itemized deduction schedule A, that was mailed to him to file with the Court, and refused to acknowledge he knew anything about an itemized deduction schedule, yet Larry has a copy of the letter that had been sent by certified mail to him to stipulate or at least acknowledge, and file with the Court as he stated he would do. Larry then tried to file with the Court the amended schedule A for itemized deductions since Blaine Holiday admitted he did not know of the Itemized Schedule A for deductions. The Court transcript wills how that Larry tried eleven (11) times to file with the

Court the Itemized Deduction schedule, and the Judge refused him this right, yet had gave the IRS the right to file his papers with the Court. At one time as the Judge is ruling against Larry, said somthing in the order of, You didn't file an itemized dedution schedule where it would have been deductable. and is not on the standard deduction schedule. Larry said at that point he had been trying to file the amended schedule A with the Court. The Judge ignored it and went on, as he seemed to be trying real hard to completely discredit Larry before the Court.

***** Larry had also presented to the Court, all the documents from past hearings and appeals process that for eight years he had been allowed the same deductions on a standard deduction schedule. The Judge stated that Larry wasn't in them Courts now, and we will not give you the allowances you have been allowed in the past your in my Court now.

***** Well everything was stacked against Larry, at least he had the transcripts so he could show what took place in the Tax Court. Turned it over to the Tax advocate from Diana DeGette office. Since the Laws the IRS administrates the Laws made by Congress that gives them the opportunity to step in and handle any wrongdoing. There was plenty of wrong doing on the part of the IRS, and the Judge and it is all on transcript. Larry just needs someone to listen to it, he expected that Congresswoman DeGette could ask for a mistrial, there was sure enough in the transcript to do that but she didn't. Instead it was forwarded to the branch office in Minnesota for consideration September 10, 2001 and that is where its been. I lost in the Tax Court, but he also has their hands tied for wrongdoing, Larry says he will win.

IT IS YOUR RIGHT TO KNOW ABOUT HAZARDOUS AND TOXIC SUBSTANCES IN YOUR WORKPLACE.

** Well that's what they tell us, but do you really have a right to know. The rights are all out there, and they are very well intended and organized, there in the law books, but the Courts will not allow them to work. You can go into Court with the best case out there spelling out all the material facts, and showing Prima Facie Evidence, and the

Courts will rule against you, unless you have a lot of money then they will let you spend it, only those with money, or one who may attract a show case attorney will ever win. Larry's case with the government where he was able to establish beyond a shadow of doubt, that pollution from the Rocky Mountain Arsenal had entered his wells, Toxic material that was made at the RMA was fingerprinted to the Land wells. Yet Larry so far lost his case in a Court of Law.

** What Larry wants to talk about next is the work place. He will explain breifly what happened to him so every on will know what a Government agency he was employed by has done to him. It was a tough situation for Larry to get started over from the terrible life threatening situation that happened to him in Colorado, but during the mid to late 70's and early 1980's even in tough times Larry managed Working for the Sheriff Departments, and operated his own trucking business, started from one bob tail and ended in 1983 with 14 Semi tractor trailer rigs. He had paid off all his losses from 72, and that is when they decided to go back to Minnesota in the latter part of 1983. Larry really didn't know what he was going to do, as the Minnesota snow flew he stayed close in to home during the winter. We ended up with 160 inches of snow, and cold, like 30 to 35 below zero, and wind chills hitting 80 to 85 below zero. Larry thought why in the world did I move back here. he knew it was cold and nasty in the Winter, but decided he would trade it for the early color of green in the spring, and the beautiful green all summer long, and the beautiful lakes with the multi colored falls, its almost worth spending about four months of winter especially for the rewards of spring, summer and fall.

** In the fall of 1983 Larry registered for classes in Real Estate with Pro Source of Minnesota. He decided he would like to learn about Real Estate and then possibly he could sell Real Estate by Auction, he thought that it would be interesting, and he wanted to learn more about investing in Real Estate. During the Winter Months Larry worked very hard, Pro Source had four levels of classes the recommend to do two course levels, then apply and take the test for licensing with the State. He had completed the two classes by December, and the State did not offer a test for licensing until February 1984, so Larry decided he would take the other two Courses in RealEstate, and then take the test. By the end of February he had now completed the other two courses and was ready to take the state Examination for license.

** Larry had heard so many stories about how rough it was to pass the test the first time around, and it kinda worried him. Well as he is taking the test he noticed some already turning the test in, and he still had a ways to go. Well Larry stretched it out and finished just as the buzzer sounded letting everyone know time was up. He turned his test in not knowing whether to feel good about it or bad about it, but he decided to be very positive and knew he was going to pass with no problem. A couple hours later the scores came out and suprising so he was at the top of the list, shown, scores either pass or fail. There were several people who didn't make it. Edina Realty met him there with a job opening for him Maplewood. Larry thanked them very much but said he had another job opportunity to consider, they said let us know when your ready.

** Look back about ten years, this reminded Larry of when he took the test for the City of Denver Sheriff Department, when he walked into the auditorium, there were 2500 people taking the test for two positions. He just about turned around and walked out, but again he stayed positive and took the test, it was a rough test, a lot of pure memory testing, what seemed Larry's worst to remember, names, people and a whole lot of arithmetic, using fractions, yes it was a timed test and he took every second they allowed, not going back and reanswering any questions, but had alotted just enough time to finish.

** Larry didn't think about it any more, and one day after the fourth of July, a letter came in the mail, City of Denver. Larry was a bit exited but didn't know what to expect, he quickly opened the letter and it said report for Oral examination on a date the following week. When Larry arrived there were about twenty other people waiting for the Oral testing. As Larry sat watching the people would go in come out and leave. There were only about five left when Larry was called in, he took a deep breath and walked through the door. Sat up straight in the chair, and across the table there was five very business like, and firm looking individuals who were ready for answers Larry learned one thing to always direct your answer to the one asking it, while only momentarily casting your eyes to the others, and to look him straight in the eyes when answering not looking up or down. When it was over they said thank you, but we would like to have you wait out front for us. Everyone else had went in and left, he wondered what was up. Then a couple more went in and left, and another went in and they ask him to wait, then the other two went in and then left. They called the two of us back in and come right out and said your hired. Wow what a feeling of accomplishment out of 2500 applicants we were the first ones hired. Absolutely breathtaking.

***** In about December of 1983 Larry had filled out an application for Whashington County Sheriff's Department, and recieved notice to report at Stillwater Highschool Auditorium to test for the position of Jailer/Correctional Officer. Larry was very excited about the position, and knew it was a position he could soon work into being a street Deputy. He was experienced and well educated and trained, in Law Enforcement and as a Correctional Officer, had his Colorado Peace Officer's license transferred to meet Minnesota standards. He had talked with some people at Lakewood Community College, who checked his background, his education, and credits toward his Associates degree in Law Enforcement. They had set him up to test through four classes to upgrade his license in Law Enforcement, for September of that year and had given him material he would need to study before the tests.

*** More good news Washington County Sheriff Department had set up an oral interview for Larry. It was set up for the ladder part of February. Larry dressed neatly, and discretely, and reported for his Oral interview. Thier were four individuals all Investigators sitting at the table, as he entered the room, They stood as he walked up to the table they each introduced themselves to him, he was then able to introduce himself to them. They started by asking a few general questions about himself, and then come the business end of the questions. With his training, education, and on the job experience, of both a Police officer and a Correctional Officer Larry was able to answer the question with first hand knowledge. The last question was how do you feel about suicide in the jails. Larry was able to answer that question with experience by his side, since his career had begun and up to that point he had prevented the suicide of twelve prisoners that were in jail, had not allowed one to be successful in committing suicide. Some were real close, but were saved with the use of CPR, and stopping the blood. Had recieved several commendations for quick action when faced with death. The interview was over and that was the end of a week, they told Larry he would hear from them. Larry was able to leave that interview feeling very positive about everthing. Then on Monday Morning about 7 a.m. the phone rang, it was the Chief Deputy from Washington County. He ask if I could make it in for a short interview, and that we would be calling your Commanding Officer at the Sheriff Department in Denver for a recommendation. The appointment was set for that very afternoon. Larry met with the Chief Deputy, who ask some very simple questions as to how do feel about the Criminal Justice System for which you have been a part of for quite somtime. Larry told him it was his choice of a career working, and helping people. He then dialed the phone to the Chief at Denver Sheriff Department. He had him on the phone asking various questions, and when he ask how would you recommend Mr. Land for a job as Correctional Officer Specialist. The Chief expressed he would endorse Mr. Land as fit, worthy, and competent, while expressing commendation. He then ask a question that took Chief Deputy by surprise when he was ask if he would ask Mr. Land to consider coming back to Denver. He will have a job here anytime, He put his hand over the phone and asked Larry would you consider going back to Denver. Larry's answer was firm and direct, "NO". He then told the Chief in Denver that Larry had made a decission to make his home in Minnesota, and thanked him for keeping his options open. That is about all it took the Chief Deputy wanted to know how soon could you begin the job. Larry said I could be ready tomorrow. He explained that he would be given a chit to purchase uniforms Tuesday, and make the calls necessary to set up a complete physical on Wednesday, and the Psychiatric Evaluation for 9:00 on Thursday. With them complete and Larry's Medical and Psychological profile in order he would have him start 6:00 A. M. on March 7 1984.

***** August 24, 1984 on a beutiful sunny day Larry had taken his 8 year old daughter Melissa for a bike ride. they rode around in the huge parking lot at the church, no cars around. Melissa road ahead of Larry about two blocks so she could throw rocks in a small lake. Larry was on his way to catch up with her down this little side street when all of a sudden a car come roaring up the street then suddenly swerved across the street striking him knocking him unconscious to the ground. He stopped for a moment and came up to Larry laying on his back half unconscious. The next thing Larry can remember the person reaching toward his head his hands slipped around his neck, he realized this guy is trying to choke him. His defenses were high as he realized the guy was straddle his right leg, So Larry used every bit of power he could muster, and kicked the guy right between his legs. He went up in the air somewhat, and was rolling in pain. Larry couldn't do much more, the guy got up and Larry though what next. Well the guy bellered out we gotcha, you little Son of a B_____. Then he got in his car an screamed backwards, and Larry thought this is it, but he missed him, and then went screaming leaving a smoke trail out of the parking lot. Larry laid there a few moments and realized he had to get help quick, his bike was bent somewhat, but the wheels were still round he straddled the bike, and started moving forward very wobbly, and here come his daughter riding toward him wondering why he was acting so strange, next thing that happened he hit his daughter on her bike falling to the ground unconscious his little daughter very frightened went to a house around the corner and told them my Dad had just fallen from his bike and was hurt so they called 911, and within a few minutes Paramedics were there, then the police came and took the daughter back home to tell her mother what had happened, and what hospital they took him to.

At the Hospital they done MRI, and EEG the EEG was a little abnormal so they kept him in the hospital, and put him on a steroid medication to help alleviate any brain swelling that may occur. He had a right broken clavical. The following day he acted as if he were delirious, didn't know who he was, and didn't know who anybody else was, didn't know his family or himself, and did not know what type work he done. He could remember bits and pieces about what had happened but couldn't remember where, and he would get angry when no one could help. After the third day they sent him home, but he was very angry, irritalbe, confused and was aggressive. so they called the hospital and told them what was taking place. They put him back in the hospital a locked area where no one can walk away. A Doctor specializing in brain disorders was called in that night he came to see him. He recognized the problem right away, and had Larry taken off the medication for brain swelling. It was a steriod that reduces swelling of the brain, their was no swelling and the medication was causing amnesia, the irritable aggressive mood swings. With in three days he was back to normal and released from the hospital. He remembered what had happened, and went back to the area, and the way that person had smoked his tires, and did the little half circle when he tried to run over me. Larry was able to call the police and they came out and could see exactly what had happened, they talked with neighbor's and got coroberating stories on what had happened, and what they had heard. All he could remember it was a car with baby blue bottom, and was a two door hardtop. There was blue paint on the bike where it was struck. At least they had a good report and not like the one they had originally written, stating that I ran into my daughter's bike fell off and struck my head. Had he died that is as far as the investigation would have ever gotten.

***** Larry could not help but think it was someone who was out to get him, could it have been someone from the Jail where he worked, and he thought no. He then remembered some calls on the phone he had got that very week stating were gonna get you, you need to keep your nose out of other peoples business. But the thing that stands out almost like putting a name on it was a telephone call he had just recieved on Wednesday from his Attorney stating were going to Washington to testify before the House Judiciary Committee, they havn't set the date but you will be heard. Then Larry gets nasty phone calls telling him he's dead, we'll get you, This wasn't the first time but it's still very scary when they call you on the phone and say you're a dead man walking. Police reports were made, and was reported to the telephone company. The telephone company had tapes of the calls and what was said, but they were kept very short, and remained untraceable. Two weeks prior to the Incident it was also reported to the police, that an army colored pick-up was seen taking photographs of the house.

***** Put it altogether in a sequence of events, (1) Larry had just closed an entire army base permanently. (2) Larry had just permanently closed Shell Chemical Company, (3) Larry had turned the situation into an international event, (4) He received a telephone call that he would be testifying before the Congress of the United States, (5) He received angry death threats for two weeks over the phone, prior to the incident. (6) He reported to the police of an army colored pick up with and individual in fatigues taking pictures of the house, and cars on two different days. (7) Then on saturday late afternoon Larry took his daughter down Edgerton st. to the Church parking lot where his daughter could learn to ride her bike without the fear of being run over. (8) This had been the fourth week he had done this at about the same time on saturday. Not thinking he had set up the perfect pattern for a stalker, (Army). (9) Then for someone to turn directly across the other lane of traffic striking him head on, Larry had done some maneuvering as the car struck him throwing him to the ground, but still conscious, he can remember this person from the car getting out, walking toward him, felt better at least the guy was going to help. "Wrong" (1) The next thing Larry remembers is the hands around his throat, and he thought damn he's squeezing hard. Larry then thought of his training and what should he do that would shock the hell right out of this person. He was bent over top of Larry perfectly strattling his right leg, everything went into slow motion, and he planted his knee with all the force he could muster right in the groin area of this person temporarily debilitating him, he got hit so hard his rear end left the ground as he flew over then to Larry's left, and just lay rolling on the ground, crying out in pain, and calling Larry a few nasty names. Larry was still half unconscious and couldn't get up, but this person managed to get up, and stated we gotcha you little son of a b____. Then he got in his car and in reverse he came squeeling backward where Larry was laying barely missing him, in fact he thinks this person thought he had run over him. Then went squeeling out into the parking lot and out toward Edgerton street. Well you have already heard the rest, now make up your mind what it is you think.

***** Larry was released to go back to work on October 10, Physically and Psychologically fit and ready for duty. What I'm about to tell you is exactly what happened at my job, to let you know how negligent some may act, and do it willfully, and with reckless intent even on the job. This happened at the Washington County Sheriff Department where I worked as a Corrections Officer. They decided the work area needed a paint job. When they started it was likened to a gas chamber, all the Investigator's and the Sheriff were gone. Every one was complaining of headaches nausea, vomiting, shakes, tremors, eye, nose, and throat, irritation, difficulty breathing, dizziness, salavation, runny nose, just to name a few. This was a small department but there were up to thirty five people complaining of being made sick by the paint. The painter's had been sneaking the paint in in unmarked containers, then one day got brave enough to bring in one of the paint buckets being used. A large black five gallon can with a large white label. In each of the upper corners of the label was a red skull and cross bones. Warning stated if you feel dizziness, nausea, vomiting, eye nose and throat irritation, and difficulty breathing you have been over exposed Get medical help immediatley. It also said Industrial Bridge Paint, 37% lead based. Must use an oxyegenated airline respirator, and your lead levels must be checked each week. This had been going on now for over two weeks, in the winter time, no windows, no doors to open and no ventilation. The jail was a tightly secured area, the fresh air returned and filtering system was working but they had it set only to pick up eight percent fresh air to mix with the hazardous air, so consequently it was getting worse every minute and every hour, it seemed to be reaching it's total saturation level. We were dry heaving, and Larry passed out on at least four occasions. We had oxygen bottle for emergency purpose and we began setting them up and using it in order to make it through a 10 hours shift.

***** Larry has already told you of the intense signs and symptoms I was unaware of what it meant when we passed out, until he started reading about what we had gone through. Passing out is the last sign and symtom before you quit breathing, and your heart stops. Larry said we could of all died of lead poinsoning, not only was it lead based but it was an epoxy type paint also, the worst, and the most radical.

***** Larry had called OSHA up several times telling him what was happening, but was afraid to tell them where. They wanted to come out immediately. The guy from OSHA then ask Larry to get a serial number from the can of paint, and he could tell him exactly what we had. Then things started to happen fast because Larry had complained to so many of the Coutny, the Health Department, the safety committee, and the Sheriff. It seem to be falling on deaf ears, except the chief Deputy who came in and was saying to another jailer

I hear we have a jailer who can't take a little bit of paint fumes, he better go find another job. Then Larry walked into the Office area and was told if you can't take a little bit of paint fumes you better go find another f________ job. Larry tried to respond and he turned and walked out. Then the painters got so brave they brought in one of the five gallon containers, and left it in a storage area. Larry got out his paper and pencil and went to work getting all the information he could find on the can including the serial number, It almost seemed as if it were all falling in place. When he got off that afternoon, went directly home to call OSHA, the man said he would be their till at least 6 p.m. Larry got home and immediately called the number and the right person was there. He gave him the information from the can, and he was assured he would call back real soon. Within the half hour OSHA called back and was really worked up, want to know where in the hell this was happening, He said that paint is not only dangerous it is deadly. Do not go back into that work environment it will kill you. That is a lead based bridge paint using oxidizers where user must be wearing an air Oxygenated Respirator, and that paint is an Industrial bridge paint not an interior paint. He said walk away, but then remember all the other, just tell us where this is happening. Larry made a deal with OSHA that he would have his Jail Administrator call him first thing in the morning. He said that is real good I'll be here at 7 a.m.

***** Larry is in disbeleif of what has taken place and how tight an employer tries to keep employees from reporting a wrongful act. He knew his career was short lived, and reprisal was the scenario, even here in Law Enforcement, at Washington County Sheriff Department, and his boss Sheriff Jim Trudeau, who is now the man who runs the Minnesota State Sheriff's association. To the Chief Deputy, who is dying of terminal lung disease, "God Knows", I wonder where he got that.

***** Larry went to work the next morning knowing exactly what type of stand he was going to take. That morning he took all the prisoners in A-block, because that was the next area to be painted. The painters were in about 7:30 a.m. with the air compressors and all, that's right they were not brushing this paint is was being sprayed with high pressure sprayers. The painters set all the equipment down outside while they looked the area over, so when Larry let them into the cell block, he turned the switch to close, locking them inside the unit and couldn't start painting. Larry informed them they were in violation of both State and Federal Law, and were tentatively under arrest. He told them he would be back to talk with them when he has had a meeting with the Sheriff, and the Jail Administrator, and not until. Breakfast in the jail was over, medications to those concerned, as Larry was called to the administrator's office. The jail administrator said what is going on I found your memo stating you had the painters locked in the jail for not supplying Right to Know information, and not posting data sheets of what they were using, and you want me to call OSHA. He said why did you call OSHA and why did you feel it was necessary. Larry explained what he had done and the fact the painter's were using a toxic and dangerous lead based bridge paint to paint our work area, OSHA has the serial numbers off the cans and stated that paint will kill you, It had been banned in the United States since 1967, and was never meant to be used in an interior. The Administrator went to the Sheriff's Office to make the call with the Sheriff on the line. They both were terribly shocked at what they heard, both of them turned white. The administrator turned to Larry and said "MY GOD" I cannot believe what they are doing as he asked Larry if he would consider letting them out of the jail with everything, and not to return and formal charges would be filed against the Facilities Department. Larry turned to him and said what about us, and it was agreed a first report of injury would be turned in.

***** There was never a first report of injury, Larry let the disgruntled painters out of jail informing them again, charges would be pressed. The Administrator apologized for what they had allowed to happen, and said the Sheriff apologizes also. There was never any charges filed for the crime committed by the Facilities Department. It was from then on a matter no one was allowed to talk about, or discuss even if it were some health malady. Larry and many others ended up with stiff and sore joints that took until finally in 1989 it was diagnosed as Arthralgia, and Myalgia from lead poisoning. A Severe and permanent disability. the other was Hypertension, causing high blood pressure and extremely bad headaches, so severe at time a person would become nauseated and possibly vomit. This was diagnosed in 1986 the year following the incident caused from lead poisoning, also a severe and permanent disability one learns to control with medication. It took nearly 5 years to reach maximum medical and Psychological improvement, the Psychological injury was caused from lead poisoning and damage it done to the Central Nervous System. The injury to the Central Nevous system here and that from the Nerve agent poisoning of 1972 and mostly reversed over several years. Never received any help from Washington County after that terrible instance of willful negligence.

***** Mr Land after clearing the full house of Congress unanimously in 1988 to sue the United States of America, It was really the Army and big corporation, however they all fall under the Authority of the United States, and when Congress gave him the right to sue, it was the right to sue the United States, whom are also protected by the Justice Department, their connections and their money.

***** During these rough and frightful times of the 1970's then then again the difficult days of the 1980's when Larry spent so much time trying to be normal once more. Years went by and the frightful obstacles seem to fade into the surroundings and weren't so complicated any more, Larry was once again able to focus on his goals. He was an entrepreneur, an inventor he loved his being able to make things happen, and he once again focused in on items to patent, and items to copyright. It was exciting for him he regained a new sense of power, and a drive toward his goals. He wrote his first book in 1988, even though it was a training manual, it was what Washington County needed, and when he introduced it to them he won the award of the month, then was in contention for the award of the year which he won. His book of knowledge grew to where the Probation Department in Washington County ask permission to use the same format for their training manual, he said sure, then the State of Minnesota wanted to use the book in their training seminars through out the State. Then he began getting requests to use the training manual in their jails all over the State, then to some other states who used the book as their training guide. Then Larry wrote a short book about suicide, and what can be done to assist and help the ones who thought that was the only way out, that book also went into the library for the Minnesota State Resource Center, as a way of controling suicides. in the jails and outside the jails.

***** Larry went out of the 1980's feeling very accomplished with his book, and some of his patents, and copyrights were falling in place. Larry has got working models of all the work he has done so far.

***** Then came February 5, 1990, as he went to work he became sick as he walked through the door. No one had said a word that they were going to paint the Jail again. There was no right to know information that is required by the Department of Labor and Industry both State and Federal, NIOSH, and OSHA. There were no data sheets placed any where in the work area as required by OSHA, and well hey we have one of the richest Counties in Minesota.

Washington County Employees gave written petitions to Sheriff Trudeau, Ed Kapler, the County Health Department, and the County Commissioners they all rejected the safety of their employees, during the 1990 toxic paint incident [O]verExposure.

There were no windows, doors were sealed, The ventilation was next to none, air return set at 92%. Employee's reported every sign and symtom up to death. Complained of eye, nose, and throat irritation, Headaches, Nausea, Dizziness, Vomiting, Body aches, Salavation, Feeling of being Drunk, Loss of consciousness, High, fatigue, whistling sound in ears, Seeing Stars, unable to focus, Blurry vision, Unable to concentrate, Short term memory loss, Ataxia, Aphasia, confusion, clumsy, Shaking, Trembling, Starving for air Difficult breathing, Sinusitis, Rhinorea, Difficulty hearing, Irritability, The world seemed to move in slow motion.

#15

EPA/OSHA

Xylene; main ingredient in the paint: Effects of Xylene on the Nervous System.

100-200 ppm.	Nausea, Headache
200-500 ppm.	Feeling "high" dizziness, weakness, irritability, vomiting slowed reaction time.
800-1,000 ppm.	giddiness, confusion, clumsiness slurred speech, loss of balance, ringing in the ears,
>10,000 ppm.	sleepiness, loss of consciousness, "Death".

Compare signs and symptoms reported, to the OSHA & EPA chart produced to show exposure & overexposure to Xylene.

Minnesota, they wouldn't be doing somthing wrong would they? He ask the Jail Administrator Captain Richard Becker what it was they were using he said oh, it's just latex paint, but Larry along with thirty five others were reporting a wide range of Health maladies, signs and symptoms. All the signs and symptoms of over-exposure to Xylene, Urethane, and Mercury. These signs and symptoms were reported verbally, and in writing to Sheriff James Trudeau, Captain Richard Becker, to the Head man of the Safety Committee Ed Kapler, who is also the head of facilities department who purchased the paint being used, and nearly 3 weeks after the Painting began the correct Data sheets were found locked in the safe at facilities, supposedly it was Ed Kapler's own safe, and to the County Health Department, who all listened and done nothing allowing the painting to continue 10 hours a day for over 30 days. This caused severe injury to the Central Nervous for Larry and five others.

The other's involved all backed away when things got hot. Larry is aware of the threats he was getting, and knows others got the same threats of losing your job, no advancement, if you continue with this. He never backed down to his employer, and when they wouldn't just stop and look into the fact the painters were using an alkyd, alkyd petroleum based epoxy paint and it was an industrial grade Exterior paint that contained high levels of mercury Bio-cide, that had been banned in the United States. He signed a complaint with OSHA, they came out and did an investigation, but it was 3 weeks before the Criminal citations were issued, by then the painting was done. The legal claims.

SECTION 1

MANUFACTURER: Mautz Paint Co. **ADDRESS:** 939 E. Washington Ave. Madison, WI. **EMERGENCY TEL.** (608) 255-1661

PRODUCT CLASS: Solvent Based (Alkyd, Alkyd-Oil, Epoxy) Paint

TRADE NAME: Industrial Enamel

MFG. CODE ID. 89-00 Line

SECTION 2 - HAZARDOUS INGREDIENTS

INGREDIENT	% WEIGHT	TLV-PPM	mg/m3	LEL	VAPOR PRESSURE
Mineral Spirits	30.2	200			
Hi Boil Naphtha	6.3	500			
Xylol (Xylene)	1.4	100			

SECTION 3 - PHYSICAL DATA

BOILING RANGE: 279-416'F **VAPOR DENSITY:** heavier lighter, than air

EVAPORATION RATE: faster slower, than ether **PERCENT VOLATILE BY VOLUME** 57.5

WEIGHT PER GALLON 9.4

SECTION 4 - FIRE AND EXPLOSION HAZARD DATA

DOT CATEGORY Flammable **FLASH POINT** 81'F **LEL**

EXTINGUISHING MEDIA Carbon Dioxide, Dry Chemical, or Foam (NFPA Class B Extinguisher)

UNUSUAL FIRE AND EXPLOSION HAZARDS: Containers should be kept tightly closed. Keep containers away from heat, open flame, electrical equip., sparks.

SPECIAL FIRE FIGHTING PROCEDURES: Water may be used to cool containers to reduce rate of burning, and resist additional ignition or explosion.

SECTION 5 - HEALTH HAZARD DATA

THRESHOLD LIMIT VALUE: SEE SECTION 2

EFFECTS OF OVEREXPOSURE Inhalation: Irritation of eyes and respiratory tract; headache, dizziness and nausea. Skin: Irritant. Reports have shown repeated and prolonged occupational overexposure to solvents can result in permanent brain and skin damage.

EMERGENCY AND FIRST AID PROCEDURES: Inhalation: Remove to fresh air. Restore breathing if necessary. Notify a physician. Eye Contact: Flush with water. Notify a physician. Skin Contact: Wash with soap and water. Contaminated clothing should be removed and washed prior to reuse.

SECTION 6 - REACTIVITY DATA

STABILITY: unstable stable **CONDITIONS TO AVOID**

INCOMPATABILITY *(Materials To Avoid)* Avoid strong Oxidizers.

HAZARDOUS DECOMPOSITION PRODUCTS: Combustion will produce Carbon Monoxide & Carbon Dioxide

HAZARDOUS POLYMERIZATION: may occur will not occur

CONDITIONS TO AVOID Heat, Open Flame, Sparks.

SECTION 7 - SPILL OR LEAK PROCEDURES

STEPS TO BE TAKEN IN CASE MATERIAL IS RELEASED OR SPILLED: Eliminate ALL sources of ignition. Ventilate area. Soak up spill with absorbent material (sawdust). Remove with grounded (nonsparking) equipment.

WASTE DISPOSAL METHOD: Incinerate under safe conditions. Disposal should be in accordance with local, state, and federal regulations.

SECTION 8 - SPECIAL PROTECTION INFORMATION

RESPIRATORY PROTECTION: Use NIOSH approved chemical cartridge respirator (TC23C) to remove solid airborne particals and solvent vapors from overspray during application:

VENTILATION: Use local exhaust to maintain TLV of most hazardous ingredient (SEE SECTION 2) below acceptable limit.

PROTECTIVE GLOVES: Yes

EYE PROTECTION: Safety glasses designed to protect eyes from liquid splash.

OTHER PROTECTIVE EQUIPMENT: Eye Bath/Safety Shower.

SECTION 9 - SPECIAL PRECAUTIONS

PRECAUTIONS TO BE TAKEN WHEN IN HANDLING AND STORING: Keep away from heat, open flame or sparks. Store in an area suitable for Flammable or Combustable Liquids.

OTHER PRECAUTIONS: Not to be taken internally.

1990 Feb. 6. Paint incident	1990 Permanent Health Maladies
1. Nausea/Vomiting	1. Chronic pain syndrome
2. Diarreaha	2. Depression
3. Shaking/Tremors/Near convulsions	3. Anxiety
4. Dizziness to stumbling/trembling	4. Toxic Hypersensitivity
5. Constant headaches	5. Amnestic Disorder
6. difficulty breathing/tightness1	6. Dysthymic Disorder
7. Hurting and aching magnified	7. Pain disorder/Physical and`1
8. Terrible Motion Sickness1	8. Confusion— Psychological
9. Impotence, from near normal to	9. Short term Memory Loss
10. Chronic Fatigue nothing	10. Moodiness/Irritibility
11. Poor Concentration/learning	11. Inability handle heat/cold
12. Short term memory loss (severe)	12. Trembling/Tremors/shaking
13. Salivation/Runny nose worsened	13. Chronic Fatigue, Ataxia
14. Irritability/Confusion	14. Amnesia, Dementia, Aphasia
15. Multi Chem Sensitivity Syndrome	15. Impotence [long term] Perm.
16. Terrible sight confusion [MCSS]	16. Poor learning ability
17. High feeling, Euphoria	17. Poor Concentration
18. Seeing things “not there”	18. very poor immediate recall
19. Terrible pain bones and joints	19. Terrible Motion Sickness
20. Depression/Anxiety, Moodiness	20. Sleep disorder
21. Inability to handle heat & cold	21. Sight Confusion
22. Hypo-glycemia	22. MCSS- Multi-Chemical-Sensitivity Syndrome
23. Sleep disorder	

started to fly the rest all backed out very quickly, leaving Larry the only one to file Workers Compensation Claim, and filed with the State of Minnesota Objections overruling their attempt to have the citations dismissed. Larry and the State Attorney General's office had numerous hearings to make sure the Criminal citations were not dismissed. This went on till mid 1993 before they were finally adjudicated. The Doctor's finally checked Larry's blood level for mercury, nearly a year after the incident the Mercury level was still significantly high and over any safe standard, it was too late for help, all they could do is work with the effects that were caused, and apparent, and very significant. Mercury takes somtimes months and even years before manifesting it's cycle. For Larry it took nearly 10 years for him to reach complete medical recovery from the effects of Mercury. The injury and damage he learned to live with, It had caused many disabilities for him. The OSHA citations were significant for the employer Washington County was fined over $2,000.00 and placed on probation to the State of Minnesota for one year, ordered to update their safety program, and the Right to Know Program, and make it available to the Court.

The Doctor's realized then, there was nothing more that could be done. Had the employer been honest about what they were using there was medication available that if quickly given would have mitigated the injury that took place dramatically. Probably would have left no disabilities. Larry dealt with the disabilities he was given all the way to 1999 when his Doctor's ask him to ask your employer for some reasonable accommodations with the injuries you suffer from this may extend your life some. Since you are an American with disabilities, you have the right to request them, so he did ask for some accommodations in July 1998, by January 1999 they took his job away because of his disabilities, without even initiating any of the Accommodations. As for the Federal Law for those with disabilities don't let it fool you, there is no protection out there for the disabled, and the Federal Courts don't take the time to even recognize what is happening. The Justice Department does nothing to support the Laws they had made. Its now just turning 2002, Larry has been totally disabled now for three years, his employer is still not paying him a dime for the injuries they purposefully caused, however it will go to trial this year, There will be at least six hearings then the trial, We all wish him luck.

If your think the situation with the Painting in 1985, and 1990, was about enough for his employer to do guess again. in 1992 November and December Larry was transferred to the new all electric and computer operated Corrections and Law Enforcement Center. Larry was in the new building in training, and to assist in setting up the new Policy and Procedure for the Department. The first week he and everyone else was coughing, eye watered, nose dripped, sore throats, lungs hurt, as the days went by it became more difficult for him to breathe. This was reported almost on a daily basis by Larry to his immediate Supervisor, but he felt it was falling on deaf ears. It continued to get worse as each day went by the construction dust problem was horrible, he said we weren't seeing the construction part too much but the construction dust was absolutely horrible, fans blowing circulating the air, and the construction dust without being filtered, The filtration system was still under construction. Larry and other's did see some people installing fiberglass, being delivered by the truck load, large rolls they were taken into the pipe chases, cutting and wrapping and packing all the plumbing areas with fiberglass. He said to his supervisor why are they all wearing respirators and we have nothing. Those who wore contacts couldn't wear them in the building. Some of us noticed little red bumps on our skin which seemed to irritate and inflame the areas.

By November 26, after Thanksgiving Larry became very ill could not breath, was taken to the emergency for Group Health. They took xray's of the lungs, and he was told he had bronchitis, with indistinct infiltrates in the lungs, it is somthing he has never had, he told the Doctor that he was also spitting up blood from the lungs. They gave him a prescription for antibiotics, and al albuteral inhaler. They had four days off over thanksgiving and with the medication it seemed to get somewhat better, his lungs and throat were very irritated and sore, his eyes were red and irritated, and he was just miserable. Well Thanksgiving was over, and it was time to go back to work.

***** The four days off for Thanksgiving made a lot of difference in how Larry was feeling, his breathing was somewhat easier, but he really dreaded going back to work, knowing the construction dust was still there.

***** After Thanksgiving Larry returned to work hoping he would get better. After the first week back to work he was again taken to Health Patners emergency. Xray's again taken showing infiltrates in both lungs, Pneumonia, and Bronchitis, the Doctor's changed his medication, and he was placed on two different inhalers for the infection and swelling, and sent back home. Back to work the following week changed medications the Doctors are confused, and perplexed as to why the medication is not working, with same inhalers increased use. What they were really doing is to help Mr. Land die.

***** This went on through out the month of December, with another four day holiday on Christmas, and another on New years, It got a little worse and a little better during this time, and probably so the people installing fiberglass were gone nearly two weeks. The Doctor's changed the medication twice more, and xray's showed infiltrates in the lower lobes of both lungs, that is where your deep breath air comes from, for Larry it was not possible by now to even get a deep breath. Back to work after the first of the year, and it got worse again back and forth to emergency at Health Partners. Larry unable to get a deep breath even when using the inhalers he was so near the point of not breathing. By January 7, 1993, after coming home from work, collapsed at his home, luckily his wife was there and called 9-11 Paramedics arrived, found him near unconsciousness, recognized the problem immediately and placed him on a Nebulizer, plus oxygen, stabelized him and was taken to the hospital emergency room. They took Xray's and kept him on oxygen, and then back to a nebulizer, He was diagnosed with Pneumonia, Bronchitis, and Asthma. He was placed on a steroid medication for the swelling, changed the antibiotics, and was kept on the Asthma Cort, and Albuteral inhalers, was sent home with days off from work.

***** Larry had asked his employer to check and see what was in the construction dust, he figured they probably wouldn't or lie about what was in the dust that was in the air. Larry had taken air samples, and dust samples just in case because he felt the problem was related to what they had been breathing. The Doctor at the Hospital had advised Larry to call the Health Department and have them check the construction dust, and the air. He called the County Health

Department, telling them about the problem he was experiencing, along with his co-workers.

***** When Larry returned to work he was immediately called on the carpet by Captain Richard Becker, (Jail Administrator), Captain Mike Johnson, (Safety Coordinator), and Captain Don McGlothlin, Sheriff's Administrator). He learned that the Health Department did make an attempt to test the air quality, but were ask to leave with all their equipment by these 3 Captains, who admitted this to Larry. They also stated they told them not to come back. As for Larry they all showed their bad emotions because he had done this, and he was even told he could be fired. Larry had given them his back to work order, with reasons for being off work, and ask them to fill out a first report of injury.

MICROSCOPIC PHOTO'S OF AIR SAMPLES TAKEN FROM WORK AREA AT WASHINGTON COUNTY SHERIFF DEPARTMENT. LARRY'S WORK AREA WAS FOUND TO BE 12 TO 15% FIBERGLASS FOR EVERY BREATH TAKEN. THIS WAS REPORTED TO HIS SUPERVISOR'S 28 TIMES, OVER A TWO AND A HALF MONTH PERIOD OF TIME.

LARRY DOES NOT SMOKE, NEVER HAD BRONCHITIS, PNEUMONIA OR ANY FOREIGN INFILTRATES IN THE LUNGS/ AND WAS TAKEN ON A VERY DESTRUCTIVE PATH IN HIS ENVIRONMENT, WHILE WORKING. THE THREE CAPTAINS AND HIS IMMEDIATE SUPERVISOR RANDY HILL DENIED HIM MEDICAL TREATMENT, WHILE HUMAN RESOURCES DENIED HIM WORKERS COMPENSATION BENEFITS, JUDY HONMYHR, AND JENEEN JOHNSON, WHO'S OFFICES WERE IN ANOTHER BUILDING. THEY ALL FAILED TO FILE THE FIRST REPORT OF INJURY AS REQUIRED BY LAW.

PLEASE NOTE ON THE NEXT PAGE, WHAT FIBERGLASS CAN DO IF ITS BEING USED IN YOUR WORK PLACE, AND YOU WILL KNOW WHAT HE HAD TO GO THROUGH. LARRY STARTED WITH ASTHMA 15%, WITHIN 2 YEARS IT WAS 30%,

Health House News Basics Victims Resources

Common symptoms of fiberglass overdose

* persistent, dry, hacking "barking" cough
* sore throat, bloody taste in throat or blood in sputum
* bloody nose
* Persistent, occasionally very severe, sinusitis and rhinitis which does not respond to common medical treatment,
* especially treatment for allergies,
* persistent, occasionally severe respiratory infections which, again, do not respond to common medical treatment,
* headaches, nausea, dizziness, insomnia, irritability, depression,
* Asthma-like breathing attacks or constant wheezing,
* other allergy-like symptoms (which do not respond to allergy treatment),
* "reactive airway disease,"
* swollen, red, watery, infected eyes,
* skin infections, ranging from mild to very severe, requiring hospitalization,

- extreme sensitivity to everyday amounts of ambient pollutants, especially: cigarette smoke, car exhaust, perfumes and colognes, some cleaning products, paints and varnishes, new paneling, cabinetry or furniture made with particle board or strand board, new cars and other new plastic, foams used for furniture and bedding, new carpeting.

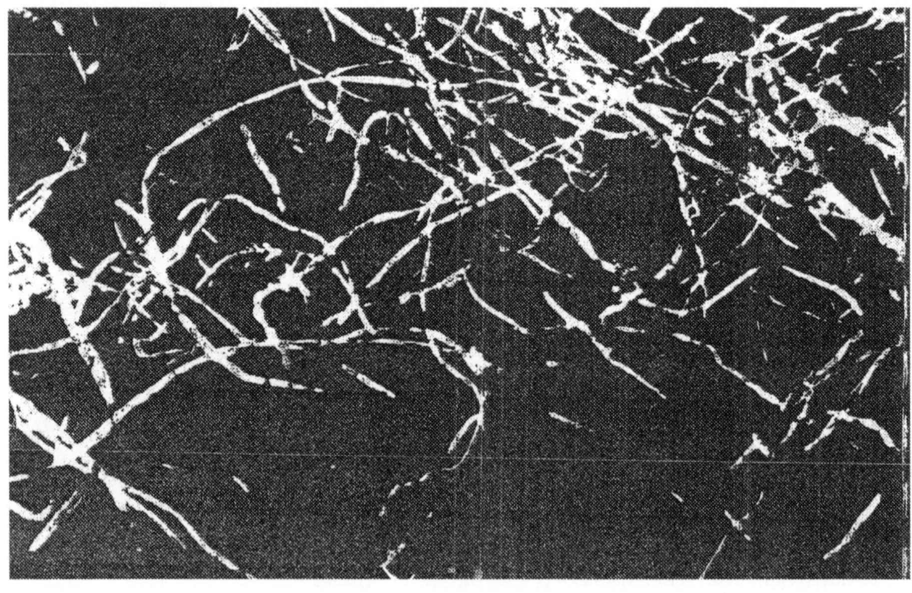

BY 1997 IT WAS 40% AND SPITTING BLOOD AS HE'S BEING TOLD THE ASTHMA HE HAD WAS PROGRESSIVE CAUSED FROM [O]VEREXPOSURE TO FIBER GLASS. BY 1998 HE WAS PLACED ON A VERY INTENSE THERAPY PROGRAM TO HELP RID THE LUNGS OF GLASS, AS IT WAS MOVING AND CAUSING MORE DAMAGE. THE THERAPY THE DOCTOR HAD PUT LARRY ON OVER TWO YEARS WAS GOOD AND SEEMED TO MOSTLY STABILIZE AND RID THE LUNGS OF THE GLASS, BUT WITH THE DAMAGE ALREADY DONE WAS 53% TOTAL PERMANENT DISABILITY CALLING FOR MANY JOB ACCOMMODATIONS UNDER THE AMERICANS WITH DISABILITIES ACT (ADA). THESE SEVERE OBSTRUCTIONS, AND RESTRICTIONS, AND TERRIBLE INJURY TO THE CNS. CENTRAL NERVOUS SYSTEM FROM THE OVEREXPOSURE TO MERCURY, AND XYLENE WERE ALL LEFT IN THE WAKE OF AN EMPLOYER WHO DIDN'T EVEN STOP TO LISTEN OR THINK ABOUT WHAT THEY WERE DOING. YET WASHINGTON COUNTY AFTER A 12 YEAR COURT BATTLE WITH LARRY HAS FINALLY ADMITTED TO THE CAUSATION. THIS IS NOT OVER YET AND WILL TAKE AT LEAST ANOTHER 6 MONTHS IN THE COURTS.

Larry tried numerous times to get hold of some one in the New York Fire Department to tell them the therapy they may consider for all their lung problems. He believes the real problem when the building came down the cement structure ground up all the insulation "FIBER GLASS" As if it were a huge grinder turning it into microscopic fibers of glass. Larry had observed samples of the dust under a microscope and wanted to be of some help to those who were, & are suffering. Even though there was some Asbestos, whenever asbestos goes into the lung it more or less stays in one place, and the lung or body has a chance to heal around that injury. With Fiber Glass still in the lungs, it will continue to do damage. Like Death by a Thousand Cuts, There is medical help that will assist your lungs to rid the fiberglass. It is understood that most of the trade center towers were in fact insulated with Fiber Glass blown onto the cement structures as insulation as they were being constructed. As the tower's came down the insulation was ground and pulverized into microscopic sharrs of glass much like a million razors in the lungs. "Or death by a thousand cuts."

Larry is Speaking from experience, he could not believe the uncaring attitude of those he talked with. Probably the Administration who was not involved with the actual dust. He was trying to share somthing good. The persistant dry hacking, barking cough. Or the sore throat blood in the sputum, or bloody nose with the lungs become sore and difficulty breathing, severe sinusitis, and rhinitis, and your not responding to medications, especially treatment for allergies. Then begins the severe respiratory infections, which does not respond to common medical treatment. Headaches, nausea, dizziness, insomnia, Irritibility and depression. Asthma like breathing attacks, and constant wheezing. reactive airway disease, swollen red, watery and infected eyes. Red spots on the skin all over then they be come infected ranging from mild to severe. Then probably an extreme sensitivity to everyday ambient pollutants, and especially cigarette smoke. You become extremely sensitive to just about everything that smells even when your trying to figure out how it could be there. Then Asthma, and other lung diseases that take your breath away, and you feel like your gonna die, severe obstructions, and severe restrictions of the lungs. Larry even tried calling the hospitals, they acted as if they didn't want to hear what he had to say, they knew it all.

Fiberglass Mutilates DNA

Accelerated cell growth likely precursor to cancer

By Robert Horowitz

Three different kinds of glass fibers were poisonous to cells and damaged DNA in studies performed by a team of doctors at NIOSH, the medical research arm of the U.S. Department of Labor. Damaged DNA can unleash a process of accelerated and even unrestrained cell growth, and the new research shows cells with fiberglass-damaged DNA exhibited these tendencies. Although the exact mechanism by which cancers and tumors grow is not yet deciphered, damaged DNA and abnormal cellular reproduction is widely thought to be the first step.

Additional research on glass fiber geno- and cytotoxicity is currently on hold because of a lack of money at NIOSH (National Institute for Occupational Safety and Health) and the Department of Labor. A final decision on whether to continue glassfiber experiments-and when-is pending. The newly-elected Republican Congress headed by Newt Gingrich, however, is threatening to completely eliminate NIOSH.

NIOSH researchers set out to answer three questions about glass fibers: whether they can introduce a transformation in the structure and form of cells, whether the induction of "morphological" changes could be related to fiber size, and whether cells thus transformed would exhibit accelerated, tumor-like (neoplastic), growth.

"The results indicate that glass fibers are capable of transforming mammalian (BALB/c-3T3) cells in vitro as a function of their physical properties and that glass-fiber-induced transformed cells possess neoplastic characteristics."

The experiment used three types of glass fibers, one microfiber from Manville, plus a microfiber and a general building insulation fiber manufactured by Owens Corning. The fibers were precision milled to lengths and widths known to cause cancer.

Prior research with cells from the lung of a Chinese hamster demonstrated the respirable portion of Manville glass fibers caused breakage and fragmentation of chromosomes. NIOSH researchers used a technique called micronucleus assay to further explore this phenomenon. "The micronucleus assay is one of the most frequently used short-term assay systems for the detection of genotoxic agents and potential carcinogens," they say.

Micronuclei are fragments of cell nucleus; oftentimes these fragments either do not have the correct number of chromosomes, or the chromosomes are damaged in some way. This does not always prevent the cell from duplicating itself, however, and passing along its incorrect genetic information.

Both Manville and Owens Corning microfibers (average diameters .2 and .18 micron respectively) caused micronucleated and multinucleated cells, with those cells given higher fiber doses exhibiting rates of chromosomal damage as much as five times higher than positive controls-cells treated with a substance known.

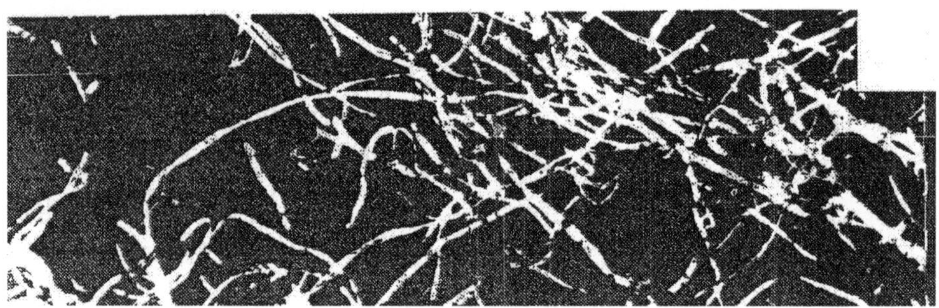

Larry says if you had felt the above, you were probably OverExposed to FiberGlass, and he wishes he could of shared his therapy program that saved his life with many other's that will undoubtedly suffer for a life time. What he knew may have played a part of making it better quicker for thousands. "No One Listened". We have Death Factory's indoor, and out door however the natural man made by terrorism act ranks among the worst.

Unless the glass is removed from the lungs it remains in there doing damage it's much like little daggers that keep thrusting into the lung tissue, causing damage each time. Then the secondary infections begin to set in, such as Bronchitis, Pneumonia, then Asthma and other lung diseases. If no one helps you will suffer for a lifetime, and how ever much it may be shortened, only friends and family will ever really know. The therapy for it is not known by many Doctors or any one else. But Larry wanted to share it with those who felt the pain, he was not allowed to get through. He knows how well it worked for him. He has spent his first year in the last ten years free of Pneumonia and Bronchitis, and has the Asthmatic illnes very much under control. Indoor air pollution, known as the Sick Building Syndrome:

According to the World Health Organization, about one-third of all new buildings worldwide are not fit for human use due to the severe indoor air pollution. Especially so if there is any construction going on in the building. There are many which can be caused by the toxic effects of indoor air polluntants. If you are suffering symptoms such as headache, dizziness, difficulty in concentration, loss of memory, severe fatigue, confusion, irritated eyes, nose, throat, and lungs sinus infections, hoarseness, difficulty breathing, wheezing, hacking, barking cough, palpitations, nausea, vomiting, diarrhea, weakness, paralysis, numbness, twitching, any disruption of the menstrual cycle. This could be caused by indoor polluntants at work or even maybe your home.

If your having problems that you have never had before and wonder why it's not going away, or you see the Doctor he gives you medication but it does absolutely nothing. your nose is dripping, and it never done that before, but now it won't go away. Take all the symptoms listed above that you feel you are having, and nothing seems to help. Keep your own personal log. 1. when is it the worst. 2. when does it seem to improve, and how much improvement, maybe some days off or a vacation, or maybe day in day out. Keep track and you will be able to figure out where and what is actually happening.

Sensitivity in people seems to vary so much, but what you need to remember we are all different what one may be allergic to will not bother the other. Not everyone entering a polluted place and breathing the same air will feel sick yet others entering that same area will have anywhere from minor to severe reaction, of fatigue, confusion, dizziness, nausea, vomiting, difficulty breathing, headache.

1992-93 Incident with Construction dust containing Fiber Glass

1. Persistent dry, hacking "barking" cough

2. Sore throat/bloody taste

3. Blood in sputum

4. Bloody nose

5. Sinusitis and rhinitis.

6. Persistent severe repiratory infections
   A. Bronchitis
   B. Pneumonia
   C. Asthma

7. Insomnia (coughing)

8. Swollen red infected eyes

9. skin rashes/infections

10. Super sensitive to dust Carbon monoxide, smoke and chemicals

11. Reactivie airways

12. Breathing attacks. Constant wheezing

13. Headaches, Nausea, Dizziness

14. Not responding to medication for allergies

15. Not responding to medication for infection

16. Emergency treatment repuired

17. Hospitalization required

18. Ambulance required

19. Continuous coughing

1. Asthma 53% disabled

2. re-ocurring bronchitis and pneumonia

3. Serious affects on the immune system

4. Immune deficiency lowered twice.

5. Coughing

6. Progressive Asthma
   A. Do not Smoke
   B. Do not Drink
   C. Never had Pneumonia
   D. Never had Bronchitis
   E. Never used drugs
   F. Never had Asthma

SICK BUILDING SYNDROME:

Sick building syndrome may have a combination of signs and symptoms you may experience when you enter your work area. It can develop within seconds or minutes, then take hours, even days, after leaving that building. perhaps a weekend goes by and you feel much better, until you report for work on Monday morning and it starts all over again.

Anderson Laboratories, "believe sick building syndrome may be primarily caused from from combined actions of a myriad of products, that may be found in the homes, schools, office buildings, warehouses. With such a huge, source of air pollutants there is probably no ventilation system that could keep the concentrations of polluntants low enough to avoid symptoms in some occupations.

A sensitivety to odors, chemicals, there are some people that have developed severe intolerance to chemicals. The sensitivity is somthing that can grow very slowly over time with exposures to this and to that one chemical may make a person very sensitive to many others. If the esposures and even overexposures continue to happen that sensitivity to one thing could be a multitude of many. Each time the reactions take place to just one or two chemicals it turns into many. Almost all of us encounter mixtures of airbourne type chemicals day after day.

Asthma:

There is a worldwide epidemic with asthma adults and the children as well. The work place is to blame, The schools are to blame, the stores are to blame. There are chemicals being used in some form or another every place we traverse. Just to walk in somewhere and to say wow this place smells so nice. In reality what your saying is wow those chemicals sure do smell nice. Occupational asthma, Lung disease, Asthma, Bronchitis, pneumonia, are so devistating, and when you can't get a breath of air, and you begin starving for air then and only then will you realize how devistatingly real it is. If everyone out there pitches in does a little bit here, little bit there, we can all do our part of making it a better world to live in. Asthma is real, and it is no fun.

1990 OVEREXPOSURE TO THE MAUTZ LINE 89-90 ALKYDE, ALKYDE, EPOXY OIL BASED EXTERIOR INDUSTRIAL ENAMEL THAT WAS USED BY FORCE FROM OUR EMPLOYER WASHINGTON COUNTY. THE MSD SHEETS

FOR THE PAINT HAD BEEN HIDDEN CONCEALED IN A SAFE BY ED KAPLER THE SAFETY MANAGER, AND FACILITIES DIRECTOR. HE WAS TOLD IN PERSON AND FACE TO FACE BY LARRY THE PAINT WAS NOT LATEX AND WHY THE HELL ARE YOU TELLING EVERYONE IT IS. HE JUST SMILED AND AS IF TO SAY GO TO HELL AND WENT ON ABOUT HIS WORK, EVEN AFTER BEING TOLD THE PAINT IS KILLING US.

LARRY ASKED SHERIFF JIM TRUDEAU BY LETTER, THAT THE PAINT WAS SO DANGEROUS, AND IT WAS KILLING THE EMPLOYEES. THE SHERIFF CHOSE TO IGNORE THE WHOLE THING, AND NO INVESTIGATION WAS EVER DONE. IT HAD BECOME A DEPARTMENTAL JOKE!

EVERY ONE WAS SAYING THE SHERIFF FINALLY FIGURED OUT HOW TO SOLVE THE OVERCROWDING PROBLEM HE HAD IN THE JAIL, "HE'S JUST GASSING ALL THE INMATES". IT BECAME KNOWN AS SHERIFF TRUDEAU'S BIGGEST BLUNDER AS HIS PERSONAL GAS CHAMBER FOR BOTH INMATES AND EMPLOYEES. LARRY HAD EVEN WRITTEN TO SHERIFF TRUDEAU, ADVISING HIM, HE AND OTHER EMPLOYEES WERE BEING INJURED, IT WAS KILLING US. THERE WAS NO RESPONSE UNTIL TWO MONTHS LATER THE INCIDENT HAD ALREADY TAKEN PLACE. THEY CAME TO ASK WHAT THEY COULD DO TO HELP. IT WAS EXPLAINED IT IS A LITTLE LATE BUT WE NEED MEDICAL HELP, AND THAT WAS IGNORED, NOTHING MORE WAS EVER SAID.

THEN IN 1994 LARRY HAD ASK FOR A REASONABLE ACCOMMODATION UNDER THE AMERICANS WITH DISABILITIES ACT, "ADA" THOSE ACCOMMODATIONS WERE GIVEN. ONE ESPECIALLY SO A LETTER WAS SENT TO ED KAPLER WHO WAS THE SAFETY DIRECTOR AND FACILITIES MANAGER. THE LETTER STATED MR. LAND WAS NOT TO BE SUBJECTED TO ANY CHEMICAL EXPOSURES AND HE WAS TO NOTIFY THE SHERIFF DEPARTMENT AT LEAST 24 HOURS IN ADVANCE OF ANY CHEMICALS BEING SPRAYED IN THE LAW ENFORCEMENT CENTER. EVERY THING SEEMED FINE UNTIL ONE SPRING DAY IN 1995 FACILITIES HAD SENT OVER THEIR CREW TO

SPRAY FUNGICIDE IN THE SHOWERS SPECIFICALLY ON LEVEL TWO WHERE LARRY WOULD BE WORKING. LARRY HAVING MULITI CHEMICAL SENSITIVITY SYNDROME HE IMMEDIATELY WENT INTO PULMONARY ARREST. OXYGEN WAS AVAILABLE AND HE HAD HIS MEDICATIONS, WAS IMMEDIATELY MOVED TO FRESH AIR BY CO-WORKERS. BREATHING WAS REESTABLISHED BUT HE WAS ILL FOR NEARLY FOUR HOURS. THE CAPTAIN NOTIFIED KAPLER THAT HE HAD LARRY'S RIGHTS UNDER ADA. HE IGNORED THIS AND THE FOLLOWING WEEK HIS PEOPLE WERE OVER DOING THE EXACT SAME THING JUST AHEAD OF LARRY COMING ON DUTY. HE NOTICED THEM IN THE AREA AND REFUSED TO REPORT FOR DUTY, & NOTIFIED THE ADMINISTRATOR, STATING WHAT IS HE TRYING TO DO MEANING ED KAPLER, SINCE HE WAS GIVEN A LETTER, AND TWO VERBAL WARNINGS. IT SEEMED RATHER ODD THE ONLY SHOWERS THAT WERE SPRAYED WITH A DEADLY FUNGICIDE WAS ON LEVEL TWO WHERE OFFICER LAND WAS ASSIGNED.

NOW THERE WERE TWO MORE WEEKS IN A ROW THE PEOPLE CAME INTO LAND'S WORK AREA ON LEVEL TWO, SAME DAY OF THE WEEK. HE ASK THEM WHO SENT YOU OVER TO DO THE SPRAYING, THEY ADMITTED IT WAS ED KAPLER THROUGH JERRY NIGORSKI THEIR IMMEDIATE SUPERVISOR, THEY WERE SENT BACK TO HIM NOT ALLOWED TO SPRAY ANYTHING, AND WERE ASKED TO TELL ED KAPLER TO FOLLOW DIRECTIVES. THE FOLLOWING WEEK HE HAD HIS PEOPLE REPORT TO WORK EARLY SO THEY COULD SPRAY LAND'S WORK AREA BEFORE HE ARRIVED. WHICH THEY DID AND IMMEDIATELY LAND WENT INTO AN MCSS, ASTHMATIC ATTACK WHICH DESTROYED HIS WHOLE DAY AND HAD TO GO HOME WITH DIFFICULTY BREATHING. KAPLER WAS AGAIN WARNED, BUT RIGHT BACK AGAIN THE FOLLOWING WEEK AND DONE EXACTLY THE SAME THING, THIS TIME LARRY WAS WARNED BY HE COWORKERS NOT TO COME INTO THE AREA. THIS SEEMED LIKE A GAME OF HARASSMENT BEING PLAYED OUT BY ED KAPLER AND HIS SUPERVISER JERRY NIGORSKI HIS OWN

STAFF FINGERED HIM AS THE WHO PERSONALLY TOLD HIS PEOPLE EXACTLY WHAT WAS EXPECTED, AND TO IGNORE LAND. THE FOLLOWING WEEK THE SAME THING AGAIN HAPPENED, AND HE WAS ASSIGNED A DIFFERENT STATION. THE EIGHTH WEEK IN A ROW IT HAPPENED AGAIN, AND AGAIN COWORKERS WARNED LARRY NOT TO REPORT FOR DUTY ON LEVEL TWO. I'M NOT SURE WHY THIS WAS DONE OTHER THAN THEY WERE PERHAPS HOPING HE WOULD STOP BREATHING, OR JUST PLAIN DIRTY HARASSMENT FOR TURNING THEM IN FOR WRONG DOING DURING THE 1990 PAINT INCIDENT WHEN THEY WERE CAUGHT DOING IMMORAL AND ILLEGAL ACTS AGAINST OTHERS THAT HAD CAUSED INJURY DURING THE 1990 PAINT INCIDENT AS WAS NOTED IN THE OSHA CITATIONS.

LARRY EVEN TRIED GETTING THE TEAMSTER'S Local 320 INVOLVED TO HELP ENFORCE SAFETY CONCERNS BEING VIOLATED. SAFETY ISSUES IN THE CONTRACT FOR EMPLOYEES SAFETY. TELLING THEM THE SHERIFF DEPARTMENT WAS VIOLATING THE CONTRACT, BOTH STATE AND FEDERAL LAW. IT WENT ON AS IF FALLING ON DEAF EARS, GRIEVANCES WERE FILED WITH THE UNION SEVENTEEN TIMES OVER. THEY STILL IGNORE THE PROBLEMS IT HAS CAUSED AND WILL NOT HELP EVEN TODAY. TEAMSTER'S IN A TIME OF NEED WAS AS USELESS AS ANYONE COULD BE, AND STILL ARE TODAY. THEY DID NOT KNOW HOW TO ENFORCE THEIR OWN CONTRACT, ALLOWING WASHINGTON COUNTY TO INJURE AND CAUSE DISABILITY TO THERE OWN EMPLOYEES. SEE SIGNS & SYMPTOMS, THE MSDS, & XYLENE DETOXIFICATION EVERYTING UP TO DEATH.

The Material Safety Data Sheets, MSDS is the empolyees only hope if the employer is using somthing they should be or are entirely not aware of what may be causing the sick building syndrome. Paints, Glues, Fiber Glass, Chemicals of any kind, Companies must supply safety data sheets for to your employer. If the employer is using somthing without telling and training the employees is an offense for which you as an employee can sign a complaint with OSHA. OSHA can issue citations, with fines, and will keep your name away from the employer so they will never know for sure who turned them in for an investigation of using toxic, or hazardous substances in the work place.

This is called the Right To Know Act. The employee has the right to know exactly what may be used in the work place, and the hazards they may face. The employer has the obligation to train the employee's what to expect, what to do if necessary, what to do if they are over exposed to the hazardous substance, and or toxic chemicals in the work place. There are products out there that can be used without emitting a hazardous, or toxic effect in the work place, also the MSDS tells the employer exactly what he must do before using the product in a workplace. Then the MSDS sheet is to be displayed in the work area that is involved as a warning to employees.

Washington County Employees gave written petitions to Sheriff Trudeau, Ed Kapler, the County Health Department, and the County Commissioners they are rejected the safety of their employees, during the 1990 toxic paint incident [O]verExposure.

There were no windows, doors were sealed, The ventilation was next to none, air return set at 92%. Employee's reported every sign and symtom up to death. Complained of eye, nose, and throat irritation, Headaches, Nausea, Dizziness, Vomiting, Body aches, Salavation, Feeling of being Drunk, Loss of consciousness, High, fatigue, whistling sound in ears, Seeing Stars, unable to focus, Blurry vision, Unable to concentrate, Short term memory loss, Ataxia, Asphasia, confusion, clumsy, Shaking, Trembling, Starving for air Difficult breathing, Sinusitis, Rhinorea, Difficulty hearing, Irritability, The world seemed to move in slow motion.

#15

EPA/OSHA

Xylene; main ingredient in the paint: Effects of
Xylene on the Nervous System.

100-200 ppm.	Nausea, Headache
200-500 ppm.	Feeling "high" dizziness, weakness, irritability, vomiting slowed reaction time.
800-1,000 ppm.	giddiness, confusion, clumsiness slurred speech, loss of balance, ringing in the ears,
>10,000 ppm.	sleepiness, loss of consciousness, "Death".

Compare signs & symptoms reported, to the OSHA & EPA chart produced to show exposure & overexposure to Xylene.

CHAPTER 16 LOOKING BACK

"HAPPINESS LIES IN THE JOY OF AQCHIEVEMENT AND THE THRILL OF CREATIVE EFFORT."

FRANKLIN ROOSEVELT

DESIRE: Desire alone is not enough. But to lack desire, means to lack a key ingredient to success. Many a talented individual failed because they lacked desire. Many victories have been snatched by the underdog all because they wanted it more. So if you desire…Intensely…and you act upon it, then everything stands within your reach…Add the imagination and reach for the stars….Seeing all possibilities, seeing all that can be done, and how it can be done, marks the power of imagination. Your imagination stands as your own personal laboratory. Here you can rehearse the possibilities, map out plans, & visualize overcoming obstacles. Imagination will then turn possibilities into reality.

Larry Land

It is the winter of 1997, soon the long sleep of winter will give way to the gradual softness of spring with the sweet chirping of the birds. The geese circling overhead in huge formations as they return to watch spring roll in. Larry makes a trip to Denver every spring, he feels privileged just to watch the return of the eagles to this area. As two majestic bird's circle above him with their massive wings outspread, ready to accept the challenges that lie ahead. Their challenge is the same as his. Simple survival, and the procreation of their species, as if a slow march toward the future.

The snow here in Colorado is now gone, the beautiful white blanket that covers the Colorado prairie, and Larry can feel mother earth beginning to stir deep within his soul he can see the ugliness and future death awaiting those who venture onto this once pristine piece of prairie.

Looking back what was once the home place that had been purchased in the early 1930's by Adam and Viola Land. They move onto an 80 acre tract of land about 6 miles south of Brighton Colorado. This land was to be their home for the next 40 years and upon which they would raise a large family of 10 children with some 35 grandchildren…they never got to know. They both died well before their time. Adam Land was a livestock breeder and a cattle buyer, (horse trader) and a dry land farmer. His children were taught animal husbandry through practical experience. Larry went on to become an expert in the field of breeding, raising, and buying and selling of livestock.

Larry was born at home on the farm March 8, 1941, exactly 8 a.m. weighed in exactly 8 pounds and 8 ounces. In the 1940's the United States was still on the edge of the depression of the 1930's when everything was still in the clutches of the drepression, along with a period of drought. The wind seemed as if it were trying to blow everything away, was called the dust bowl days. The government was giving away trees of all kinds to the farmers to plant anywhere, up and down fence lines, up and down along the roadways. cattle were laying dead everywhere, some farm houses were actually buried under mounds of sand.

In 1942, the Army constructed the Rocky Mountain Arsenal. on its most northern boundary was one mile south of the Land property. Although its mission was a military secret, its primary purpose was the manufacturing of war gases. Information made public in the late 1970's, reflect that is manufactured over 800 tons of mustard agents. From the beginning, the Arsenal utilized an unlined pit, Basin A, to dsipose of unwanted chemicals.

Such chemicals would seep into the shallow aquifer and would then be flushed under the surrounding lands. This disposal process was initiated without studies as to where the ground water would do and what effect it might have on the surrounding property owners. At the end of WWII, the army dumped all of the mustard agents it had manufactured into basin A. These mustard agents continued to seep into the ground water for the next 50 years.

Following the end of WWII, the Army leased the Arsenal facilities to a series of chemical manufacturers who commenced the manufacture of insecticides at the arsenal. The first disaster was when they were trying to manufacture DDT. It was a living nightmare for everyone even those who tried to make it. It was finally banned in the United States of America because of it's deadly consequences to everyone. This manufacturing facility was ultimately acquired by Shell Chemical Company, a Subsidiary division of Shell Oil Company, (Foreign Owned). They manufactured the Organo-phosphate insecticides and nemogen, The organophosphate insecticides include dieldrin, aldrin, endrin, which are a class on nerve agent insecticides which have been banned in the United States of America. Nemagon was an agent used in agriculture which could cause impotency in men. Because of proprietary interests, Shell was nearly as secretive about its activities as was the Army. Information released during and following 1975 reflected that dieldrin, aldrin, nemagon and other agents and byproducts had been introcued into the ground water by disposal into Basin A and leaking lines leading into Basin A. Millions of gallons was purposefully leaking into the underground water stratus, Both Shell Chemical Company, and the United States Army were aware it was leaking, into the water stratus that would eventually move to the North and Northwest where it eventually polluted & contaminated fifty to sixty thousand acres of land off the arsenal property with airborne chemicals and the underground water.

Then beginning in 1953, the Army constructed facilities at the Arsenal to manufacture GB nerve gas. GB nerve gas is an organophosphate compound of the greatest potency. A single drop applied to the skin is sufficient to kill a man within the hour if left untreated. Again this activity was a military secret at the time. Information supplied by the Army during discovery on a later congressional reference, reflected that the process for making GB agents was highly exothermic—i.e. generated lots of heat. Large quantities of water were used to cool the process. This caused a rise in the water table which was a primary factor resulting in injury to the Lands in the 1970's. The process to make nerve gas resulted in a liquid consisting primarily of the nerve gas and a compound known as DIMP. (diisopropylmethylphosphonate). Distillation was used to separate the two compounds. Distillation is an inefficient means of separation and necessarily resulted in nerve gas entering the ground water at the same time in association with the DIMP. Some batches of nerve gas could not reach military specifications. These batches were treated in sodium hydroxide. This was an enhanced chemical reaction that created sodim fluoride, a very toxic substance, in itself. Nearly that of GB nerve agent, which as also discharged into basin A & also entered the ground waters.

Sodium Flouride is a very deadly form of fluoride, while calcium the fluoride is not yet the tests used to discover fouride in a water sample, may show safe levels of fluoride but does not distinguish whether it is Sodium Flouride, or Calcium Flouride. In effect if they were to check your water and discover the fluoride concentration was safe, your water would kill you in moments if it were sodium fluoride. Sodium fluoride is a man made fluoride and calcium is a natural fluoride.

Because of the large amounts of water entering Basin A, it was quickly enlarged, and the enlargement was found insufficient to contain all the wastes generated by the Army and Shell Oil Company. Basins C, D, and E, were constructed to contain the excess wastes. Each of these Basins were unlined. Basin C was called the "STRAW" basin because it was the most permeable of all of the basins. It had been constructed right into the upper water stratus, so what ever was put in Basic C disappeared very rapidly. Army Memos on Basin C, if they wanted to get rid of somthing quick they would put it in basin C. Then over the years since the 1950's 60's, 70's, and right into the 1980's, it was used by them as a flushing mechanism to flush the chemicals from beneath the Arsenal out into the water aquifers off the Arsenal, and therefore slowly poisoning every one the the North and Northwest of the Arsenal.

The year 1954 was an unusually dry year. Because of the unusually dry conditions farmers to the west of the Arsenal, whose lands were irrigated, began using ground water to supplement irrigation water from the Platte River. Corps irrigated with ground water would die. Animals consuming the ground water would become ill. Subsequent studies showed that the source of the ground water was from the Arsenal. Although all precise figures have been lost, it is estimated that the Army paid over $600,000 to adjacent landowners for the pollution although the exact chemicals involved were never identified. Timing of the movement of ground water would suggest that such contamination was the result of the manufacture and disposal of mustard agents. With the mixture of the various chemicals and a natural process in the waste ponds somehow 2-4-D was formed in the well water off the Arsenal. 2-4-D was not in itself manufactured at the arsenal.

This contamination resulted in a report authored by Dr. Graham Walton of the Robert A. Taft Sanitary Engineering Center, U.S. Department of Health, Education & Welfare, Cincinnati, Ohio. This report recommended a testing program to identify the chemicals involved & a testing program to define the boundaries of contamination. The study also recommended and education program to that adjoining landowners could be advised of the existence of the pollution. The study also recommended a program requiring special practices in completing water wells in the area to protect against ground water contamination. The recommendations of this report were never implemented. Compliance with the recommendations of this report would have prevented the later tragedy that happened to the Land Family.

One of the results of the 1950's pollution was a study by the Corp of engineers to build a containment system similar to the one later constructed in the 1980's. This system was abandoned as too expensive and modified to a decision to build a lined basin covering 93 acres and known as Basin F. Basin F was constructed in 1957. Basin F began leaking while it was being filled for the first time. It was drained, and repaired & refilled. It was not long before the Army suspected that Basin F was leaking again. The solution this time was to classify the information as secret so that the fact of leaking was concealed from the Colorado Health Department the the United States Department of Health and Welfare, The United States Congress, and the public. Such information did not become public until the 1990's when Basin F was abandoned.

Basin F was designed for a maximum of 180 million gallons of toxic waste. It was designed with a life expectancy of 10 years, the black asphalt liner was 3/8 of an inch thick blown onto the surface. The Army was warned in 1965 that the Basin F was leaking, they at that time drained it down and replaced some of the higher surface with more blown asphalt. then immediately filled it to the maximum. By 1969 the Army was warned by the Armies own Health and Hygeine agency, and the United States Department of Health and Welfare. They were quite frankly told to stop using basin F as it was without doubt Leaking. It was inspected and then confirmed by two Generals from the Pentagon it was in fact leaking and large parts of the liner had been completely eroded and had disintagrated. The General's said this we can't have the public find out, so they had it immediately classified, and then continued using Basin F for an additional 15 years. Remember when I had said it was designed to hold no more than 180 million gallons of toxic waste. During the 1970's they through away the plans the basin was designed to hold, and drew up their own plans they knew it was leaking out across the

Country side endangering peoples lives. In the later part of the 1970's Shell Oil Company knew their days in Denver were numbered, by 1977 and 1978 they doubled, and tripled their production, pushing the limits of waste in basin F to 200 Million gallons, by 1979 and 1980 they had now quadrupled their production of deadly pesticides, filling Basin F. to the maximum point of over flow, 250 million gallons of waste, and yes they were aware the Basin was leaking, so did the Pentagon.

Despite all its leaks, Basin F was not adequate to contain all of the liquids being generated on the Arsenal. During the 1960's The Army then installed a spray raft on Basin F to spray wastes into the air to encourage evaporation. This also resulted in mustard agents & possibly nerve agents being sprayed into the air. The Land family was affected by this program. Jim Land testified how they would awaken in the morning choking from the fumes from the Arsenal The cause of such fumes was classified secret and unknown to the public. This program resulted in the Land family's initial contact with chemical agents and increased their sensitivity to subsequent exposures.

A hydrologic study releaed by the USGS in 1975, contains information supporting a conclusion that the contamination from the nerve gas manufacturing reached the free aquifer underlying land adjacent to the Land property in the late 1960's. It was about this time that the corrosive nature of the contaminated ground water consumed the casing on the Land's original 800 foot well and a substitute 400 foot well was drilled. It was about this time that Adam Land began complaining about a chemical smell to the dishes. It was at this time when his diabetes became difficult to control and he underwent personality changes. pesticides, filling Basin F. to the maximum point of over flow, 250 million gallons of waste, and yes they were aware the Basin was leaking, so did the Pentagon.

Despite all its leaks, Basin F was not adequate to contain all of the liquids being generated on the Arsenal. During the 1960's The Army then installed a spray raft on Basin F to spray wastes into the air to encourage evaporation. This also resulted in mustard agents & possibly nerve agents being sprayed into the air. The Land family was affected by this program. Jim Land testified how they would awaken in the morning choking from the fumes from the Arsenal. The cause of such fumes was classified secret and unknown to the public. This program resulted in the Land family's initial contact with chemical agents and increased their sensitivity to subsequent exposures.

A hydrologic study releaed by the USGS in 1975, contains information supporting a conclusion that the contamination from the nerve gas manufacturing reached the free aquifer underlying land adjacent to the Land property in the late 1960's. It was about this time that the corrosive nature of the contaminated ground water consumed the casing on the Land's original 800 foot well and a substitute 400 foot well was drilled. It was about this time that Adam Land began complaining about a chemical smell to the dishes. It was at this time when his diabetes became difficult to control and he underwent personality changes, & under went several surgeries for colitis problems, along with problems of digesting food.

By this time, Jim Land and Lois had married and he and his wife lived on the north end of the 80 acre tract. Jim had been in partnership with his father in a large farming operation. Adam Land became so difficult that Jim was finally forced to dissolve the partnership. Then Adam's wife was also forced to dissolve their marriage of nearly 50 years.

By 1969, Larry Land had moved to the Box Elder Creek Ranch which he and his wife were farming, on shares and building a large cattle herd. He had many hundreds of acres of ranch and pasture Land he was leasing, there and several grassland ranches in the mountains that over looked the city of Golden, called the Flat Top Mountain Range. Larry had become very successful in raising livestock. He was also a Biologist, a micro biologist, self taught, who had his own special breeding program with cattle. He had what was known as the widest of wide outcrosses for cattle ever before known. He already had 20 of the out crosses, some were crossed with Holstein, Black Angus, Santa-Gertrudis, Black Baldies, and Herford. The whole idea was through genetics to have a calf that was smaller than the ordinary therefore making it much easier for the heifers and cows that would be calving, and to cut any death loss at birth nearly 100%. Also the change in the genetics bred out most illnesses calves would normally suffer. Larry had already proved his point there, he had the first 20 head that were much smaller at birth, and immune to most diseases. These cattle now were in the range of 12 to 20 months old.

This giving Larry a much better idea of the growth. At that point in time the figures showed, once all ordinary cattle had topped their weight class, the breeds were somewhat smaller, but then they kept growing where others quit, then within 6 months caught up with the others in weight, and still growing, then with an additional 4 to 6 months they would out weigh the others by 400 pounds or more. As for the feed and quality It seemed the breeds could actually thrive where the others would not.

This was just one of the biggest setbacks Larry had ever experienced in his life. This was one of the hopes and dreams he had built into and around his life, and then was actually able to do it, and prove he had genetically engineered into cattle breeding somthing never before done in history, cattle that would thrive on less feed, or feed of lesser quality, His genetics in breeding actually bred many diseases suffered in the livestock industry out. He had an animal smaller at birth, saving much death loss there, and then an animal that would outgrow others.

By the fall of 1971, he had purchased 75 acres of the 80 acre tract and began moving to the family farm. It was the spring of 1972 when he began moving his 150 heard of cattle that were left of nearly 900 at the Box Elder ranch over to his ranch where he was born and raised. His dreams seemed to be coming true. He worked hard thru the winter of 71 to construct the corrals, barns, and lean to's, so every thing would be in order to build his 1000 head heard of Holstein Dairy cattle and a place he could continue working on the Genetic Engineering System that was meant to be a "Directed Alteration" in the livestock we know today, Larry had already successfully altered the genetic process by gene splicing within his breeding and genetic engineering program. He had already set up leases on several thousand acre's of pasture he would be able take advantage of almost year round.

Using the old 800 foot well, and the existing 400 foot well, Larry then realized the two existing wells would be inadequate, since the 800 ft. well kept filling with sand from the 35 to 60 levels where the casing had been eaten away by chemicals. Not knowing the chemicals were coming from the Arsenal in the shallower water and then getting into and polluting the deeper wells. The first part of april the cattle Larry had already moved onto the farm, some of the smaller animals seemed not to thrive. Larry was already having trouble with the deep well, so he decided to have a new well drilled to sustain more livestock. Larry had begun to question the quality of the present water. The new well he had drilled further to the east and to a depth of 320 feets. When he began pumping the water, it was red in color, he ask the well man what is this. He said it is just iron, we would be pumping the well for 48 hours straight in order to determine where the static level of the water would be, and how much water it was producing. then called the State Health, and Tri-County Health Departments who are required to do tests on the water and wells to approve or disaprove and let you know if your water is safe to drink. So it was pumped for forty eight hours and was still running red, The well man said go ahead and pump it for 72 hours, and it should be cleared up by then. The reports came back positive from the State Health and the Tri County Health Departments, stating the water was good potable drinking water, meaning there was no toxins, or harmful bacteria in the water samples, I ask again then why in hell is the water running red. The well driller, and the Health departments both claimed it was just iron and would not be harmful at all. During this time Larry had gotten in several hundered head of livestock, the count was now up to 635 head he then put the breaks on any more until he got everything sorted out, most of the calves were eight weeks upto twenty months old. The smaller ones showed signs of not thriving. By th 15th of April Larry decided to go ahead and turn on the water since he was told by the Colorado State Health Department, and the Tri-County Health Departments that the water was OK. He trusted their expertise in this matter, after all he even had the Chief Chemist from the State Bill Dunn out to examine the water who said it is good potable drinking water. Larry thought if someone were using a red dye or somthing to trace leaks from the Arsenal Bill Dunn would know since he was the man in charge of making sure no chemicals leaked from the Arsenal.

Larry turned the water into the animals, and he began to see their ears drooping, eyes red and watery, they seem to lap at the water, hair became scruffy and small splotches of hair was falling out leaving small dime and quarter size bare spots, They began to get sores in their noses, and the mouth, then the diarreha began, and the weaker and smaller calves began going into convulsions, the legs would be going in the same rythem as the eyes no matter what you done they were gone. "That was the last throws of life caused from nerve gas poisoning". The vetrinarian knew they were being poinsoned but how, he began checking the feed, changing the feed putting them in different pens then knew the only thing the cattle had common with each other was the water. Their teeth began turning black as they would try eating the day their teeth began falling out, and the cattle began to vomit, Dr. Scott said my God I've never seen anything like this in my lifetime. The State Health Department was called back out, and stated no it cannot be the water, Bill Dunn said, "there is no way on God's green earth contaminated water could be coming from the Arsenal." "It was later found out that he knew exactly what was happening, where it was coming from and why the water was running red". Yet he said there was nothing wrong here even when he seen the cattle dying, dead, and the heifer's aborting their calves. Yes around 70 head of heifers were with calf, about the seventh and eighth month of pregnancy. They all began aborting their calves, within 5 days they had all aborted right while the State Health People were observing first hand and could not put a finger on it. Larry believes they knew first hand exactly what was happening because it was the State Department of Health that tried their best to conflict what Larry was giving to the news media. Trying their best to make it all look as if it were his fault, mismanagement with livestock. Larry had just came from the Box Elder Farm where he had been doing the same thing with nearly 2000 head of livestock with a 3 year death loss of only 1 percent. which is considered "excellent". The Colorado Department of Health kept saying the cattle were dying of hemmoragicsepticemia which is a contagious form of pneumonia, or shipping fever. As they were witnesses to necropsies done by Dr. Sctott. (Necropsy is same as an autopsy). All they seen was the lungs destroyed, They didn't even bother to put into their eqation, that the kidneys, the Liver, the Thymus gland, the Heart, the eyes, were totally destroyed, & that spelled out somthing very toxic was taking place and the immune system was totally overwhelmed then and only then this secondary infection Pneumonia that only takes over once the immune system had become overwhelmed, and medication was of no help. Larry had managed and took care of over two thousand head the exact same way just a year before, he knew best as Larry was considered an expert in animal husbandry, ("a scientiffic control of the production and management of livestock"). This he had done already when he was out on the BoxElder Farm and only had a one percent loss with over two thousand head of livestock over three year period of time. This was considered excellent since 3.5% to 4% is considered average when raising livestock. This what he had accomplished out on the BoxElder farms was a prelude of what was planned when he moved to the home place.

Dr. Scott, the attending veterinarian had soon become convinced that the water was the problem. He finally convinced Larry of this fact & they moved the remaining herd to another one of Dr. Scotts client's property. Upon reaching the new property and new water the remaining animals ceased becoming ill, and began to recover. The only thing separating the dairy animals on this other farm was a barbed wire fence, as his cattle become exposed to the Land animals there was no contagion passed on. That was because there was no contagious problem in the first place, but the people from the State Health still would not buy it, even though none of the contagion they talked about passed on. Most of the remaining animals were young, and were some of the last brought in. Although they ceased to die, they also ceased to grow. Eventually, they were sold in their stunted condition, at a great loss to the Land's.

About six months later, Jim Land began having the similar problems with his herd of cattle. He lived ½ mile further to the North from Larry's place. He quickly began hauling water to his cattle, and the problems ceased. The expense of hauling water to his cattle forced him to sell his herd.

By now the human members of the family of this Horrible tragjedy were having many prolems. Initially it was thought to be fatigue, with a lot of gastric disorders, Nausea, Vomiting, Headaches, Eye, Nose, Throat, and Lung Irritation, with difficulty breathing, aching and hurting joints, Forgetfulness, Shaking, Gettiness, Dizziness. At a Congressional reference, Dr. Tietlebaum, a Denver toxicologist, who had been working with the Land's expressed the opinion that the humans in this case were suffering the same problems as the cattle that had died. Even down to the terrible rashes on their bodies, and the small splotches of hair missing. The Administrative people from the State Health Department seen this also first hand, as well as what they seen with the livestock, but they done absolutely nothing, they were just plain skeptical while they all knew there was somthing wrong. The striking similarity between the symtoms of the land family and the so-called Gulf War Syndrome is very much the same.

No matter how many times the State Health Department had been out to the farm, between 1972, and 1976, they had tested the well water and found Dimp. at levels that were two to three times the states level, they found dieldrin, and endrin in the Land wells at lethal levels, and even then the State Health Department responsibility to distribute bottled water to those affected, everyone got it but the Lands. The State Health Department found toxins from the Arsenal in our wells yet failed to notify us, and we didn't find out until 1992 nearly 20 years after the fact. as part of the trial had finallly begun. The State Health Department at no time after discovering that the Land wells were in fact polluted from the Arsenal did they ever say to us don't bathe in the water either as the toxins absorb through the skin, and can kill you, nor did they ever offer bottled water. The Land wells were the first one's in the area that were in fact found to be contaminated from the Arsenal, actually fingerprinted to the Land wells.

Three generations of the Land family were affected by their consumption of this water. Because of the financial devastation and the loss of the cattle, the Land family had to rely upon public health assistance. Initail conclusions at medical clinics at the University of Colorado were never followed through, or never followed up on. There is evidence that the health problems are extending down into the fourth generation suggesting that some of the adverse effects were and are genetic.

Larry Land's appeals to the Army for help fell on deaf ears. The Army initially denied and continues to deny any responsibility. The only water testing during the poisoning episode was performed by Tri-County Heath, and the State Health Department's, and all such tests were solely search for bacteria. This never even dawned on Tri-County or the State Health Departments that somthing was wrong when the water was completely void of any bacteria. With the toxins in the water there could not have been any bacteria good or bad, and the fact the water ran so red in color. Despite the absence of testing during the period of poisoning, there is substantial circumstantial evidence that nerve gas was at least one of the poisons in the water. Subsequent testing indicated a strong probability that sodium fluoride was also present. Larry Land continued to do water testing, His efforts finally resulted in tests done by the EPA. These efforts eventually led to the Army's admission to the public of the existence of Dimp which as found in the groundwater on property adjoining the Land property at concentrations of 2 to 3,000 ppm. Dimp does not occur other than in association with the manufacture of nerve gas.

In 1972, the procedures for testing water for unknown organic substances was few and far between, and extremely expensive. One expert hired by Mr. Land stated that the cost would have been in the millions of dollars. In 1975, the Army began a testing program. It's budget of about $5,000,000.00 million dollars for the first year was considered inadequate by the person supervising the program. After 25 years of testing, new compounds are still being discovered on the Arsenal in the vicinity of Basin F.

Mr. Land being a self taught Micro Biologist decided to take a sample of the red colored water, and very carefully placed it under his high power microscopes, just to see if the red colored material was in fact micro fine particles of iron. Larry took one look and came away shocked. The red material was perfect red colored spheres. He immediately took it to the Brighton Labs to see if they would know what this was. They took one look and said my God that looks like lon-Exchange Resin. So Larry asked what that was and what would it have been used for. lon exchange resin is used to clean remove certain chemicals and metals from a finished product to make it pure, and refined. They inturn directed Larry to certain chemist with a laboratory in Denver. He took the sample and placed it under his microscope, and immediately stated that's exactly what you have is lon Exchange Resin. This man was an Army Scientist that had been retired. He began explaining the different types of lon Exchange resin and how it is used to clean up batches manufactured and to purify the finished product. What you have here is a very fine grade of lon Exchange resin of the 300 to 400 mesh size, he said that will go where ever water will travel. It is used by the Army as a tracer if they suspect a leak in the system. lon Exchange Resin is non biodegradable, will last forever, cannot be destroyed by acid, or corrosive materials, it becomes just that what it may be attached to, always bright red and cognizable. In your case just North of the Rocky Mountain Arsenal, They have many leaks of very toxic materials and have used that material to trace those leaks. He said I'm surprised someone hasn't told you already, because they are working hard to determine where the toxins are leaking.

Counsel for the Land family advised them that the government could not be successfully sued because of the discretionary function exception to the Federal Tort Claims Act and the Doctrine of Sovereign Immunity. In 1988, with the assistance of then Congressman Hank Brown, the House of Representatives passed a resolution referring the Land problem to the Federal Court of Claims for it's advise. The Court of Claims are the Fact Finders in a Congressional Reference. They work on three levels of discovery, Legal Claims, Equitable Claims, and Gratuitis Claim. The Equitable claim and the Legal Claims are handled by the Court, who would make all determination. The Gratuitis Claim is turned back over to Congress with all the facts and findings of the panel, and it is then up to Congress or the Congressmen and or Senators of the State of Colorado would then Place the amount on the Bill setting in the files of the Full House of Congress, HR. A Bill 816 with the necessary amount of funds to be appropriated that would settle all claims brought against the United States Of America by Larry D. Land et al.

The Tenth Circuit Court of Appeals has summarized the law of the State of Colorado relating to water pollution as allowing two causes of action. The first is a traditional cause of action and is known as a "circumstantial case". An example of a circumstantial case typically involved cattle which suddenly became sick and started dying when drinking water from a contamination source. The proof of causation was sufficient if the cattle then ceased to become ill when they were removed to a new and non-polluted source of water. The second cause of action is the toxic tort case. A toxic tort case is proven by testing showing the presence of a chemical known to have a certain effect upon life.

In the Congressional Reference, the Land family presented proof that was based upon a circumstantial case. The Army presented a defense based upon the theory of toxic tort that the concentrations of Dimp was insufficient to prove causation. Judge Robinson ignored the the plaitiff's circumstantial proof and ruled that the testimony of the veterinarian Dr. Scott and the toxicologist, Dr. Tietlebaum, was inadequate because they could not state with certainty based upon chemical tests what compound actually caused the poisonings. The limited holding of Judge Robinson was that the proof was inadequate to show that concentrations of Dimp of less than 500 ppm were sufficient to prove that it was the causation of the losses of cattle and injuries to the Lands. Even though the Judge was totally aware that anything above 8ppb is the law of the State of Colorado and anything above that standard would have caused losses of the cattle and the injuries to the Lands. Even though he was to be hearing the case under Colorado Law he deliberately through that part out and used the EPA standard of 500 ppm. Even this conclusion is subject to suspicion since the testimony of the government's experts was based upon test the results conducted by the government on samples obtained by fraud and several years after the poisoning episode under conditions that were entirely different than those encountered during the poisoning episode. Test of ground waters adjoining the Land property showing DIMP with concentrations well in excess of 500 ppm were ignored.

Subsequent to the Congressional Reference various residents of Irondale, Colorado sued both the government and Shell Oil Company for the damages resulting from contamination of the air during the closure of and clean up of Basin F in the Late 1980's and early 1990's. It had been predicted by counsel for the Lands, The Tenth Circuit Court of Appeals held that the government was not liable for damages resulting from its activities at the Rocky Mountain Arsenal because of the Doctrine of Sovereign Immunity since such activities were not subject to claim under the Tort Claims Act because of the exception for discretionary functions. In the same case, the Tenth Circuit Court of Appeals found that the activities on the Rocky Mountain Arsaenal were Ultrahazardous and applied the Doctrine of Strict Liability to Shell Oil Company. Based upon this case, it is apparent that the sole remedy to the Land family for relief for the conduct of the United States Government for damages that resulted from its activities at the Rocky Mountain Arsenal is a private bill by the Colorado Congressmen for the appropriation from the Congress of the United States to be placed in the blank space on HR A Bill-816. The decission of the Court was made when they awarded a Gratuites Award the amount is to be awarded by appropriation from Congress.

It has now been setting in the files of the Untied States Congress for action, by the Colorado Congressional members or the Senate. That was placed there for some type action by Colorado, since that is where it began. It would now be their resonsibility to act on Lands behalf so the award from the Court is appropriated. In 1997, Larry began to contact all the Senators and Congressmen in Colorado. This is May 2002 and the Bill is still not been acted upon. We have ran into a stone wall, and none have made a move to assist getting this back before the full house of Congress.

1. Senator Ben nighbhorse Campbell, Denver, Pueblo, Grand-Junction.
2. Senator Wayne Allard, Greeley, Denver Colorado Springs, Pueblo.
3. Congressperson Diana Degette 1st District Denver, Westminster.
4. Congressman Bob Schaffer——-4th District Greeley, Ft. Collins, Sterling.
5. Congressman Scott Mcginnis—-3rd District Pueblo, Durango.
6. Congressman Mark Udall 2nd District, Westminister, Washington D.C.

Larry Says', he is a believer, We the people I know helped our Senators, and we helped our Congressmen, most don't even remember after their elected, and some don't even remember what it is there suppose to do. That could be why I'm still waiting the Courts award because it was up to Congress and they somehow forgot how to do it.

I somtimes wish Hank Brown was still our Congressman, because I know for a fact I would not be sitting here waiting for Congress to do the Job they were suppose to do 5 years ago. If Hank would have been here it would've been handled 5 years ago. That's just the kind a person he was always ready to do what was needed.

MARTIN MARIETTA CORPORATION, A Maryland Corporation, and city and County of Denver Colorado. Acting through the Board of Water Commissioners, Defendants.

TCE in water being treated at the Kassler operations. Kassler was immediately closed due to toxic contamination of the water and because of that contamination Kassler has not reopened.

Between April and June of 1985, the State of Colorado issued two compliance orders to Martin charging that the waste management units at the facility in Waterton have violated and continue to violate the requirements of the CHWA Colorado Hazardous Waste Act, and its implementing regulations and that Martin Marietta's Management of hazardous wastes at the facility have caused significant contamination of the ground water.... Then again similarly, in July, 1985 The Environmental Protection Agency (EPA), issued a cease and desist order to Martin for unpermitted discharges of hydrazine wastewater into Brush Creek from Waterton. May, 1986, Martin and the State of Colorado entered into a compliance order which required Martin to bring Waterton into compliance with the CHWA and to pay $1, million in civil penalties.

Plaintiffs, and all the residents of a SouthWest Denver suburb known as Friendly Hills allege that they have suffered a variety of injuries which were caused by Martin's improper handling & hazardous waste practices as Waterton. Plaintiffs assert that Martin improperly and with calous disregard for others discharged hydrazines and other hazardous waste into Brush Creek and allowing it to enter into the ground water system, that these contaminants infiltrated Kassler water treatment systems, and that this contaminated water was then distributed to their homes through the City and County of Denver's water distribution system.

Plaintiffs' also did allege that this contaminated water caused twelve primary injuries as well as related injuries (Cancer). Specifically, four of the plaintiffs are children who developed cancer, one of whom developed leukemia. One plaintiff is an adult who suffers from kidney cancer, five of the plaintiffs are children who suffer from seizure disorders, and two of the plaintiffs are children with birth defects of the heart. All these are horrfying injuries that can be caused by hydrazines, and (TCE). Yet through the Court, and again large Corporations having what they have done totally dismissed. (IS WRONG) (The Victims at the beginning and Victims at the ending..........

The Community had brought toxic tort action against the missile manufacturer Martin Marietta alleging injuries, due to contaminated water. On defendants motion for summary judgment the District Court, Weinshienk, J, held that the community failed to present Prima Facie case of causation.

Judge Weinshienk, "states the Community that had alleged contaminated drinking water due to missile manufacturer's imporper handling of toxic substances failed to show that its members were exposed to contaminants at levels sufficient to cause alleged injuries, as was necessary to state toxic tort claims against a manufacturer".

The Communities direct evidence of exposure proved only that it was possible that members were exposed and the community had submitted virtually no circumstantial evidence that would create evidence linking of exposure. Under Colorado Law a case like this can be presented, or defended in a manner of circumstantial law or under Tort law, It then seems to be up to the Judge which law he will choose to make his decission. They usually lean heavy toward the Tort law, Larry says because, "it is the law the Government and large corporations will use, usually gives them a broad scheme, & power with very little morality in releasing the Government, and large Coporations from any responsibility, to the victim. Even though they may be morally, and with just cause entirely responsible for injury and destruction they have caused. When there are several chemicals involved, The Defendant will lock in on one chemical most likely to have not caused any harm or would be totally impossible to prove what harm it may have done even though that chemical was a marker, a FINGERPRINT as to where other chemicals may have come from that did the harm Making it vitually impossible for the victims to move foreward with there case to a Jury. The Judge then blocks it completely, and will not allow it to go any further, even though the victim may have a perfect circumstantial case, and would with out doubt win with a jury. The Judge blocks it, then will dismiss it under Tort Claim Law not allowing it to move forward to a Jury. The Judge's in most Federal cases will dismiss the case against any Government agency, or large corporations before it ever gets to a jury. The cases that are dismissed by the Judges are clearly done to bar the victims of having their right to tell their story to a Jury before the Court.

The Court calls this Preliminary Evidencial determination as to whether methodology employed by an expert is of the type normally relied upon by experts in the field before expert's testimony may be presented to a jury. Larry Says, "this is a ploy for the judge to hear expert testimony from the witnesses in the case, he then must figure out how, when, why, and what, method s will be used by the Court to discredit the Expert witnesses in the case." "In essence the Judge first locks in on maybe the chemical that was being used as a tracer of other substance involved, The Defendant's Attorny's will lock on that very same chemical and because it is not one the experts had testified too. The Judge then locks in on Toxic Tort, that forces you to prove that all the injury came from this one chemical. Unlike Circumstantial which would allow a person to name any of many chemicals and to prove part of the injuries may have been caused by each one. He then forces the victims to prove that chemical caused the injury or else. When they cannot, the Judge unleashes fury upon the Expert witnesses by destroying their credibility, and the credibility of the victims, He dismisses the case under the Tort Claims Act." "Now the victim is suddenly again a victim of The Judge the legal system & the Tort Claims Act. is a most difficult Hurdle or Obstacle, the Judge will put in the pathway to a jury to stop those who have been victims of wrongdoing by the Government. As the victim tries to clear that hurdle, It seems that the Government, or Corporation works in harmony with the Judge to throw out all Common Sense, or Circumstantial evidence in favor of the Tort Claims Act, therfore making that hurdle impossible to ever get your case to a Jury. Once the Judge strikes the hammer, your case is forever dead". You have appeals, "Larry says no matter how you may prepare your case for appeals, they won't even look at it. (They will take your money though). No matter how many error's the Judge may have made, does not sway the appeals Court, In most cases the only thing they may look at is the Judge's de-cission, saying that is a good sound decission and that is the end of your appeal.".

Then there is a matter of issues of which factors should have been considered and what impact each should have been given when deriving decay coefficient were issues to be resolved by a jury in toxic tort action against Martin the Missile manufacturer in which the community alleged there were injuries due to contaminated water. Issue of whether the water sample was taken from liquid or sludge was for the jury in toxic tort suit in which community alleged that its drinking water was contaminated by the missile manufacturer.

Even if the community that brought a toxic tort action had been able to prove exposure to contaminants through the drinking water by direct evidence, the community would have been required to submit epidemiological evidence in support of their causation contentions.

The matters brought before the Court are the defendants' motions for summary judgment. An evidentiary hearing was held on these motions, from July 9, to July 13, 1990 Jurisdiction was proper pursuant to 42 U.S.C. s 1983 and 28 U.S.C. s 1331. The Court has reviewed the motions the briefs supporting and opposing the motions, the testimony & exhibits presented at the evidentiary hearing, the opinions submitted by the expert witnesses appointed by the Court Pursuant to Fed. R Evid. 706, and the pertinent case law, and is now prepared to rule. The District Court, in the State of Colorado, Weinshienk, J., held that the community failed to present a prima facie case of causation. "ORDER OF DISMISSAL GRANTED"

PROCEDURAL HISTORY AS THE COURTS VIEWED IT:

Plaintiffs initiated this action in January, 1987. and In their Complaint, Plaintiffs have set forth numerous claims sounding in tort as well as based on 42 U.S.C. s 1983. This litigation has been long and arduous. Almost every discovery request was hotly contested. At a hearing held in February, 1990, the Court expressed frustration with the uncooperative tendencies of all parties. The Court also initiated a discussion concerning how to make significant progress in this protracted litigation. Both plaintiffs and defendants were of the opinion that something other than a full trial on all claims and all issues might be desirable, since a full jury trial would probably take between six and nine months to complete. Plaintiffs suggested that a "test" trial, in which a full jury trial would be held on all of the claims of three of four of the plaintiffs, would be most expedient. Defendants suggested that the most efficient way to proceed would be to hold an "issue" trial, in which only evidence concerning causation would be presented to a jury.

After several status conferences and considerable discussion, it was determined that the most efficient procedure would be to hold a series of summary judgment proceedings. The centerpiece of these proceedings would be an evidentiary summary judgment hearing at which plaintiffs would present their prima facie case of causation as if they were presenting the case to the jury at trial. If the Court determined that plaintiffs had met their prima facie burden, then the Court would schedule a test trial in which all of the claims of one or two representative plantiffs would be tried to a jury.

The Court determined that the first of these summary judgment proceedings should address plaintiffs's 1983 claim. On April 26, 1990, the Court denied said motions for summary judgment after oral argument.

[1] At the April 26 hearing, the Court also granted defendants' motion for court appointed experts. The Court was informed that one of the primary issues of the evidentiary hearing would be the admissibility of the plaintiffs' expert witnesses' opinions. A court must make a preliminary determination as to whether the methodology employed by an expert is of a type normally relied upon by experts in that field before the expert's opinion may be presented to a jury.

This determination necessarily includes an evaluation "prusuant to Rule 104(a) whether particular underlying data is of a kind that is reasonably relied upon by experts in the particular field in reaching conclusions. In order to assist with this evaluation, the Court appointed three expert witnesses; One a Doctor and geochemist with expertise in chemical hydrogeology; and a Doctor a physician with expertise in environmental health and epidemiology, and a Doctor specializing as a Toxicologist.

The Court at that point instructed the Plaintiffs that their burden at the July evidentiary hearing was to present a prima facie case that contaminants reached plaintiffs' taps in quantities sufficient to cause the injuries they have alleged. The Court was to utilize a summary judgment standard in order to determine whether plaintiffs satisfied their burden of proof. The issue was whether a reasonable juror could conclude that contaminants from Martin reached plaintiffs' taps in quantities sufficient to cause the injuries they have alleged. If so, then the motions for summary judgment must be denied. If not, then said motions must be granted.

In essence, this standard actually "mirrors the standard for a directed verdict under Federal Rules of Law. THe threshold inquiry in this case requires much more extensive factual development than has traditionally been permitted in a summary judgment hearing but comports with the Supreme Court's directions for addressing motions for a summary judgment.

A CLOSER LOOK AT THE EVIDENCE

We will be looking at the evidence in the Plaintiffs' Prima Facie Case. The evidence plaintiffs presented at the July hearing can be classified in four distinct categories: 1. Factual evidence of contamination at Waterton., 2. The expert testimony regarding the levels of contamination at Waterton and the proportion of these contaminants that were transported to Kasslser (chemical fate and transport testimony); 3. expert testimony regarding the amount of Kassler water that was actually received by the Friendly Hills residents, and the levels of contaminants that were in this water (water distribution testimony): 4. Expert testimony regarding the probable cause of plaintiffs' primary injuries (medical causation testimony).

FACTUAL EVIDENCE OF CONTAMINATION

Plaintiffs submitted thousands of pages of documents which confirmed the poor hazardous waste management paractices at Waterton. They presented the testimony of a Doctor, and an expert in industrial hygiene, who testified about numerous unacceptable practices. Plaintiffs also then presented the testimony of an investgator from the Colorado Department of Health (CDH) who had spearheaded the 1985 Martin investigations. His testimony included his comments that on every visit he made to Waterton in 1985, and 1986, he observed improper waste discharges from various locations.

CHEMICAL FATE, AND TRANSPORT OF CHEMICALS

Plaintiffs' chemical fate and transport evidence was the foundation of their presentation. Plaintiffs presented the testimony of another expert in chemistry and environmental sciences, and yet another who was an expert in environemental engineering, Hydrology and pollution transport. These Doctors, along with that of a civil engineer and ground water hydrologist, jointly authored the plaintiffs' chemical fate and transport report.

It was the primary role in this process for the expert in chemistry and in environmental sciences, to calculate an initial "load" factor, which would quantify the amount of hydrazine contaminants that entered Brush Creek from the wastewater treatment pond at Waterton known as Pond T8-A. The Doctor concluded that 100 parts per million (ppm), was an appropriate and the initial load factor. This factor represents a total concentration of hydrazines and NDMA. He derived this factor from one sample that had been taken from the pond in 1985. These contaminants were found in an analysis of this sample in a concentration of 160 ppm. He then testified that 100 ppm concentration was a conservative factor to employ in the chemical fate and transport analysis, given the 160 ppm sample. All the conclusions reached in plaintiffs' chemical fate, transport and water distribution analyses were based on this 100 ppm load factor.

Then the good Doctor also addressed the improper waste treatment processes relied on at Waterton. It was the Doctor's opinion that there were additional sources of hazardous waste at Waterton other than from pond T8-A but that these sources were unquantifiable. Plaintiffs continually criticized Martin's almost complete lack of documentation concerning hazardous waste at Waterton and its treatment or lack of treatment. Again the Doctor stongly concurred with this criticism.

The good Doctor, an expert in environmental engineering, hydrology and pollution transport testified, that these contaminants reached Kassler in concentrations ranging from 16 to 88 parts per billion (ppb). He then explained the chemical transport model he and the Doctor specializing in Civil Engineering and ground water transport had constructed and the methods he employed in deriving the decay coefficient. Then again the Doctor developed a model in which he attempted to depict the pathway the contaminants traveled from Pond T8-A to Kassler. With this model, the Doctor attempted to calculate the rate at which the contaminants would disintegrate. This involved determining what the applicable "decay coefficient" should be. Then He relied on the published work on decay and coefficient to further develop his model. Then both Doctor's did at that point concur and stated that there were other contaminant pathways from Waterton to Kassler that the model did not attempt to quantify due to the lack of available data.

WATER DISTRIBUTION

Plaintiffs submitted the expert report of two Doctor's who specialize in water distribution as evidence that contaminants traveled from Kassler to the plaintiffs' homes in Friendly Hills. It was uncontested that the only periods of time in which contaminants might have reached the plaintiffs' homes were from June, 1977 through December 1980, & from June, 1982 through February 16, 1983, it was also uncontested that during these periods, Friendly Hills residents never received more that 10 percent of their water from Kassler.

Then they testified that, during the exposure periods & assuming that the contaminants concentrations at Kassler was either 88 ppb or 16ppb, contaminants reached Friendly Hills residents in concentrations above 0.0ppb on 78 percent of the days. These Doctor's reports had relied totally on the Chemical Fate and Transport reports that had drawn a conclusion to say that contaminants had reached Kassler in concentrations ranging from 16ppb to 88 ppb.

MEDICAL CAUSATION

A Doctor of Veterinarian Medicine, with expertise in chemical carcinogenesis and toxicology, testified that he has been engaged in extensive hydrazine research. It was his opinion, based upon his own research as well as a review of the literature, that hyerazines are highly carcinogenic, as they cause cancer in many different animal species as well as in multiple organs within any given species. The Doctor declined to render any opinion as to whether hydrazines are a potential cause of seizure disorders or birth defects. He also declined to render any opinion regarding NDMA or its effects.

Another expert a Doctor in genetic toxicology with emphasis on cancer, teratology and transmissible genetic damage, testified that animal studies have confirmed that hydrazines are among the most potent of the known carcinogens. He then opined that the plaintiffs' injuries were consistent with known effects of hydrazines and NDMA to a reasonable degree of scientific probability.

Then a Physican with expertise in clinical medicine who specializes in workplace & environmental medicine, then testified that plaintiffs' injuries were consistent with hydrazines exposure. She expressed her opinion that there was no other probable cause of plaintiffs' injuries; therefore, the contaminants from Waterton probably caused the alleged injuries in her analysis, The Doctor presumed exposure to the contaminants. She made this presumption based on the expert reports from all the other Doctor's.

RESPONSE FROM THE DEFENDANT

The defendants then attacked the plaintiffs' allegations and their expert witnesses opinions at every juncture. As an initial matter, defendants also contended that there was no evidence of contamination at Waterton or at Kassler. Defendants asserted that Martin had promulgated an extensive standard operating procedures for hazardous waste treatment at Waterton. They argued that there was no evidence supporting plaintiffs' contention that these procedures were not followed.

Defendant then called upon five different expert witnesses to attack the methodology employed by the Doctors who gave expert Opinions and testimony regarding, Chemistry, along with Environmental Sciences, then Environmental Engineering, Hydrology, and Pollution Transport, and the Civil Engineer and ground water Hydrologist. These expert witnesses did harshly criticize plaintiffs' chemical fate and transport report as follows:…1. It was not scientifically acceptable to base a chemical fate and transport contamination concentration conclusion on only one data point……. 2. The sample from which the date point was derived was a sludge sample rather than a liquid sample as plaintiffs' asserted. There fore, the contamination concentrations in that sample were not representative…….3. The decay coefficient that had been selected was inadequate because it was derived from laboratory experiments that were not applicable to the Brush Creek environment. Furthermore, he failed to differentiate between the four separate chemicals involved in the analysis, and he failed to account for non-laboratory decay mechanisms…….4. The Brush Creek flow rates used in the chemical fate and transport model were not a true representation of the observed flow rate of Brush Creek……5. The contaminants, if impermissibly discharged were discharged on an intermittent basis, yet the report concluded that contaminant concentrations of 16 to 88 ppb were constantly in the water arriving at Kassler for the 11 year period of model attempts to depict. In addition defendants' expert witnesses criticized several of the techniques relied upon by Doctors' in their water distribution analysis, and their evidence in how the contaminants traveled from Kassler to the homes of the plaintiffs'

Defendants also presented a strong rebuttal to plaintiffs' medical causation evidence. This was done by a Doctor of pediatric oncology. Who testified regarding the four cases of cancer in children. He stated that the four cases were not causally related and that there was no identifiable cause of the cancers. He expressed the opinion that in the absence of such indicia of causation, the only means of determining causation is by epidemiological studies.

Two Doctors for the defendants in this case, both physicians with expertise in toxicology, both testified that methods employed by plaintiffs' medical causation experts were not scientifically acceptable and that there was no evidence on which to base a causation conclusion in this case. Then came a Doctor who's expertise in oncology, biostatistics and epidemiology. He testified that he had conducted an epidemiological study on the difference of the cancers in the children in Friendly Hills. It was his expert conclusion that the difference between the expected and observed incidence rates was not statistcally significant. In other words, it was this Doctor's opinion that the four cases of childhood cancer was within the expected range for the Friendly Hills Community.

Plaintiffs then presented the testimony of a Doctor, and physician in Public Health Physician with expertise in environmental medicine, Public Health, and Epidemiology, and that of an epidemiologist, they had then discussed the testimony findings and conclusions by the defendant's. Plaintiffs' then presented testimony that it would have been difficult even at best, if not impossible, to perform a meaningful community based epidemiological study in a small community like Friendly Hills. He also opined that since it was clear that Friendly Hills residents were exposed to the hydrazines contaminants, The Defendant's testimony in Oncology, and biostatistics and in epidemiology, and their findings that the childhood cancer incidence rate in Friendly Hills was approximately twice what one would expect to be significant. The Plaintiffs' then testified that the rate being twice what was expected, "Is Very Significant" even if a statistician would not conclude it is significant.

DISCUSSION ON THE APPLICABLE LEGAL STANDARDS

In order to resolve the pending motions for summary judgment, the Court must determine whether a reasonable juror could conclude that contaminants from Martin reached plaintiffs' taps in quantities sufficient to cause the alleged injuries. Because this being a one time remedy, for the defendants, they must establish beyond a reasonable doubt that they are entitled to summary judgment. All of the claims plaintiffs assert sound in tort. In order to prevail on a tort claim, a plaintiff must prove by a preponderance of the evidence that a defendant committed an act which did cause an injury to the plaintiffs'. "Under Colorado law, and event is the proximate cause of another's [injury] if in the natural and probable sequence of things, it produced the claimed injury. It is an event without which the injury could not have occurred. In order to prove causation in the toxic tort setting, plaintiffs must prove that defendants caused plaintiffs' exposure to toxic contamination and that the exposure caused, or contributed to, plaintiffs injuries. Therefore in order to establish a causal connection between the contamination at Waterton and plaintiffs' injuries, it was necessary to show facts and circumstances which indicated with a reasonable probability that the injuries resulted from, or were caused by the Waterton contamination.

Certain issues were clearly established at the hearing. Despite the defendants' assertions to the contrary, the evidence was overwhelming that there was massive contamination at Waterton. The evidence of the observed discharges and the discovery of TCE in the Kassler water supply constituted ample evidence that Martin's hazardoud waste standard operating procedures were not followed, Its hazardous waste treatment system was mechanically flawed, or some combination of both. The existence of plaintiffs' injuries, other than the alleged seizure type disorders, was uncontested. However, proof that Martin committed reprehensible acts coupled with evidence of injury is not enough to prevail on a tort claim. Plaintiffs must prove that the reprehensible acts caused or increased the likelihood of, the alleged injuries.

There were two ways, that plaintiffs could have esablished their prima facie case of causation. First, plaintiffs could have shown that contaminants traveled from Waterton to Friendly Hills and that plaintiffs were exposed to these contaminants at levels sufficient to cause the alleged injuries. A second method of proof would have been for plaintiffs to have shown that the injuries in Friendly Hills were such that is was probable that there was exposure to the contaminants at levels sufficient to have caused the alleged injuries. Plaintiffs relied primarily on the former.

The focus on the remaining portion of this opinion will be on the sufficiency of the evidence plaintiffs presented and the methodologies employed by plaintiffs' expert witnesses. If these expert witnesses based their conclusions on underlying data not normally relied upon by other experts in their particular disciplines, then those conclusions will not be considered in evaluating the sufficiency of plaintiffs' evidence because evidence of such conclusions may not be presented to a jury.

DISCUSSION: SINGLE DATA POINT CONTROVERSY

Plaintiffs submitted the expert reports in chemistry, while included environmental sciences, environmental engineering, hydrology, and pollution transport. Other's in civil engineering, ground water hydrology called specifically the report on chemical fate and transport. The Judge in reading these reports together it seems, noted they postulate that hydrazine waste water containing a constant concentration of hydrazine and NDMA of 100 ppm was discharged into Brush Creek on a regular basis over an 11 year period. Further, the reports concluded that these toxins traveled from pond T8-A at Waterton to Kassler and arrived at Kassler in concentrations ranging from 16 to 88 ppb, and that the contaminants were frequently delivered in measurable quantites to Friendly Hills residents.

Expert witnesses testifying on behalf of defendants attacked many of the methodological techniques employed by the authors of each of the expert reports. Most of these alleged methodological, was seen as flaws, however, they must be considered by a jury when evaluating the weight to be given to the opinions, rather than by the Court in making any preliminary assessment concerning the admissibility of a given opinion. For example defendants strongly criticized the factors used by the Plaintiffs' had considered, and then failed to consider when deriving the decay coefficient.

Defendants' expert witnesses opined that among other factors, no reasonable expert would fail to account for chemical volatilization when deriving a decay coefficient. Asserting that he did account for chemical volatilization, while the Plaintiffs' expressed their opinion that this effect would be negligible. Therefore, he did not factor this effect into the derivation of the decay coefficient. In deriving the decay coefficient, again the Plaintiff's testimony expressed that it was their opinion that this effect would be negligible. Therefore, he did not factor this effect into the derivation of the decay coefficient. They then determined what factors did in fact contribute to the contaminants' decay coefficient, and then they attempted to calculate the effect the identified factors would have on the contaminants' rate of decay. The issues of which factors should have been considered and what impact each should have been given in deriving a decay coefficient are issues to be resolved by a jury.

One alleged flaw does not raise the factual question of what weight an opinion should be given but instead raises the legal issue of whether any weight whatsoever may be accorded the opinion. That flaw is plaintiffs' reliance upon a single data point in formulating their scientific conclusion. All of the Plaintiffs' evidence of exposure rests upon this data point because it is the starting point for plaintiffs' expert witnesses' analyses. If the data point is removed, the expert witnesses have no basis on which to reach any conclusion other that that it was possible that the plaintiffs were exposed to these contaminants.

As an initial matter, defendants contested plaintiffs' assumption that the sample was a liquid sample taken from Pond T8-A. While defendants that the sample was sludge sample taken from the floor of Pond T8-A. If the sample was of sludge, then the concentration of measured contaminants would be much greater than the concentration of any contaminated wastewater that was discharged. The factual determination of whether the sample was taken from liquid or sludge is one for a jury.

The dispute, however, is illustrative of the danger of relying upon only one sample to formulate conclusions about what the "normal" contaminants concentration of Pond T8-A was during the 11-year exposure period. Larry says, "The wastewater being discharged at the time may have been much greater than any concentration in the sludge, or the samples that were taken at either Kassler or Waterton because of the enormous amount of dillutioin". The dispute illuminates one potential source of sampling error. Simply put, no one has any idea to whether this sample is representative of the "normal" contaminants concentration. The Court appointed expert witness in geochemistry and hydrogeology, wrote in her report to the Court that

"[I]t is unsound scientific practice to select one concentration measured at a single location and point in time and apply it to describe continuous released of contaminants over an 11-year period of time.

The Plaintiffs then argued that defendants may not fail to keep proper records and then assert that plaintiffs' prima facie case must fail due to a lack of data. Plaintiffs relied on other cases in support of their assertion that quantification of the exposure level is not required under these circumstances. The Courts in the other cases did not hold that the absence of data permits the inference that there was an exposure. In those cases, the courts found ample circumstantial evidence of exposure and then refused to require precise quantification of the level of exposure. In much contrast, the direct evidence in this case is insufficient to support the contention that it is probable that there was an exposure. Plaintiffs' chemical fate and transport and water distribution evidence, when viewed in its best light, proves at most that paintiffs were exposed possibly to contaminants from Waterton. Mere possibilities or conjecture cannot establish a probability. Therefore, this Court condcluded that no reasonable juror could find, based on these expert opinions, that it was probable that plaintiffs were exposed to the Waterton contaminants.

MEDICAL CAUSATION AND THE EPIDEMIOLOGY CONTROVERSY

Plaintiffs contend that if defendants prevail on their argument that the plaintiffs cannot prove there case because of insufficient data, then plaintiffs have no means of meeting their required burdens and defendants become the beneficiaries of their own impermissible failure to maintain the proper records. This contention is unfounded. Even in the absence of direct evidence, plaintiffs in toxic tort cases can present circumstantial evidence in order to show that it is probable that they were esposed to contaminants. The two primary forms of cirmcumstantial evidence in a toxic tort case are etiological evidence and epidemiological evidence. For example, plaintiffs could have attempted to show that their injuries exhibited lesions or other indications that hydrazines of NDMA were a potential cause of the injuries. Plaintiffs also could have attempted to show that there was an abnormally high incidence rate of the injuries they have alleged in

Friendly Hills compared to other similarly situated communities. Although not required, such evidence might support an inference that it was probable that they were exposed to Waterton Contaminants.

Plaintiffs made no attempt to present such circumstantial evidence on the exposure issue, however. Instead, plaintiffs relied upon their chemical fate and transport and water distribution reports as direct evidence of the exposure. Plaintiffs' medical causation expert witnesses all presumed the exposure and testified that plaintiffs' injuries were consistent with such an exposure. Plaintiffs discussed animal study experiments which revealed that hydrazines are highly carcinogenic in many species and in many of the organs within a species. Others testfied that plaintiffs' broad array of injuries could all have been caused by Waterton contaminants, in essence because hydrazines have the alleged ability to cause almost any cancer or birth defect in a human being. Plaintiffs presented no evidence that the etiology of any of their injuries revealed that hydrazines caused those or any injuries. Instead, plaintiffs submitted evidence in the form of expert witness opinion that hydrazine causation of their injuries is not inconsistent with the known etiology of their injuries. This evidence, however does not assist plaintiffs in their effort to prove that they were exposed to the Waterton Contaminants. It is useful only after plaintiffs have met their burden on the exposure issue by some other means.

There was another potential means of proving an exposure it is by epidemiological studies. The parties were strongly divided in their views in regards to the necessity for corroborative epidemiological evidence in a toxic tort case. This argument focused not on the utility of epidemiological evidence as circumstantial evidence of exposure but on whether epidemiological evidence was required to show that the Waterton Contaminants have the ability to cause plaitiffs' injuries.

Plaintiffs asserted that epidemiological studies are not required as proof in order to establish a contamination-injury, cause and effect relationship in a toxic tort case. The cases here that plaintiffs cite involve various individuals' and claims that they were exposed to a toxin that caused their alleged injuries. These alleged exposures were only of an individual and not of an identifiable group or community. In such an instance, it is impossible to submit corroborative epidemiological evidence because it is impossible to identify a group or community which has been exposed to the alleged carcinogen, and an epidemiological study therefore cannot be conducted. The cases defendants cite involve a known exposure of an entire community once again to an alleged carcinogen. In those cases, the courts found plaintiffs' causation evidence insufficient because they were unable to submit epidemiological proof of the alleged cause and effect relationship. All of the cases cited, when read together, support an approach to toxic tort cases that requires submission of epidemiological evidence to establish causation in cases where collection of such evidenc is possible. Collection of such evidence is possible in situations where an identifiable exposure population is large enough to perform a meaningful epidemiological studies.

In this case, even if plaintiffs had been able to prove exposure by their direct evidence, they would have been required to submit epidemiological evidence in support of their causation contentions. Such a submission would have been necessary because this is a case where it is alleged that an entire community has been exposed to the contaminants. Under these circumstances, it is possible to perform an epidemiological study. Therefore, such evidence should have been submitted; Otherwise and without epidemiological studies the plaintiffs fail to meet their causation burdens.

The Court, however, will not base the ultimate resolution of any of the motions for summary judgment on this ground. The Doctor's epidemiological study, asserted that the Friendly Hills community was not large enough to perform a meaningful epidemiological study. They assserted, in effect, that this case is much more akin to the cases in which only individuals have been exposed that to the cases in which an identifiable population has been exposed. In such cases, a court should not require any submission of epidemiological evidence.

The problem with this argument, even if the Court accepts it, is that it will eliminate the plaintiffs' ability to submit circumstantial evidence that the injury incidence rates in Friendly Hills are abnormally high. This, in effect, eliminates the only other means available that plaintiffs have to prove exposure. The expert medical witnesses for the Plaintiffs' had already attempted overcoming this by opining that the other Doctor's finding that the number of childhood cancers in Friendly Hills is approximately twice what one would expect is significant. This opinion does not assist plaitiffs in their effort to prove that the other eight primary injuries they have experienced are causally related to Waterton contaminants.

Plaintiffs allege that the Waterton contaminants caused four childhood cancers. According to one of the expert witnesses, while they expected at the time the number of childhood cancers in Friendly Hills was approximately two, twice that level to four cases at the high end of the what one could even see as the expected range but still not outside that expected range in that community. Some communities with the demographics of Friendly Hills will have no cases of childhood cancers, some will have one, and some will have two, others may have three still others four and so on. One of the Doctor's testified that in order to be considered statistically significant in this situation, Friendly Hills would have to experience and eight fold increase in its childhood cancer incidence rate. He testified that he believed this standard to be too strict. However, the Fifth Circuit, when addressing a similar argument, stated that "we find….the Brocks' failure to present statistically significant epidemiological studies that prove that Benedectin causes limb reduction defects to be fatal to their case." The fifth Circuit then went on to encourage disctrict Judges faced with similar issues to be "especially vigilant in scrutinizing the basis, the reasoning, and statistical significance of the studies presented by both sides.

In any case, plaintiffs primarily argued that the Defendants expert was incomplete and inconclusive on this issue. Plaintiffs did not attempt to perform an epidemiological study of their own because plaintiffs viewed such a study as futile. Freindly Hills, in the opinion of other of Plaintiffs' expert witnesses, that Friendly Hills was too small in population to yeild any meaningful results in an epidemiological study under these circumstances. Plaintiffs did not attempt to rely on epidemiological evidence as circumstantial evidence of exposure. Therefore the Defendants' in this case then corroborated in their expert reports, that there was no epidemiological evidence which can be considered as a basis for satisfying plaintiffs' prima facie burden in this case.

CONCLUSION

Platintiffs' burden at the evidentiary summary judgment hearing was to present a prima facie case of causation. In order to accomplish this, they had an obligation to show that they were exposed to Waterton contaminants at levels sufficient to cause the injuries they have alleged. Plaintiffs' direct evidence of exposure proved only that it was possible that they were exposed. They submitted virtually no circumstantial evidence of any exposure. There evidence was insufficient and did not meet their prima facie burden. No reasonable juror could conclude that it was probable the plaintiffs were exposed to Waterton contaminants.

DECISSION

Therefore, the Court will GRANT defendants' motions for summary judgment and will dismiss this action. Accordingly, it is further ORDERED that defendants' motions for summary judgment are granted.

It is FURTHER ORDERED that the complaint and cause of action are dismissed with prejudice, the parties to pay their own costs and attorneys fees.

It is FURTHER ORDERED that all other pending motions are moot.

POLLUTION COVERUP BY MARTIN MARIETTA CORPORATION, IS NOW ALLEGED BY TWO SEPARATE AGENCIES.

At the same time the Courts are dismissing the actions against Martin Marietta Corp. for the pollution injuries that were felt caused by them at Freiendly Hills subdivision near Denver Colorado after the contaminated water had been treated at the Waterton water treatment center. Larry's thoughts, "The residents were forced to prove much more than humanly possible in a Court that seemed so Biased toward ordinary people trying to show a crime endangering the lives of their children was taking place. All because a large corporation was allowing rocket fuel to escape into the water where children would soon be drinking and bathing in unknown amounts of one of the deadliest chemicals known to mankind Hydrazine, and as I now understand by our sources that also Aerozine was being used, while carelessly and reklessly allowing it to escape into the water that these unsuspecting children would be drinking and bathing in a toxic material that absorbs through the skin. Of course again a large corporate business who was doing business with and for the United

States Government, as well. As I followed the case I'm sure there was only thoughts that were given to how much did they drink, and never a thought given as to how much had they absorbed through the skin while taking a warm bath".

Federal agents and the Colorado attorney general's office are investigating a possible coverup of environmental crimes by the Martin Marietta Corp., the Denver Water Board and the Colorado Deparment of Health. Colorado's Attorney General Duane Woodard confirmed that his office is conducting such an investigation. The separate federal probe was revealed this week by sources familiar with the inquiry. It is being spearheaded by Denver-based agents of the FBI and the U.S. Environmental Protection Agency's National Enforcement Investigation Center, said the sources, who spoke on condition of anonymity.

Martin Marietta employes 12,000 workers at its plant near Waterton in Jefferson County. Their company uses hazardous chemicals such as solvents and rocket fuel "Hydrazine" in the manufacture and testing of military rockets and other aerospace equipment.

Woodard said his staff is investigating allegations that Martin Marietta, the water board & the health department covered up the illegal dumping of contaminants into streams near the Kassler water treatment plant. Those streams led into the Denver drinking water supplies until 1986. The water board formerly operated the Kassler water treatment plant near the Martin Marietta facility. The plant is now shut down.

Allegations that the health department ignored violations of health and environmental laws at Martin Marietta also are being probed, Woodard said. "I assure you if there's anything there, we'll find it," Woodard said. He declined to give details of his investigations.

In a seperate federal probe, FBI and EPA agents were investigating many possible violations of the Clean Water Act and of federal laws prohibiting false statements, the sources said. Alleged violations of those laws also are being investigated in the continuing federal probe of the Rocky Flats Nuclear Weapons Plant.

However, unlike the Rocky Flats probe, Federal agents involved in the Martin Marietta Investigation have not themselves accused any person or organization of committing a crime. Also, the Martin Marietta inquiry so far is low key, in sharp contrast to the widely publicized Rocky Flats investigation. The targets of the federal probe denied knowing of its existence. "We have no knowledge of a federal crimninal investigation and have not been contacted by the (State) attorney general's office about an investigation," said judy Stowell, spokeswoman for Martin Marietta.

"We have no indication here of that "federal investigation" going on said Denver Water board spokesman Ed Ruetz. State Health Department Director Dr. Tom Vernon said he knew of the state probe of his agency, but not of any federal investigation. He denied the health department was involved in any wrongdoing. The FBI and EPA would not comment on whether they are conducting such an investigation. However, federal agents were observed this week by the Denver Post interviewing a person with knowledge of Martin Marietta's pollution record.

Woodard said he was unaware of the federal probe. "But we didn't know about the Rocky Flats investigation until everybody else did," he added, referring to the fact that federal agents kept their probe of the weapons plant secret for two years before revealing it last month. Woodard said his office received an executive order from Governor Roy Romer on July 14, giving him full authority to conduct a criminal investigation into a possible cover up of environmental crimes by the Martin Marietta Corp.

The executive order was needed because the investigation crosss jurisdictions of local District Attorneys, Woodard reported. He said the order protects his staff from having any evidence suppressed by a Judge at any future trial. The executive order also will open doors at Martin Marietta previously closed to his investigators, Woodard reported. Woodard has assigned two people to the case, including Bob Sexton, considered one of the state's top Criminal investigators.

An investigation by Woodard's very similar previously was halted last year after two attorneys there determined the State's statute of limitations has expired on certain alleged crimes under investigation. But investigators in the earlier probe didn't target the Health Department and didn't have access to certain records that will now be available because of Governor Romer's order, Woodard said.

The Attorney general had earlier predicted the new investigation could take up to six months maybe more. Woodard's investigators already have interviewed several Denver Water Board employees, including Jack Dice, who is the director of water quality systems.

Environmental activists....inclucluding Adrienne Anderson, western director for the National toxics campaign...have accused Dice of supervising the destruction of records related to alleged Martin Marietta pollution of public drinking water. Dice has said records at Kassler were routinely destroyed as a housekeeping measure, but denied the destruction was to cover up any crimes. Martin Marietta and the water board were already targets of a civil suit alleging that pollution from the company's plant did cause illnesses and deaths in a nearby neighborhood. It is unclear now whether those specific allegations are being investigated in the state and federal crimiinal probes.

In 1987, 13 families living in the Friendly Hills subdivision brought a $130 million lawsuit agains Martin Marietta, the Denver Water Board and the U. S. government. They charged that pollution had seeped from the Martin Marietta facility and from adjacent Air Force land and did in fact enter their drinking water at the Kassler treatment plant. The Friendly Hills lawsuit is in a pretrial phase called discovery, during which the two sides exhange information and quesiton possible witnesses under oath. Trial date is yet to be set. Larry says "It can be seen very clearly why the investigation by both the State and Federal Agency's is taking place, There are lies, and there are damn lies, and when the government can't get by with that they make statistics. That's exactly why governmental agencies have now been ordered to make this a statistic, and as you can see in the case where young children were dying of cancer birth defects, seizures, in Friendly Hills, a rate twice the national average. November 9, 1990 Thier case was made "Just Another government statistic".

Thanks to The Denver Post, Environmental writer Thomas Graf, picture by Bruce Gaut.

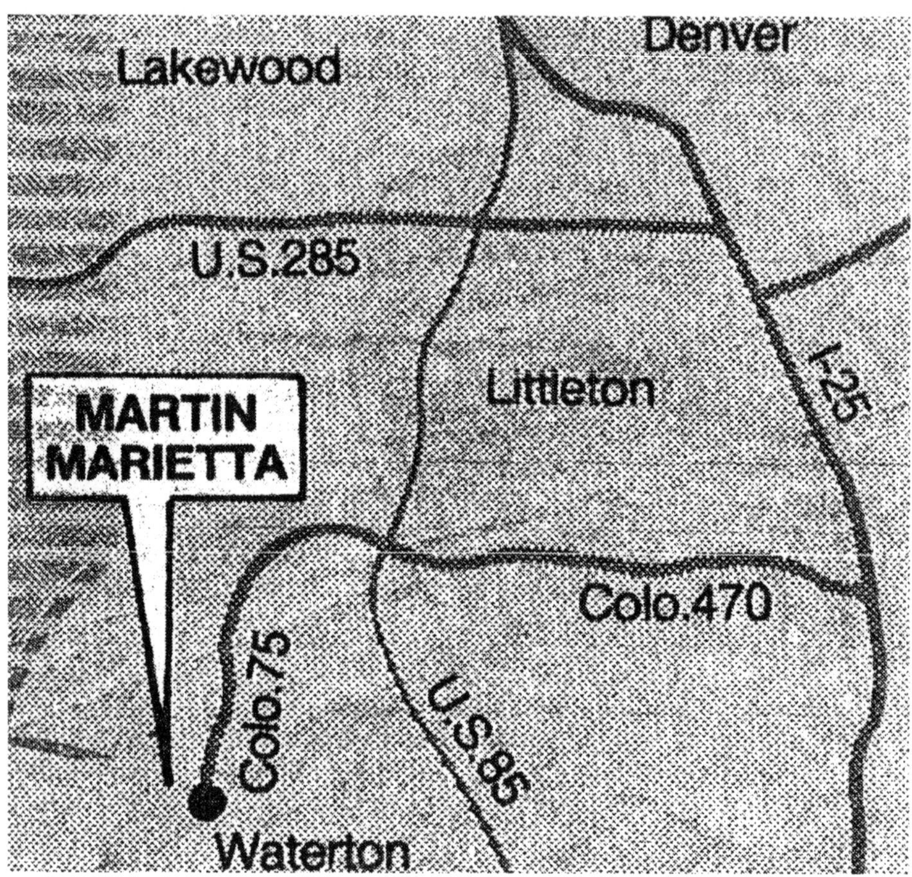

CHAPTER 17

Learning from still another, "DEATH FACTORY IN OUR MIDST" ROCKY FLATS, A well known Death Factory near Denver Colorado located in one of its surrounding Communities. It is on the United States list of the most polluted sites on earth.

THE UNITED STATES KNEW FIRST HAND THAT BETTER WAS POSSIBLE AT ROCKY FLATS, BUT THEY CHOSE "GOOD" AND THAT WAS NOT ENOUGH. FOR ALL THE PAIN AND SUFFERING CAUSED BY THEIR MISTAKES AT THE ROCKY FLATS NUCLEAR WEAPONS MANUFACTURING PLANT.

Larry D. Land

See Page 355 AS pictured, a Plutonium glove box exhaust system picture by Alexander A. Papesh Rocky Mountain News service. Their source was Rocky Flats management.

Pollution debris in the ducts accounted for up to 20% of the plutonium. The dirt and debris had nearly clogged all air ducts at the Nuclear Weapons Plant. Other materials found in the dust and debris were, Florine, Magnesium, Calcium, Silicon, Iron, and aluminum, and water waste, in some if not most exhaust lines depending on the process That fed the air lines. This information was according to Joe Mellon of the EG&G INC. which does operate the Jefferson County Facility for the Department of Energy.

Researchers identified radioactive materials and chemicals released between 1952 and 1989 from the former Rocky Flats Nuclear Weapons Plant as part of the nine-year Historical public Exposures studies of Rocky Flats. Purpose of the studies was, to determine the potential cancer risks to residents living in surrounding communities now, and then resulting from contaminant pollution released off the site. This Plant which is owned by the Department of Energy (DOE), It would appear is now trying to blend in with Society by calling it the Rocky Flats Environmental Technology Site. Very much like it's Sister just to the north of Denver. The Rocky Mountain Arsenal which went from a manufacturing disaster that can never be cleaned up, too The Rocky Mountain Arsenal National wildlife refuge. These animals are not natural residents of the arsenal, they are brought In and kept behind 8 foot chain link fences more likened to a Concentration Camp, they have regular daily pick up and clean up of all dead animals then just introduce new animals on an as needed basis to make it look as if it is a pristine wildlife habitat, when in fact the Rocky Mountain Arsenal is known as the most contaminated place on earth, and larger than the CherNoble nuclear incident in Russia. They fumigated all the prairie dogs with cyanide, because they were so contaminated they were killing all the Eagles that were being brought in to what is now called the Bald Eagle Management Area. Pesticide, Herbicide, Nerve agents contamination found in the Arsenal Wildlife. Many of the animals were contaminated with pesticides and other materials once manufactured and produced at any one of many facilities of death at the Rocky Mountain Arsenal, including Shell Oil Co., while they are claiming others are being electrocuted by non insulated power lines, and I find that absurd they are being killed by the pollution it's just another one of their damn lies, from Army and Shell Officials and until it can simply be called just another statistic.

The Army struck a deal with the Air Force at Buckley Air National Guard, to trap and move hundreds even thousand's of the healthy prairie dogs from Buckley Air Base to the Arsenal since they were healthy and so numerous. What they had done was actually an on going program, and a way for the Government to admit what was killing the Bald Eagle and bringing them near total extinction was death and deletion from pesticides. (so they say), "Records Show" Jeff Edson of The Colorado Department of Health states, "It's obvious that past contamination at the arsenal is having an adverse effect on some wildlife species," (if not all), Larry says "All animals and birds that come in contact with toxic chemicals at the Rocky Mountain Arsenal in any form have adverse effects, Immediate, long-term, and to all future generations and into extinction.

The former Pesticide, Herbicide, and deadly chemical weapons plants, was turned into a so called wildlife refuge, by an act of Congress Larry says, "that is just a blinder or cover up to the public, they won't tell you this land can never be clean again, will never be suitable for building, never a suitable for human habitat, and millions of tons of deadly spoiled earth being entombed for history. It was once said if they took a cubic yard of contaminated soil placing it end to end it would reach from there to China. Then word is out it will take until at least the year 2060 to have it maybe clean enough it can be turned back over to the State of Colorado. Over twenty years ago the federal government for health reasons had banned and outlawed certain pesticides, that are still showing up in most of the animals tested by the U.S. Army today even after the former pesticide and chemical weapons plant was turned into a wildlife refuge.

The Colorado Division of Wildlife spokesman Dave Weber stated, "all the wildlife is not keeling over every day out there," The real question is more at, How healthy is the wildlife? Are they able to live normal life spans? At least the Bald Eagles were nearly and totally on the edge of extinction because of the chemicals at the Arsenal, (RMA). Songbirds, ducks, hawks and owls have been found dead at the arsenal all showing elevated levels of contamination.

An Army study found that 426 to 686 terrestrial birds, mammals, plants and insects tested at the arsenal from 1988 to 1990 were contaminated with a toxic pesticide called Dieldrin. In the same three year period, the chemical also was found in 145 of 269 aquatic plants, birds and fish tested at the arsenal. Dieldrin was banned as a suspect carcinogen by the Environmental Protection Agency in 1974. The only place in America that manufactured dieldrin was Shell Oil Co.'s plant in Colorado.

Looking back at the activities at Rocky Flats, which included the processing, purifying, machining, and preparing plutonium for the manufacture of pits, the key component of what was known as "Triggers" for nuclear weapons. The plant also manufactured other weapons parts using Uranium, Beryllium, Stainless Steel and other materials. The parts were then shipped to Texas where the preparation and assembly of the nuclear weapons was done. In addition plutonium was recycled and reprocessed, from the retired warheads. Rocky Flats Plutonium operations were suspended in late 1989 and did not resume. The DOE had officially ended nuclear weapons production at the plant in 1992.

Rocky Flats is located just 16 miles North-Northwest of downtown Denver Colorado. It sits in Northern Jefferson County. The site is located on a 350-acre partial of land, zoned Industrial area, and is surrounded by 6,100 acres of open space, called the "buffer Zone". This so called buffer zone was not any part of the Rocky Flats Manufacturing plant. My question to them is who did they advise that their Land was considered a buffer zone to a Military Installation that manufactured some of the Worlds most hazardous, and toxic materials in the World.

In an effort in June of 1990 to clean the ducts soon met with a decission to removing some before processing operations could resume at the plant, Officials said yesterday. At least ten air ducts in three different buildings will be cleaned out with the use of drills and vacuum cleaners or to be removed, will probably last till three to four months. DOE had shut down operations late in 1989 because of recurring safety problems and concerns about plutonium building up in the 40-year old ductwork. An Independent scientific inspection last summer found one pipe containing enough plutonium to potentially create a constant and uncontrolled release of radiation and "Saw indications of other places where there may be undesirable buildups," said Roger Mattson, the head of the team.

A computerized camera used measured a half-inch in diameter was being made to snake through the miles of piping in the plant's processing buildings, identifying problem areas.

One video of some of the camera's work—-was shown yesterday at a public hearing—revealed ducts nearly blocked by salt and plutonium deposits that look much like stalagmites growing on the bottom of the pipe. A tiny drill is snaked into the ducts to chip away at the buildup and open the pipe. That residue contained 1% to 2% Plutonium Mellon said. Another pipe held only piles of Plutonium-contaminated dust that were easily vacuumed away. Studies that followed showed that these piles contained 14% to 20% pure plutonium.

At the time the debris that had been removed was far less than a pound——-and will be stored in a protective "Glovebox" that will be saved for reprocessing to reclaim the valuable plutonium, Mellon said.

The Department of Energy Nuclear Weapons Plants are more serious than the DOE realized, and will take many more years to solve than the DOE believes according to a Congressional report. Hazardous waste and contamination problems the Department of Energy Nuclear Weapons Plants very simply were not prepared. "THE TRUTH IS THAT LONG TERM SAFETY WAS NEVER FIGURED INTO THE EQUATION, NOR WAS THE BILLIONS OF TAXPAYER DOLLARS IT WILL NOW COST FOR CLEAN UP.

"Science and the consequences of putting the cart before the horse" The public cannot easily accept DOE assertions that contamination at the plants poses no serious public health threats because the agency 'hasn't by no means' done a thorough investigation that would give them the knowledge to reach such a conclusion. "Once again they are putting the cart before the horse."

The DOE release of their report on the future of the Nuclear Weapons Complex, is concise, after the release of OTA, Office of Technology Assessment, findings closely mirror the DOE report, which very simply states Rocky Flats should be shut down immediately and it's operations moved elsewhere. The reports and findings of "OTA" Office of Technology Assessment were at the request of the Senate Armed Services Committee, Congress, who researched hazardous waste and contamination problems at the Rocky Flats Plant near Denver and the 14 other branches of the nations Nuclear Weapons Complex's of operations. OTA reported just prior to their report, there were deficiencies in nearly every aspect of the Energy Departments efforts to address the problems. I believe this says it all for the rest of the Department of Energy's 14 other branch's or complexes across the United States, maybe next to your home, your children's schools.

While they were praising EG&G's work in cleaning the ducts and in the company's overall safety improvements at the plant often plagued by human error, lax management and a poorly trained work force. "If they had a poorly trained work force, then they had poorly trained management first", and that finger is pointed all the way to the Department of energy.

DOE, states Safety is soft Science with a lot of attitude involved, and that is probably true, but when management ignores Equipment error's, It's malfunction's and over burden the equipment then blame it on human error, or maybe equipment that just needed cleaning and/or repair, management blames it on the most convenient wouldn't you know A poorly trained work force, and force employees to process materials that can not be Made with safety in mind, nor do they know how to take apart the science they have created, nor the waste product left in it's wake. It was management and it goes all the way to the top of DOE, and any other Government or Corporate giant that become another Death Factory Amongst us all.

DOE's stated goal—-is to clean up all weapons sites within 30 years...... That is totally unfounded, since it is not based on an adequate assessment on how much work is going to be required. Then it's not a matter of how clean the site should be, "It's how clean is the State of Colorado going to demand it be cleaned, and are the right technologies going to be available and at what cost. The report states it will take decades to restore only some sites to their original conditions, and there are others that are beyond any hope of ever being restored, straight out of the reports from OTA Project.

Colorado's former Governor Roy Romer, and Secretary of the DOE, Admiral James Watkins signed an agreement in principle June 1989 to provide the State of Colorado with the funding for Health-Related studies and other oversight activities of Rocky Flats. The Colorado State Department of Public Health invited national scientific experts on the environment to assist in designing the program for research. To increase public accountability, it was recommended by the experts, to do the research in two completely different forums by two different contractors.

Phase I will be the Historical Public Studies, a dose reconstruction and toxicological review, that began in 1990. With ChemRisk, a division of McLaren/Hart Environmental Engineering, who conducted Phase I. Radiological Assessments Corporation will be conducting Phase II, a toxicity assessment and risk characterization, from 1992 to 1999.

A recent report that listed plutonium tainted soil at Rocky Flats as one of the particularly troublesome clean up problems that faces DOE today. However DOE "hasn't done much clean up today", as explained by Johnson at a News Conference. "They will be spending $3 (billion) to 4 (billion) a year "on clean up", At this point there hasn't been a whole lot accomplished at least as of yet." As for Health risks to the General Public, that may exist have not yet been discovered, at least not anything that could be pointed directly at and say here is the problem for today. However it was added neither DOE nor anyone else has researched that question well enough to answer it definitively.

"There has been no imminent hazards", added OTA senior analyst Tara O'toole. "There will be a lot of stuff out there that sounds just awful", however it may not threaten public health. Larry says please note, "what she really meant was "If may not threaten public health", or maybe "It may threaten public health." She said DOE still must develop a certain process for assessing which contaminated areas pose a danger and which do not. In a floor speech, Senator Tim Wirth, D-Colorado, called the OTA report "A stinging indictment". and "a vindication of the warnings that prominent scientists and leading environmental groups have been issuing for years."

The National Resources Defense Council attorney Dan Reicher agreed, saying, "This reflects many of the points we have been making over the years about the environmental problems in the nuclear weapons complex, and I think it puts a government stamp of approval on many of the points that we've been making. "Even the DOE claimed to agree in principle with the OTA's conclusion. "The bulk of the information that was contained in the report did provide strong support for" what Energy Secretary James Watkins had stated on numerous occasions. "The fact this was a problem of enormous proportions and will require yet unavailable technology and trained personnel to resolve it. This report is an independent confirmation of that fact".

What happened at Rockwell Internationals operations at Rocky Flats, but also under the control of DOE. United States Department of Energy, is not something that just started, their problems began back in the 1960's early 70's when protestor's were trying to alert the nation to what was happening there, but they were hauled off to jail.

By 1985 a project engineer blew the first whistle. Shortly thereafter the Federal Bureau of Investigation, visited the shop where the items were made. It was said the practice was ignored by officials of Rockwell International Corp., which operates the plant for the Department of Energy. More than $1 million in gadgets, trinkets and more elaborate items,,,, including jewelry, a wine still and a grandfather clock.... These were items made over a 17 year period of time at the Rocky Flats nuclear weapons plant....

The practice was in violation of federal conflict of interest regulations. Some of the goods were paid for by skimming government contracts, or through a violation of federal purchasing policies. Thousands of plaques, jewelry, minted medallions and gem-like stones that were manufactured in the shop ended up in offices and homes of DOE, managers and employees, and on the desks of corporate officials doing business with Rockwell.

A Federal grand jury failed to issue indictments in the cased, but the DOE inspector generals office in Washington D. C. assures everyone the investigation will continue. "The extent of the fraud, theft and abuse of government funds and facilities in this case was only limited by "One's own imagination", said J. David Navarette, 49 who for ten years was principle project engineer in Building 881, known as "The Model Shop", where the activity occurred.

Abuses in the building 881, were all a part of Rockwell's Futures Systems Department, and were carried out with full knowledge of Warren L. Rooker, who ran the shop for 23 years, 17 years at Rocky Flats and six years prior at the now defunct South Albuquerque Works in New Mexico.

An elaborate suspended Oak and birch spiral staircase had been made for Rooker himself. Which had been taken out of the plant piece at a time in a false bottom suitcase. At the same time he Rooker also authorized a Wine Still, and a Grandfather Clock. This clock the Craftsmen had built from scratch and taking nearly 15 years at a cost exceeding $10,000.00 The reason given for reporting the activities by the project engineer, of building 881 is when he was ordered by Rooker to design a set of plans for Rooker's Planned retirement home near Farmington New Mexico. "I had it" Navarette said. At the time Rooker left his post in June 1985 he declined to discuss the matter. However his Attorney Joseph Saint-Veltri, had implied knowledge of the activity went higher into the company, and into government ranks as well.

Rockwell Security Officials had been questioning the project long before the FBI probe began, and were told all that was necessary to keep it under wrap. Also made at the shop were gold plated phalluses. Donald A. DeGasperi, who worked at the shop back in the early 70's, "referred to each as the famous golden monk." He said they were 10-inch statues of monks in the shape of phalluses which were sent to Rooker's friends. Jack E. Door, plant and general manager 1981 thru 1986, was aware of the work on plaques, he had received several, but insisted they were of little value.

More protest have become just another part of Rocky Flats Job:

When Dominick Sanchini learned again there would be another demonstration at the Rocky Flats nuclear weapons plant just West of Denver in Jefferson County. He felt he had to be there, his feeling were very strong when he reiterated "You have to respect what the protestors are doing. Sanchini recently spent around six hours outside the plant mingling with the demonstrators in the blazing August sun.

It was an especially unusual afternoon for Mr. Sanchini. For him as General Manager of Rockwell International's operations here at Rocky Flats, he is usually accustomed to spending his days inside the plant. However he wanted to see, and was interested in seeing how the police and the Flats' own security force were able to handle the demonstration. These demonstrations have been taking place known to me taking place since the early 70's and through most of the 80's and 90's. What Mr. Sanchini was really trying to do is figure out what it is that motivated the protestors'. "Why were these young people so dedicated and strong with their intentions, emotion, and dedication. There were times some would strap themselves to the tracks leading into the plant with trains coming toward them. They had chains available, but I'm not sure whether they were ever used.

Sanchini, who has been general manager of the plant for just over a year now has accepted dealing with the demonstrators, and just another adjustment he has been forced to make as part of his new job. As he searches out the general feeling.... out there doing it Mr. Sanchini does not speak of the demonstrators in an angry tone, but feeling both curiosity an bewilderment with a pinch of admiration for sure. Mr. Sanchini a 34 year veteran with Rockwell, Stated "I never heard of a protest" he said. His life here at Rocky Flats has been drastically changed. Rocky Flats is a critical link in the government's nuclear weapons productions and operations. The plant make plutonium "Triggers" that fit inside warheads and bombs. The triggers are set to create the first in a chain of lightning quick reactions that denotes a nuclear weapon.

Rockwell also over-hauls older nuclear weapons, where plutonium tends to degrade over time, and the material must be reunified every 10 years or so. There are about 6,200 people working here at Rocky Flats, 5,400 of them are Rockwell employees, the rest are subcontractors. Rockwell took over the operation from Dow Chemical in 1975. Running the facility has not made Rockwell popular in the eyes of a lot of people. It is a large defense contractor that feels it has an obligation to provide the government the managerial services it needs. "Some look at it as good or bad depending what sides your on."

Rockwell has an operation and does carry a threat to its corporate images. Larry Says, Much like the image of the Rocky Mountain Arsenal, and Shell Chemical Corporation who had an operation of deceit, fraud, and malicious negligent wrong doing all the way to the pentagon, who in 1960's realized the public was going to find out that basin F had been Leaking and polluting the over 40,000 acres off the Arsenal, The General at the time said God we can't let the public find this out. So they classified it completely, as they continued to pollute and ravage the Country side for over twenty more years to end up with the most polluted piece of land on earth, while it remained classified until 1990. Follow this story very close and I will bring you proof of the second most polluted piece of Land on earth Rocky Flats with land that can never be redeemed.

Because Rocky Flats handles immediate and imminent super toxic dangers working with radioactive materials, there is always the danger, for example, of an accidental release of radiation into the environment. Such an accident could be disastrous for the environment. Mr. Sanchini States it is a "HIGH RISK" operation and they risk their reputation. "PLEASE NOTE HE DIDN'T SAY ONE WORD OF THE THOUSANDS IF NOT MILLIONS OF LIVES MAY ALSO BE AT STAKE."

What your seeing here is the stainless steel box, right, is the remote control for the room, Rear, in the plutonium-recovery building, as shown inside Rocky Flats. Rockwell always worried about the Corporate image being at stake, and of course the almighty "BUCK". The Department of Energy has told Rocky Flats it's business as usual for the foreseeable future at the plant. Even a move toward disarmament would keep the plant busy. Yet officials decry lack of "TOOLS" for clean up the pollution Rocky Flats Nuclear Weapons Plant. Health Officials are warning. While Rockwell is asking for increased budgets, and more tools for the safety of the general public. But the Department of Energy has an unrelenting grip over their ears and are not hearing the Pleas for help, they cover their eyes so they can pretend not to see what is taking place, and cover their mouth because they don't want to talk about it.

In order to clean up the pollution here at Rocky Flats nuclear weapons plant, and about 20 other federal facilities in Colorado alone will take strengthened laws and increased budgets, Health Officials warn. The citizens need an assurance that Rocky Flats is being operated safely and properly, but quite frankly, they are crying for more tools, "The tool box is empty", said Thomas Looby, assistant director of the Colorado Department of Health. "We need money, groundwater and drinking water standards, and a mehanism to Apply standards to air emissions also".

Looby, who heads the health and environmental protection, made his comments quite clearly at a legislative committee meeting regarding the States role in any clean up of Rocky Flats. The Chairman of the Legislative committee William D. Breckenridge wants to first look at what other states are doing faced with similar situations, and how effective those laws are in health and safety, and environmental concerns.

Larry says, "It is very clear Mr. Looby realizes they have a critical, and significant problem on their hands and headed to the critical stage". Looby couldn't provide figures at that meeting just how much money will be needed to enforce pollution clean up, but said quite specific recommendations would be presented at the committee's final meeting in October. "DOE has assumed responsibility, but that doesn't mean all the issues have been addressed as fully as we'd like," "I don't think we have a good handle on all the air emissions from that facility.", And it is said by our source we have no good handle on the soil, and water contamination at this site.

Rockwell International Corporation, manages Rocky Flats for the Department of Energy. Manufacturing plutonium triggers for the nations nuclear weapons stockpile. In the process, the plant generates tons of radioactive and chemical waste. Because there are other operations not always talked about. Up until last year, the state had no say over how those wastes were treated or disposed of. During the 30 years of operations, spills, fires and waste disposal have contaminated air, soil and groundwater at the facility 16 miles northwest of Denver.

In addition, the plant's day-to-day operations spew small amounts of radioactive elements into the air. The only air emissions the state of Colorado can regulate are those that will come from the proposed mixed-waste incinerator. Looby states, "We need some mechanism to apply standards to emissions not covered by hazardous-waste programs." and said also, "but again that takes resource." Resources is evidently something the State has never had.

Rocky Flats is one of more than 20 Federal facilities in the state with significant pollution problems, and the State of Colorado has no enforceable groundwater regulations. Efforts last year to develop strict controls failed after heavy opposition from industry and mining groups. Federal Officials have been very blunt in saying, State Officials have absolutely no authority to condemn property or order pollution cleanups. Information forwarded by Mr. Looby, Colorado Department of Health.

Late in 1988 Governor Roy Romer got the red carpet, and white glove tour of Rocky Flats, and was able to see and inspect what they wanted him to see, or what was staged for him to see on his investigative tour just ahead of the FBI crack down on the entire plant. He inspected storage areas and processing buildings, and despite all their obvious problems looming out, Governor Roy Romer declared the facility safe, while in the air ducts of the Plant there is enough plutonium to make 9 Atomic Bombs, and this is lying there just waiting to happen. 9 atomic bombs would make Denver and suburbs disappear, and lay waste to the entire State of Colorado. Governor Roy Romer is in a position to talk harsh and straight to the point in safety. And yet has no real idea of the immensity of the "TIME BOMB". The components of 9 nuclear bombs sitting untouched and unaccounted for, A nuclear disaster waiting to happen, and as the governor stated “A Ticking Time Bomb”.

But the Governor, who described the problem of nuclear waste storage as a time bomb with a “CLOCK TICKING”, and demanded the Department of Energy be held accountable, and responsible for the monster they created, and responsible for quickly developing long-term solutions. Governor Romer angrily declared during a press conference on the from lawn of the Governor’s Mansion. “I’m saying to the federal government: Get of your duff. Let’s go. Let’s get a solution.” and he says “I’m warning everybody in sight.”

Romer, who met with several Rocky Flats employees during the morning tour, also reiterated his resolve to halt the plant’s operations sometime next spring if it exceeds its mandated limit of 1,601 cubic ards of radioactive waste material. Romer said, “My objective here is very clear in making the citizens of Denver safe.” “Secondly (my objective) is to save 6,000 jobs“.

The Governors office is “hopeful” DOE officials will be able to present some alternatives at their meeting in Salt Lake City on Dec. 16 with Governor Romer, Governor’s Cecil Andrus of Idaho, and Garrey Carruthers of New Mexico. Governor Andrus Idaho quickly banned any further shipments of radioactive waste to the Idaho National Engineering Laboratory, near Idaho Falls, where hazardous materials have been stored now for two decades.

His ban came during repeated delays in the opening of the Waste Isolation Pilot Plant, which would be a permanent disposal site right near Carlsbad, N.M. It will not be made available to receive radioactive waste at the earliest maybe June. Governor Romer was even suggesting to the President, to summon all western Governors, and ask that each would accept a portion of the radioactive waste now stored at Rocky Flats at least until June.

Governor Romer said he would be willing to meet with the western Governors, and explain to them “We have 15 (box) cars filled with radioactive waste”. “Each (governor) would take 3 and that would solve the problem at Rocky Flats. Romer stated “he would even suggest to the president of the United States, to put us all in a room, and say OK now show your patriotism”. Larry says “This is not a matter of patriotism, It is a matter of Government Negligence at the highest level. They should be responsible to dispose of what they make.

Governor Romer said, "we've got a real problem out there", "It's been an area that the federal government has neglected. I just want to solve Denver's problem quickly." He examined two of the most contaminated sites within the facility. He praised the plan by Rocky Flats Officials to spend an estimated $20 million to clean up some areas of the plant, for some its too late. One of the most important parts of his visit would be a briefing on the dangers inherent in the "Maximum credible accident".... An unlikely situation that might arise, for instance, from a plane crashing into the storage area and starting a fire......

Governor Romer: States, "If it can't meet [Safety] Criteria....We will close it".... As an FBI Spy Plane caught the Flats in the Act. Affidavit claims illegal burning, and polluting of drinking water.

FLATS OFFICIALS TACKLES

TACKLES BALANCING ACT

FLATS SAFETY REPORT

PESSIMISTIC AS BEST.

NUCLEAR PLANT POLLUTION

IS GREAT ENOUGH TO CAUSE

CANCER, JOE ADMITS

FBI says spy plane
Caught Flats in act

Affidavit claims illegal burning, and pollution of drinking water Page 17 picture of Rocky Flats, FBI caught Flats in the Act. Agents seize nuclear triggers intended to be shipped to Iraq.

Agents seize nuclear triggers
Intended to be shipped to Iraq

TROUBLE AT ROCKY FLATS

Rockwell ad invited tourists
Before agents raided plant

FLATS STILL NOT SAFE ENOUGH, NEW BLDGS. AT FLATS ARE REJECTED

PLUTONIUM OPERATIONS MAY NEVER RESUME AT FACILITY.

PLUTONIUM MAY TAINT DUCTS FOREVER.

FBI says spy plane caught Flats in act

Affidavit claims illegal burning, polluting of drinking water

Agents seize nuclear triggers intended to be shipped to Iraq

TROUBLE AT ROCKY FLATS

Rockwell ad invited tourists before agents raided plant

FLATS STILL NOT SAFE ENOUGH, NEW BLDGS. AT FLATS ARE REJECTED

NUCLEAR MYSTRY AT ROCKY FLATS NUCLEAR WEAPONS PLANT, IN JEFFERSON COUNTY NEAR DENVER COLORADO.

Investigations in the past always revealed shabby magement, freaky accidents with dangerous pollution. The FBI has sent in 75 FBI agents, they are currently looking for proof of deceit and fraud in the disposal and incineration of plutonium laden wastes. What has puzzled most environmental Officials, and caught them all by surprise when they found unexpectedly even for trouble prone Rocky Flats. Traces of radioactive strontium and cesium that a nuclear chain reaction could produce….even though there is no nuclear reactor at the site. The Environmental Protection Agency has demanded a study to determine now the mysterious isotopes got to Rocky Flats.

DOE'S Deputy Secretary Henson Moore, states "Our first priority in the operation of any of our facilities. "Last week as the FBI invaded Rocky Flats owned by the Department of Energy, announced that State Environmental Officials would henceforth be allowed to inspect its plants and that if any safety violations were uncovered the operators would be fired. In addition State Safety Teams will conduct surprised aerial surveillances. DOE'S CREDICILITY AT ROCKY FLATS TAKES A HUGE NOSE DIVE: ROCKWELL INTERNATIONAL IS FEELING THE HEAT OF THEIR ILLEGAL BURNS OF CHEMICAL WASTES. THERE IS A GUARANTEE THIS INVESTIGATION WILL NOT BE OVER TILL IT'S OVER.

This episode is another reminder of the scandalous way's safety is routinely monitored "OR NOT MONITORED" at the plant. The FBI raid on Rocky Flats may or may not uncover environmental crimes however it was felt necessary at the highest levels. Rocky Flats in a sense is likened to a HenHouse under the Supervision of the "FOX".1. The Department of Energy is the agency that runs the facility meeting production quotas. While at the same time, 2. acts as a self appointed monitor of safety, 1 and 2 are two different responsibilities that don't always meet in the middle.

This is a criminal probe and is a reminder that Congress is wrong to shield nuclear weapons-plant contractors such as Rockwell International Corp. in this case…. That cushy deal is hardly an incentive to be extra careful. Where there is substantial Ultra Hazardous liability exposure. Federal Officials say that despite the probe, there is no immediate safety threat to Rocky Flats employees or the nearby residents. They admit they would be more comfortable if the charges were something like embezzling of funds. The actual allegations here aren't nearly that benign; They include concealment of environmental contamination, and pollution, false certification of reports and unpermitted treatment of waste……the kinds of high level crimes that could literally ultra poison their victims….

Governor Roy Romer should have been quietly told much earlier about the year-long Investigation…. He had a right to complain about not being informed until this week; and he should keep up the pressure to make sure he's not frozen out again, especially with the seriousness of the allegations…… Larry Says, "by not informing the Governor, and keeping him in the —-dark—- is the same as keeping the public in the dark and that is exactly how DOE, worked in this case along with Rockwell International Corp. ("MUCH LIKE MUSHROOMS, KEEP THEM IN THE DARK AND FEED THEM ALL THE BULL SH_T YOU CAN.")

The reasons why the FBI and the Environmental Protection Agency didn't even tell top Energy Department Officials about the probe until it was well under way is more understandable. Investigators may not be brimming with confidence in the agency's ability to police itself. The energy officials were taking their duties more seriously to be sure in recent years. However the best of intentions will never dispel the fact that they have a definite conflict of interest here at Rocky Flats with DOE, and Rockwell International Corp.

There are a dual set of tasks for Rockwell International, to manage one being the health and safety duties……and on the other hand they oversee an important national security function…. Rocky Flats makes plutonium triggers for nuclear bombs, it's their responsibility in keeping the production moving. That's where the potential for friction, and it is a fact that could undermine public confidence in the entire nuclear-weapons production process. Health and safety monitoring should be vested in an agency with no stake in the production process, or writing rosey reports. Whatever the outcome of the probe at Rocky Flats, This should resinate in Congress to restructure the entire oversight system at all nuclear-weapons plants everywhere.

Just prior to the FBI raid on the plant Rockwell had invited tourists. It all began with an ad in the Broomfield Enterprise, "You're invited to visit the Rocky Flats Plant,“ then further stated “Learn about national defense missions, and working restoration programs, environmental protection and radiation controls.” This was a full quarter page ad signed “Rockwell International…… Where science gets down to business.”

These tours have been offered for several years. Officials said the tours is a way and method for the company in explaining to neighboring communities how the nuclear weapons plant operates….and the reasons why. The tours were planned by Rockwell International who manages the plant for the Department of Energy long before Federal Agents raided the plant. “The ad was placed before the investigation,” “It wasn't in response to the investigation. The ad just sort of jumped out at people, “No one can remember them doing any advertising like this. It's a bit ironic,” Any tours always seemed to be geared toward the 6000 member workforce and their families, and not for the general public. Rocky Flats tours are restricted, its tour schedule to one Tuesday per month thru the summer only.

Two of the most popular tours in town, are the Denver Mint, smack downtown Denver, and then you have Adolph Coors brewery in Golden, one could almost call it Coors Town since they even have their own police force. Actually, “tour” at Rocky Flats is really a misnomer since much of what goes on at the plant is classified. Tourists will never be allowed to visit or even peek inside Building 771. It is the main plutonium processing facility which is the focus of the investigation, and where nuclear weapons triggers are made. If there is a visit it is most always confined to the visitors center. “We do show them a movie and some demonstrations.” says a spokesperson.

AGENTS SEIZE ROCKY FLATS RECORDS

Countless FBI agents and Environmental Protection Agency Officials swept through the Rocky Flats Nuclear Weapons Plant on Tuesday, seizing record as part of a year long Criminal Investigation. Our sources tell us the agents suspected the plant operator, Rockwell International Corp., may have been illegally burning chemical wastes, releasing deadly dioxins and possibly other toxins into the air.

Colorado State Health Officials by the end of the week joined in the hunt, and citing Rockwell with 25 violations of state hazardous waste handling laws. Incinerators used at the plant are used to reduce plutonium contaminated items to ash. This allowing the recycling of the radioactive material. While it may have a purpose for the Plutonium, and the recycling of the radioactive material. Burning these chemicals would release dioxins, which have been linked to birth defects and cancer. Both State and Federal Officials, still maintain that there is no immediate health danger from continued plant operations. This is just a short scenario that has just touched the tip of an iceberg at Rocky Flats. It will show exactly how far out of touch with reality the Government has been with industry and their own agencies who are all claiming to be the United States, and cannot be touched Law enforcement continues trying to do the right thing when some government officials tries lying to them about what is right and wrong. Law enforcement is out there continually trying to enforce the rules and laws for Health and safety of the people. But with corrupt government officials and the corporations they work with it is a strenuous and arduous job in separation of deceit, fraud, Reckless negligence, and misreprsentation. Very simply put separation of "Wrong from Right". The bottom line is they know whats right and they chose to do it wrong then scream foul when they get caught.

Beginning with the FBI raid that had just occurred, it was largely a psychological warfare operation to the recent revelation of finding a significant amount of plutonium in the plant's air ducts. Then how interesting it is to hear that the U.S. Department of Energy wants to modernize the Rocky Flats nuclear weaons plant just N.W. of Denver, eight miles south of Boulder. The federal objective was to make it appear that the plant can be run "safely" and to divert public attention away from the main issue of permanent closure.

The underlying principle is that no one seemed to have noticed that Rocky Flats, running or not, is a federal disaster in and of itself. With modernization, safety and shutdown all aside the very existence of this plant is by far the greatest threat posed upon Coloradoans, and will remain so in the years, decades and millenniums ahead, as this sick legacy of nuclear stupidity begins to take on its true human environmental and economic costs. Right now during and until that time, one can surmise that it will be "business as usual" for the newly modernized Rocky Flats Nuclear Production Facility…despite the incredible number of Health, Safety, and Environmental violations of its past and it's current records. They haven't yet learned how to handle plutonium or its waste.

Barrells corroded and leaked plutonium-contaminated oil unto soil. Weeds grew through the badly corroded bottoms of these barrels.

DOE TAKES NOSE DIVE

RAID ON ROCKY FLATS STRAINS AGENCY'S CREDIBILITY

EPA's Investigation unit as it entered Rocky Flats Weapons Plant.

Dozens of FBI agents and Environmental Protection Agency Officials swept silently through the Rocky Flats Nuclear Weapons Plant, as part of a year long investigation seizing records, Information collected indicates illegal burning of chemical wastes, that are releasing deadly dioxins amongst other toxins into the air. State Department of Health citing numerous hazardous waste handling laws violated.

EPA Investigations team from Denver arriving at Rocky Flats.

Raid on Rocky Flats strains agency's credibility, DOE takes nose dive.

In the enormous ballroom of a Denver Hotel las week, Speaker after Speaker bashed the Department of Energy, as the Officials sat and listened impassively. The main subject was a plan by DOE to start trucking toxic wastes by this fall from Rocky Flats Weapons Plant in Golden to a new storage facility in the salt caverns beneath the New Mexico desert.

As the Department of Energy were being bashed, across town at that very moment, more than 70 FBI agents were serving operation…"DESERT GLOW"…. A criminal investigation of illegal dumping and storage of toxic materials at Rocky Flats…on the DOE., their credibility already sagging, they flat out took a nose dive. However operation DESERT GLOW …. Was apparently limited to Rocky Flats Operations….This was the Bush's administrations chance in saving DOE from persistent and tenacious congressional assaults. The speakers were very skeptical of DOE's updated study of the environmental impacts of the Waste Isolation Pilot Project (WIPP) and the transport of radioactive and hazardous wastes there.

One source close to DOE said, "If you have everyone trying to murder an agency, maybe the way you save it is to effectively commit suicide." The premise is not to allow DOE to go down, or to commit suicide, but rather it would be to allow the Bush Administration, and the Congress to clean up the agency. It would be a dual effort, and opportunity for the government to save DOE, and clean it up, to where they will have the Health and Safety of Americans and their future in mind everywhere. This would be intended to help America first and then from experience show the World how to clean it up. Let's not forget what is happening elsewhere in the World also has an affect on America.

Since Larry was close to the Ronald Reagan years, and Congress he believes that the Reagan administration with Operation Desert Glow, was extending a hand to Congress to work closer as a Government in managing their own Government, and help organize DOE. Members of Congress, including Rep. David Skaggs, D…. Co. … have slowly been turning up the heat on DOE and its problem – plagued nuclear weapons plants for more than a year. A whole string of measures to force independent policing of DOE have already been approved or are in the works, and operation "DESERT GLOW" just placed it in priority time, and something DOE can no longer say when "WE GET A ROUND TO IT".

Skaggs is proposing a measure to beef up independent oversight, arguing that a bill Congress passed last year setting up an outside review board wasn't adequate. Congress had already set up it's own oversight panel. In the face of the growing congressional agitation, then DOE secretary John Harrington's last year set up another panel to review internal safety. Then Doe Secretary James Watkins was accused by congressional leaders in his attempts to undermine safety concerns by reorganizing the internal safety operations. Of course Wakins denied the charges leveled at him, and asked Congress to stop creating committees to oversee his management efforts at Rocky Flats, and in fact other nuclear weapons plants. It seems to be "very clear Watkins was not running a tight ship." and he expects a hands off approach by congress. His inability to deal with the scrutiny he's now facing what Larry see's as his, "Illegal grip on DOE." This way both Congress and the Bush administration can clear up a problem with DOE, and it's management team who can't be trusted. Their management skills are and have been in the past atrocious, It seems as if everything regarding safety has been left on the back burner.

Operations Desert Glow, has actually stepped up the campaign on Capital Hill with Congress in charge. Skaggs purportedly used the incident as ammunition in the argument for his proposal to beef up outside monitoring. Larry Says, "Scagg's bill is the best we've had and about 10 to 15 years too late." He and other Congressional leaders are pointing to the operation as still further evidence, that DOE has not been able to handle the job, and cannot be trusted to manage the Country's nuclear weapons plants without some oversight. The Investigation at Rocky Flats, means that every aspect of the weapons complex is under review and/or the penalty phase for safety violations. DOE silently had WIPP on schedule for September, until Operation Desert Glow.... Which is expected to last at least for another six months.... Public at this time has dealt a severe blow to DOE's efforts to get the atomic waste storage program in New Mexico (WIPP) started.

Doe Officials were still trying last week at the Denver hearing, insisting they would still be able to meet the opening date. However the skeptical environmentalists, and Officials on the fast track ...... Environmental Impact Update. to whom they had to convince, and then the simultaneous raid at Rocky Flats was surely a failure of their sales pitch. This week DOE Officials continue their public hearings on WIPP in New Mexico. It was expected the agency would drop any discussions for September opening date......as they felt the bashing they took in Denver last week was a mere scrape compared to the (POST OPERATION) Desert Glow review they will get.

AGENTS SEIZE NUCLEAR TRIGGERS INTENDED FOR SHIPPMENT TO IRAQ

U.S. and BRITISH agents in London seized 40 Iraq bound nuclear triggers that had been illegally smuggled out of Southern California. There were six people arrested two were Iraqi nationals. A cargo hanger at Heathrow Aiport culminated in an extraordinary sting operation. It all began in San Diego 18 months ago, the operation was a sting said a Department source. The U.S. Customs Commissioner Carol Hallett stated, "We've stopped some very serious business from going forward". Hallett and British Officials would not discuss the operation in any detail. Was this the final stage for Iraq to complete their atomic weapons capability.

President Bush at the White House stated "that arrests underscore his deep commitment and concern about the issue of nuclear proliferation in the Middle East." The President added: "We again call upon nuclear suppliers to exercise special restraint in providing materials related to the development of nuclear, chemical and biological weapons and intermediate-range missles in the volatile area. The United States furthermore urges all states in the area to adhere to the non-proliferation treaty. Iraq is a signatory of that treaty. It has been made very clear that our views on nuclear proliferation, for which we have made very clear on several occasions. President Bush's concerns were echoed "Loud and Clear" by all experts in nuclear proliferation.

"This just ponts out the shortcomings in the international system, which isn't going to catch them every time", said Deborah Holland, issues director for the Nuclear Control Institute, based in Washington. It is an organization that studies terrorism and nuclear proliferation.

ROCKY FLATS IS A FEDERAL DISASTER

DEADLY PLUTONIUM: Even in minuscule quantities, of plutonium can cause cancer and chromosome damage if ingested or inhaled, all scientists agree. Because plutonium remains highly radioactive for 24,000 years, if it lodges in an internal organ it can bombard surrounding cells with deadly alpha radiation for a lifetime.

Twenty-eight kilograms of plutonium, if dispersed in a fine dust into the air and spread by winds throughout the state, it could cause a catastrophic increase in cancer risk for everyone in Colorado and beyond. However, the plutonium discovered in ducts at Rocky Flats has been contained by filters and has not been released into the atmosphere, officials stressed.

Plutonium in Flats air ducts, enough dust to make at least 6 Atom Bombs, and possibly more….

The Rocky Flats nuclear weapons plant may remain in business for at least twenty more years, and cleaning up after it will take much longer than that as stated by Energy Secretary James Watkins. A report from the President has what remains the objective to close Rocky Flats, Watkins told the house nuclear facilities panel that moving plutonium operations out of the Denver-area would take time. How much time "Twenty Years", He states "I couldn't move it in less than 20 years," and then it might be difficult finding a state willing to accept the operation.

The environmental restoration could go well beyond 20 years. That is the same scenario they gave for the Rocky Mountain Arsenal's clean up. Sister to Rocky Flats in the production of "WEAPONS OF MASS DESTRUCTION AND DEATH". In 1983 they had anticipated the environmental restoration would be 10 to 15 years before it would be ready to transfer back to society, at a cost of maybe a billion dollars. The Untied States marched Shell Oil into Court to see to it they paid their fair share. That fair share was based on only a Billion dollars, and Shell was out of the loop. So far the cost is running into several $ billions of dollars, and it is the taxpayers who are footing 100% of that cost. They have now estimated it will take until 1060 to finish the clean up. ("Maybe") It is 18,000 acres now known as the most contaminated place on earth, and can never be transferred back to society.

"It is definitely good news that this secretary of the DOE has embraced the (Flats Phase Out). Aspect of the 2010 report." said Skaggs, a Boulder Democrat whose district includes Rocky Flats.

Technicians who handle the deadly plutonium using gloves attached to arm coverings in the sealed units known as glove boxes. The 60 pounds of plutonium did not escape into the outside environment, but some of it did pass through intermediate filtration designed to block its progress through the duct.

The plutonium in a slurry form, allowing tiny bits of solid radioactive matter mixed with a liquid solvent, and other material used in the trigger manufacturing process. Bill Reimbach a spokesman for plant operator EG&G INC. said "We think the figure of 28 kilograms is pretty solid." There are about 2.2 Pounds per kilogram. "We have a lot more measurements to do before we can restart." A spokewoman admitted some plutonium may be left in the ducts after operations are restarted. The Department of Energy which owns the plant, says clean-up of the ducts will begin before plutonium operations are resumed.

"The pipes will be cleaned before we restart" said Catherine Kaliniak of the Department of Energy." Im not sure if all the activity will be completed prior to restart. But the Energy Secretary won't order restart if there are any safety questions. To prevent the reoccurrence of this problem in the future, the whole system at the plant needs to be changed. A member of the Rocky Flats Environmental Monitoring Council who was the first to learn of the plutonium discovery stated a wake up call for prevention.

Melinda Kassen, attorney for the Environmental Defense fund, and a monitoring council members, stated "Operations should not be restarted until they can make sure they're not going to have those releases in the future." Such a move would delay the restart at Rocky Flats, could be for months beyond the scheduled date that was set for June.

Catherine Kaliniak had said the plutonium problem can be solved by better monitoring systems and doesn't require them to totally rework the plutonium handling system at the plant. The latest revelations from Rocky Flats dismayed just short of shock Colorado leaders who said this latest discovery surely calls into question the plans to restart plutonium operations by June.

Tim Worth said the risks at Rocky Flats are much greater than what was reported by DOE, and that he was deeply disturbed by the report that the potential for a criticality accidents are real. The Indication that 28 kilograms of plutonium have found their way into the ventilation system are truly disturbing and deserve prompt and unequivocal responses from the DOE. U.S Rep. David Skaggs said this discovery surely illustrates that a lot more work needs to be done before any decisions are made for resuming Operations. "This is a very serious matter, and a review of all criticality standards, and procedures needs to be done as part of any restart.

If your wondering what is criticality? Let me explain, the fact they have 28 kilograms or 62 pounds of raw Plutonium just sitting in the ventilation ducts. One way to look at it is that "It is an accident waiting to happen. Enough has gathered in the air ducts if it were to get wet would cause an uncontrollable reaction (criticality). It's much like having at least and possibly of more than 6 ticking Atom time bombs, sitting on the hill just northwest of Denver, a very short distance to Golden, Boulder, and all the suburbs of Denver. Safety is saying to DOE to stop and clean it up, but DOE is pushing to leave safety out to dry, and to continue operations at Rocky Flats.

Yes up to 62 pounds of radioactive plutonium will remain stuck in the air ducts because DOE is pushing in putting Safety on the back burner so operations can continue. They say they have safety on their mind and its true they just haven't found it yet. And the plant officials fully intend to resume operations as soon as possible. In spite of all safety recommendations to clean up all the nearly 62 pounds or plutonium in the ventilation ducts, It really makes one wonder whose air are they ventilating. Even with Congress very involved in this whole mess DOE continues to push their weight around. Plant Officials say they will clean up only about 12 pounds of this deadly plutonium that have clogged at least 10 ventilation pipes before resuming some operations. That means that about 80% will remain stuck in the ventilation piping and more daily if operations are to restart.

Plutonium processing needed to make bomb parts at Rocky Flats was shut down late last year. This was done after years of investigations of wrongfull and negligent actions in regards to the safety and health of employees, and Denver area residents. When they were found in over 30 violations by the Colorado Department of Health of polluting the air, water, and soil. Illegal dumping, Illegal shipments of radioactive waste, illegal burning, what else. DOE agreed to remove all of the plutonium from the ducts before resuming operations, but there's no estimate on how much we think we can remove. They say they are going to do those ducts with safety concerns first. First they say were gonna clean the ducts, then they keep pushing to let it ride so they can once again resume production.

At this point they have agreed not to use the 10 air ducts in buildings 707 and 771, Nelson said, "The 10 air ducts in buildings 707 and 771 will not resume until the plutonium is removed, however other operations will restart as soon as DOE approves". Melissa Kassen of the Environmental monitoring Council was skeptical. During the internal assessment, finding nearly 61 pounds or Radio

Active Plutonium in the nearly 4000 feet of the air vents, a build-up that occurred over the nearly 40 year history of Rocky Flats.

Keeping the above paragraph in mind, Plant Officials yesterday said "their study was incomplete, but the amount found thus far did not represent any threat to workers, the nearby residents or the environment, and it was spread over a vast area in several buildings at the facility". Then in their next breath EG&G Inc., the company running the plant, conducted the inventory after federal inspectors last year estimated up to 15 kilograms may be in the pipes, then they discovered there is at least 28 kilograms of plutonium in the pipes and this could create what is known as a (criticality) a very significant, dangerous, and uncontrolled nuclear reaction that would release radiation, putting employees and the public in harms way.

Melinda Kassen of the Environmental Defense Fund and a member of the Rocky Flats Environmental Monitoring Council stated and mincing no words, "I don't care whether there are little bits of it throughout a whole bunch of buildings…….That misses the point…… which is that there is more than six bombs' worth of plutonium in the air ducts". "They have two problems…… they don't know how to get it out of there and that they're going to have a bit of a problem saying they can run a safe operation without changing the equipment which allowed this stuff to get in those ducts in the first place". Please note, these are called ventilation tubes, what in the world was they ventilating.

Plant Officials still have no plan for the removal of the plutonium, changing the filters or redesigning the "glove boxes" where the employees work with the radioactive elements. It has been reported that employees often poked holes in them rather than change them, in order to get the work done faster. This allowed plutonium to escape into the ventilation ducts that which would vent the contaminated air to the out side, for unsuspecting residents were being contaminated with radioactive plutonium. Rocky Flats, is just 16 miles from Denver, about 4 miles to Golden and Boulder. The Plant produces plutonium triggers that detonate nuclear weapons and recycles plutonium from old weapons. EG&G says "We want to go through and complete our measurements using very sensitive equipment and then decide what we will do from there. We want to be absolutely sure of what we have before we have a plan to deal with it."

SAFETY PROBLEMS BEDEVIL EFFORTS TO CLEAN UP THE ROCKY FLATS NUCLEAR WEAPONS PLANT NEAR DENVER.

Most of the Jefferson County plant has been shut down since September because of chronic safety problems. Rocky Flats uses several other radio active elements, along with Uranium, and Plutonium to make detonators for nuclear weapons. Cancers, brain tumors and reproductive disorders have been linked to exposures to plutonium. There has now been a heightened focus on safety due to the increased radiation problems here at the plant.

RADIATION EXPOSURES PLAGUE ROCKY FLATS SAYS DOE' REPORT.

The seventh contamination report occurring at Rocky Flats in less than six weeks. Four of those occurred in building 771. This is a sharp increase in contamination incidents that have occurred in buildings specific to plutonium. Doe says elimination of these contamination incidents is the essential key to the safe resumption of operations, especially the plutonium buildings. It was observed in working room 427 two people were working without their full face respirators on even though the room was posted.

In September 1988 this was part of the basis of an off normal event that shutdown building 771, and now one and a half years later these types of events are still occurring at building 771 Rocky Flats. It would appear the lessons learned from the closure of building 771 in 1988, were not implemented, or adhered to during the subsequent restart regarding radiological protection across the Rocky Flats site. One would imagine that EG&G's better management policy to correct the OSHA deficiencies prior to a Department of Labor-OSHA inspection at Rocky Flats.

- Procedures designed to prevent an accidental nuclear reaction are not always followed.

- Radiation monitoring equipment occasionally does not work.

- Fire protections in some cases is most inadequate.

- Warning signs to tell workers to wear respirators are not properly posted.

- This was the same problem that forced a partial plant shutdown last year.

Internal reports from the Department of Energy have shown that eight Rocky Flats workers were exposed to radioactive plutonium just in the past two months.......Is this a sign of employee's failing to follow safety directions, or is it beleaguered failure of safety equipment at Rocky Flats??????

DOE Washington Headquarters.... Outlined a variety of safety violations that could stall the plans to restart the weapons parts production this year. Echo'd in the new report's criticisms, raised several times over were Safety, and Inadequate equipment. Although slight exposures and low-level contamination problems that may have gone unnoticed, are being reported for the first time. It's a reflection on the upgraded equiment and safety programs, along with better training and more intense routine surveys, says EG&G Inc., spokesman Jeff Schwartz, Yes it will get better, we have a commitment to excellence and a clean and safe work environment.......

Two of the plutonium exposure incidents that occurred, when two employees were working in the containment room without respirators. Still another, two employees wearing full protective cothing improperly removal of their suits contaminated their skin and hair. Any exposure from plutonium, or other radioactive materials are considered serious but manageable. EG&G says it is very disturbing because it's an indication that it doesn't matter if you change the managing faces at the top the results are like looking in a mirror.

EG&G took over operations after DOE had dumped Rockwell International Corp. because of chronic safety violations, that led the the contamination of many square miles North West of Denver, some of the polluted sites are so bad they have been listed as the most contaminated sites on earth, much like the Sister Plant Rocky Mountain Arsenal. There are places to polluted they will never be returned to the public. Rockwell not only endangered it's employees, it has endangered the public for years to come. EG&G with new upper management have vowed to run the plant safe, however mid-level and lower level managers are the same people that were responsible for the day to day plant Operations that led to unsafe working conditions, and to the general public.

Although DOE had dumped Rockwell International leaving one to believe the FBI's investigation of the facility would ring down criminal violation against Rockwell Int. There were a lot of written citations, as I understand were no more than a slap on the wrist for Rockwell……for all the lives they endangered and the billions of dollars it will cost the taxpayers who are left to pick up the tab, not only for the clean up but the continued operations by EG&G while the clean up will be forever in progress at least until the year 2060, even then it won't be done.

The Department of Energy has been in the business of relieving the private contractors, like Rockwell who managed one of the nations nuclear weapons plants from any financial accountability for their actions, and are able to walk away with millions or billions in severance packages. Larry say's "while the new start up programs will cost billions to clean up and start all over again. IT'S ALL AT YOUR EXPENSE".

DOE inspector General John Layton, in a recently released report and in Congressional testimony earlier their week, criticized the department's policy of indemnifying the private companies that run the department's 17 nuclear weapons production facilities under contract to DOE. It was Layton's recommendation the department revise its policy and contracts to ensure private contractors assume some financial responsibility.

What was and is the rational behind DOE and other Government agencies including the military in protecting its' weapons contractors from financial loss. They were paid for doing what really should not have been done in overstepping safety of the workers and those in various parts of this nation that may or did live in close proximity to these lethal Factory's of death, and they are all across this nation and the world. This has been in the past, now and the future, criticized by environmentalists, many members of Congress who were left considering reauthorization of the Price Anderson Act, twas the laws that governs the nuclear weapons and commercial nuclear power industries. This law requires DOE to indemnify its weapons contractors from the "risk of public liability for a nuclear incident, without regard to how substantial that risk may be." They are also protected against any bodily injury, sickness, disease, death, or property damage that may result from the release of any radioactive or non-radioactive but hazardous materials.

DOE agreed with critics, and Congress without them being in a position as to indemnify the contractors, "The risks would be so great the Department of Energy wouldn't be able to run the plants. Larry says, "What they are saying is the contractor's no matter how shabbily they run the plant, with little to no regard for human life, either working in the plant or those who live near the plant, and to get production to the maximum, they throw away all the safety rules as they have at Rocky Flats now and in the past. They talk of safety but really when they talk of production, safety goes out the window."" At least until they are caught like by the FBI, or the Environmental Protection agency, or your State Department of Health all very critical of safety violations to the workers and the general public, they issue hundreds of citations that were merely filed in the round circular file because they were totally unenforceable." About every four to five years when mismanagement, or downright malicious, and reckless negligence surfaces DOE or the Military are there to as you can see change the top management who then promises they will follow the safety guidelines for a safe operation, or at least until they are caught. Let's all remember another thing this is all done by contract, and the severance packages are in the millions, even billions. These are the things you never hear about. The Contractors for DOE are indemnified for doing what DOE couldn't do. They support the contractors then give them production demands that cannot be met, and that is when safety at the plants goes right out the window.

At that time a provision was added to the law allowing DOE to levy fines and penalties against top contractors who may willfully violate department regulations, but really the affect of this provision meant nothing and was merely negated by the contracts, completely indemnifying the companies. Even as a Federal Grand Jury in Denver is investigating allegations of criminal misconduct on the part of Rocky Flats Officials in connection with a massive disregard for safety and environmental laws. As for Rockwell International the private contractor that formerly operated the nuclear weapons plant just northwest of Denver. DOE. Voided their contract after Rockwell filed a lawsuit against the government contending the company shouldn't be held responsible because they did not meet federal environmental regulations.

LOOKING BACK FOR COMPARISON

Then there is an unknown law the United States has used many times over when they are sued in a Congressional Action in order to have a case, it must be heard under the Tucker act which allows a person to sue the United States of America in the Congressional Court of Claims, and the only way to that Court is with a bill that has been passed by the Senate or House of Congress. When they have clearly violated every law, every rule and everything considered safe. This Law is called "THE DISCRETIONARY FUNCTION"

This is heard by the Court of Claims on claims based on the exercise of a performance, or failure to exercise or perform a discretionalry function or duty on the part of a Federal Agency or Government Employee are not cognizable under the FTCA. The Court of claims will hear a case where there are legal, or equitable claims. The purpose of this law is to protect the Government from liability that would seriously handicap efficient Governmental operations. "Whether or not the discretion is abused is immaterial". United States V. 331 f.2d 498 (10th cir. 1964).

If you have sued and there were concrete evidence of Governmental Wrong Doing, the Court of claims must decide on Legal, or Equitable claims and even when you have proven there is Government Liability, the Court will take away your rights to any further legal or Equitable claim for which they must make a decission. They take it back out of the hands of the Court as fact finders, and back to Congress using full force of all evidence gathered clearly showing the Government is left standing holding the smoking gun. It's given back to Congress as an award from the Court, and for Congress to appropriate the funds to settle all claims full and complete. Overall something like this may take years. As in my case over 30 years and in 1997 an award was given to the Land's by the Court of Claims and was placed in the Congressional files in February 1997 for Congress to act.

That is exactly where it has been with Senator's Wayne Allard, and Ben NightHorse Campbell, turning their back on the bill, something that had been past before their time, and they have let it be known they don't do private legislation. Then their were Congresswoman Diane Degette, Congressman McGinnes,

Congressman Bob Schaffer, there is a bill sitting in the files awaiting them to appropriate the funds to make the Lands whole again. Past Congressman Brown who is now the president of a Colorado University in Greeley Colorado, was notified, since it was him who passed the bill in 1988 for the Lands to have their day in Court of Claims hoping he would have some input on the Decission being made but there was none.

LOOKING BACK FOR COMPARISON

ANOTHER CASE AGAINST A GOVERNMENT ENTITY, MARTIN MARIETTA PLANT IN ANOTHER JEFFCO SUBDIVISION IN COLORADO KNOWN AS FRIENDLY HILLS SUBDIVISION, JUST DOWN WIND OR ROCKY FLATS.

According to Court records Martin Marietta Plant had contaminated the water used by the Kassler water treatment plant. The records indicate two medical experts have opined that eight children who lived in Friendly Hills Subdivision developed cancer, and other illnesses, and in fact died because they drank water treated at the kassler plant. These Doctor's reports had been filed in Federal Court to support a lawsuit by residents suing, the Denver water board and Martin Marietta, when material had leaked from the Martin Marietta Plant that were both hazardous and toxic and capable of causing cancer and other illnesses to those who drank the water. According to Judy Stowell a spokesperson for Martin Marietta Company said "we have stated all along that it is our belief that none of the plaintiffs in this lawsuit have suffered any kind of injuries as a result of anything Martin Marietta has done, as far as polluting the ground water or anthing else, and we continue to believe that".

Dr. David M. Ozonoff and Dr. Janet Sherman, "attributed all of these ailments to the children's exposure to water containing carcinogens, toxic chemicals and other hazardous contaminants". Both Ozonoff and Sherman, "said the cancers were unlikely to be due to genetic factors and that the children had no exposure to pesticides, radiation, or other carcinogens that could have triggered the malignancies".

"There is no other more plausible cause of these cases of cancer than their exposure to water contaminated with……Carcinogens" Ozonoff said. The medical reports reviewed the medical histories of eight children. Four were diagnosed with cancer between the ages of 3 months and 15 years......two developed brain cancer......one kidney cancer......and one leukemia......two have already died......Three of the other children developed recurrent seizures,...... and one boy was born with a major heart defect that required surgery.......

DEPARTMENT OF ENERGY SPECIAL STATUS FOSTERS ARROGANCE

If the charges that are being probed at Rocky Flats, and against Rockwell International are true. Did the plant managers illegally burn and dump hazardous wastes and did conspire to cover up that fact??? “”YES YES”” It should come as no surprise. The Energy Department and it’s contractors have been granted privileged status in our legal system for years. Why should anybody be shocked with DOE managers came to believe they had a license to toot and flout the environmental dictates of mere mortals in Congress???

DOE has forever acted like a law unto itself when it came to hazardous-waste management at it’s nuclear facilities in particular, the agency argues it isn’t covered by the resource conservation and Recovery Act (RCRA), the nation’s principal solid-waste management legislation. The Environmental Protection Agency can’t sue DOE or its contractors over RCRA transgressions, (Resource Consveration and Recovery Act. Which is the nations principle solid-waste management legislation. Colorado along with many other states hare handicapped in trying to enforce their own laws, many of which parallel EPA rules, on all Energy Department Facilities. Even though the ACT to control DOE Facilities were drafted by key lawmakers who may challenge the agencies interpretation, the executive branch has always taken DOE’s side (Right or Wrong).

“Sovereign Immunity” is the fancy term for all agencies of the United States and a coddled status. This arrangement amounts to an assault on the rights of the American people. In this case most notably, their right to protection from hazardous materials handled by the government. Rep. Dennis E. Eckart, D. Ohio, introduced Legislation that would right this wrong by making it clear that RCRA applies to DOE as well as it would to private polluters. It was Eckart’s desire to give all the states the power to regulate and levy fines for hazardous-waste violations at all DOE plants and make it easier for the EPA to take civil action.

The real danger lies in not giving states and the EPA more freedom to hold DOE's mind on their business, and Safety has always taken back seat. Recent events here at Rocky Flats alone would underscore the crimes ignoring safety in other atomic energy plants across this nation. There is a need in every one of them, for all the watchdogs that can be mustered. There are critics that believe the states would abuse that authority, and they would be an insurgance of legal actions with DOE, on even slight pretexts, and the states already enjoy similar enforcement rights in the areas under a variety of federal environmental regulations that have much less teeth.

Will it be an ominous legacy we leave for our children who will inherit, debt, pollution and poverty. Have we already failed our children. Leaving to them the environmental disasters today that will cause catrastrophies tomorrow. Experts already warning of everything from radon radiation and ozone depletions to to disappearing landfills, and hazardous waste disposal now going on in just about every state of the union. "These wastes the most toxic known, along with the nuclear wastes, that cannot be returned to normal will now pollute our world forever.

Coloradoans are especially sensitive to environmental issue; Our world-class mountains streams and air are already seriously contaminated with by a variety of sources: radium dumps, nerve gas dumps, pesticide dumps, the Lowry Landfill, The Martin Marietta rocket manufacturing and testing range, The Rocky Mountain Arsenal Chemical Weapons Plant, Shell Oil Company with the most toxic for Pesticides, and Herbicides, that have fouled the air and water just north of Denver for over fifty years and counting, then we have Rocky Flats. Rocky Mountain Arsenal, and Rocky Flats sister munitions plants, are listed as the most contaminated places on earth. Thru mismanagement over the years they have created a monster that will never go away, and will be sitting there to hand down to our children's, children and their children.

Rocky Flats and the other 48 nuclear weapons facilities will cost billions, if not trillions to only begin the clean-up there are areas that cannot be cleaned up, and in some cases are being entombed. Environmentalists protest because the jobs of cleaning up can now never Be completed. With new ecological disasters just awaiting a time to happen. "THE TWO HEADED MONSTER KEEPS GETTING BIGGER AND BIGGER". As for the Rocky Flats nuclear weapons plant, they have promised a lot of clean up, and as they clean up, they are creating more waste that will only surpass the dilemma they have now, since what they are doing is manufacturing more as they try cleaning up the impossible. "But they keep it all under wrap by saying were cleaning it up.

WHAT COMES AROUND GOES AROUND AND ITS JUST GETTING STARTED

"MULTI CHEMICAL SENSITIVITY SYNDROME"

A whiff of perfume or anothers cologn, or after shave, maybe just a breeze from the neighbor's lawn can and will leave chemically sensitive people speechless, shaking, tremors, weak and fatigued, nauseas loss of balance, headaches, difficulty breathing, people are becoming increasingly vocal about it. They complain that they're getting sick from everday exposure to even the common levels of chemicals that may be in a home, or office, or environment.

Multi Chemical Sensitivity Syndrome can leave sufferers with headaches, burning, nose throat lungs, nausea, difficulty breathing, a totally paralyzing disorientation that can render one incapable of speaking. Those in the insurance field, medical and chemistry industries are skeptical think the misery is all in the victims head. The condition still has a ways to go to convince those in the Insurance and chemical industries, although the medical industry has come a long way in recognizing it as a very serious dilemma. Scientists insists that society is seeing the first casualties of the proliferation of chemicals. Since World War II, annual U. S. production of synthetic organic chemicals has exploded from less than 1 billion pounds to over 273 billion by 1988, and to over 325 billion by 2003.

Sensitivity to multiple chemicals possibly is just an incident away, a smell of perfume, cigarette smoke, cologn, household sprays, insecticides, herbicides, maybe a spray at work, or the smell of industrial chemicals in the air we breathe, even carbon monoxide or even the water we drink or the food we eat. Something one day will trigger the incident you will never be able to forget. The chemical "MONSTER" is everywhere, it may be seeping into your home, seeping into your work place, seeping right out of the streets. "IT HAS GOTTEN OUT OF CONTROL".

Whats happening with this today "May well be the canaries" Claudia Miller stated, who is a University of Texas Health Science Center allergist and immunologist who co-authored a study on the malady for the Department of Health of New Jersey, which has the nation's second biggest chemical industry behind Texas. "We don't understand the mechanisms," she said "It may be neurological or immunological, then again it may be biochemical." What ever the mechanism, the effect can be dramatic, say experts for those who suffer this overwhelming biochemical sensitivity are the victims.

The illness can often begin with a single exposure, such as the installation of a new carpet in the office, in your home, or maybe an accidental release of some toxic materials from a sewage plant……or some chemical manufacturing plant, It will overwhelm the body into a permanent hypersensitive state. What follows is a bewildering array of symptoms and a search for help and a diagnosis for treatment.

The things you should try to avoid, fumes from pumping your own gas, carbon monoxide, glues for carpets, paints, building materials, pesticides, fumigants, herbicides, in some cases even cigarette smoke, any chemical smell that is out of the ordinary, will trigger a mild to severe episode…… Some sufferers find breathing relief with use of adrenaline, while others react to drugs. How wide spread is this It is running rampit around the World and nobody knows what to do about it or really don't care. "AGAIN YOU OR YOUR CHILDREN ARE THE VICTIMS". According to the New Jersey report, the National Academy of Sciences suggested that about 15 to 20% of the population has heightened sensitivity to chemicals. Many of the victims face intense and almost over whelming symptoms with the exploding headaches, vomiting, tremors, and inability to breathe.

I'll tell you a quick short story of what happened to me, in 1972 we were overexposed to Sarin, and GB nerve agents as we lived within on mile of the Chemical Weapons Plan the Rocky Mountain Arsenal and Shell chemical Company who manufactured pesticides and herbicides…. We had noticed the boys were waking up in the mornings with their eyes matted shut, and choking, the Doctors didn't know nor did they care, it got much worse we lost over 650 head of livestock within two week period of time, everyone was out there no one could figure it out, then it was noted the only thing we had in common with the cattle was water, the water was contaminated. Sure enough Nerve Gas fingerprinted to the Rocky Mountain Arsenal along with Pesticides from Shell Oil Co. were found in the water, and in the air we were breathing. The livestock did not survive, All of us except for one of the boys survived, eventually we moved to St. Paul Minnesota only denial from the Army and the Justice Department.

We were getting our life back together again when my employer Washington County Sheriff Department they decided to paint our work area using a Lead based Bridge Paint that had been banned in the United States by EPA in 1967. They overexposed at least 35 employees to this over 21 days, before OSHA got them stopped. Larry ended up being very sensitive to various chemicals, perfumes, household items, ending up with three major health maladies, 1. Arthralgia, 2. Myalgia, and 3. Hyper tension, to the maximum.

Then in 1990 the same employer Washington County Sheriff Department here in Minnesota used an Industrial grade Exterior enamel, Mautz line 89-00 Alkyd, Alkyd, containing Mercury, Xylene, and Urethane, paint. This paint caused every sign and symptom to overexposure to its contents except for death, within the first 15 minutes for 10 hours a day for 30 days before the could be stopped, OSHA was called in and issued Severe type citations to the employer. These are criminal citations, Washington County tried every thing under the sun for nearly 3 years to get out of them. They were found liable by the State of Minnesota and placed on probation to the State, as well as fined nearly $2,000.00.

That set Larry up completely for Multi Chemical Sensitivity Syndrome, causing a lot of injury to the Central Nervous System, causing him disabilities that eventually cost him his job after 15 years. The paint line 89-00 Mautz Alkyd, Alkyd, had one ingredient that was not on the label, and had been banned in the United States because it was so dangerous. The name was MERCURY, nearly the same as a nerve agent. Larry's blood level was found after over 10 months still 450% higher than normal, it was like a burning time bomb going off in Larry until he was totally disabled in 1999.

As you can see by now, I'm Totally disabled but still trying to warn the World of the types of things one must watch for even in your daily life. Even to the places it may take you in the Work force. What Washington County done to myself and many other employees is inexcusable. They knew Right from Wrong, and it was them who chose to do it wrong then try ignoring the consequences. I feel with telling the story I have told you may prevent a terrible disaster in your life or to your children, then I have done my job.

The physical basis to the disorder by some credible organizations have given at least some credence to a physical basis for the disorder. Just to name a few, found in the most toxic and deadly nerve agents, and nuclear waste, are Silver, Mercury, Lead, Berrillium, Phenol, Plutonium, deadly Calcium, Sodium, Flouride, Boron, Ytrium, Gasoline, Xylene Benzine, Touline, Chlorine, Bleach, These are all materials that could cause death if wrongfully used, Some are used in every day life in buildings, construction, Yes we even have enough materials under the ordinary sink in the home to mix and make deadly nerve agent, in some cases high explosives. Getting back to the thought on nerve agents and how all these items were used as such or waste. An injury to the central nervous system must first be a physical injury……. WE look at the injury nerve gas causes to the brain such as Sarin, G B Nerve agent, B Z nerve agents, V X nerve agents, all cause physical injury to the Central Nervous System, if you managed to clear yourself of the gas, shower, and what ever else you may try to lessen the effets, the one part you will never rid yourself of the injury caused to the Brain, You will always have psychiatric problems from then on, and those of you that have been through this do know what I'm saying.

The first place to be injured is the Neuro- Transmitter in the brain millions of neurons being transmitted within the brain day and night. The nerve gases I listed above and all the other things such as gasoline, Carbon Monoxide, the Mercury, Lead, Silver, Flouride, bleach, Sodium, All these can cause problems with the Neuro Transmitter in your brain or the brain of your child. Once the Neuro Transmitter in the brain is affected even if you live you will have affected your psyche. No matter how small the amount of Neuro transmitters that have been affected the brain has now lost those neurons that protected us against certain chemicals, and the more neurons missing the more you will have what is called the Multi Chemical Sensitivity Syndrome.

This reminds me of the United States Army in the 1950's, and 60's they were flouting to the Country about the new bomb that could be dropped over a sleeping city, and no one would ever wake up. It would destroy anything living, and leave the city Intact with no or very little structural damage. As a young person I often wondered what in the world this may be, it was called the NEU-TRAN bomb. Much later I had put two and two together since nerve agents at that time was a big secret, One the United States had won from the Germans in 1938 from a scientist who defected to the United States, bringing with him the secrets of G-B nerve agent, Sarin nerve agent, VX- nerve agent, and possibly B-Z nerve agent that caused hallucinations in the victim before death.

The United States had developed the Honest John rockets to deliver the blow, any where in the world. Each would hold up to 360 small nerve agent bombs, each able to wipe out every living thing within 4 square miles, leaving all structures intact. When moving in on the target area the rocket was set to explode above the center of any city, disbursing the highly charged grapefruit sized bomblets loaded with nerve agent in every direction each having it's own highly charged explosive that would vaporize the nerve agent allowing each unit to cover four square miles instead of 1 square mile four hundred percent more of total obliteration of life. The deadly cloud of nerve agent hovering over the entire city. The nerve agent being vaporized became so potent that just one particle, so small you would not be able to see it with the naked eye, it has no odor it has no smell.

The exact reasoning behind the name NEU-TRAN Bomb is very simple when we think of how the brain works. The NEURO:::::::::::TRANSMITTER nerve gas like sarin, GB, Or VX the deadliest of all, act as blockers at the most susceptible part of the body, the brain is the most vulnerable. Where it then blocks the Neuron's:::I:::I::I:ITransmitted and with this you have death within seconds. Even if you are a victim that somehow survives you will always be that victim with Psycho-motor problems for life. This is truly a horrible thought but it's the truth and has been available since the 1938. These are nerve agents, exactly as stated they enter the body by nerves and nerve endings anywhere in the body traveling by nerve it is more profound than anything in the World. Except for Mercury which is a main ingredient for nerve agents and also will enter the body via the nerves, that is why it has been banned for use in anything from the United States.

This was supposedly the most powerful weapon on earth, it was called silent death and it was really scary when you were led to believe something so destructive actually existed. they started talking of entire cities like New York, Chicago, Las Angeles, San-Francisco, Denver, Dallas, Houston, both Minneapolis and St.Paul, Atlanta, Kansas City, one bomb and there gone. Let me tell you this was really scary after having seen what the United States had accomplished in Japan Nagasaki, and Hiroshima. I used to think my God where will this end. The super power of the United States seemed unending, I can remember air planes, the ground would just rumble, all you could see was airplanes in –V- type formations from Horizon, to Horizon.

Then as a child in 1945 my dad had told all us kids to watch our for any balloons going over or landing, and we were watching the sky from morning to night, then we walked the country roads, and out across the pature land looking for balloons. We finally found our first, it was a large balloon with a canister below, we did what the Army had said don't touch it just call them. Wouldn't ya know it seemed as if we had the whole damn army there wanting the balloon, so we grabbed our bikes and rode up the rode with the Army following, this made me feel as if I were going into battle. We got there at the place and they encircled the entire area, and a special team and trucks went out to retrieve it. They claimed it was an experimental balloon and they needed the data untouched they said keep up the good work.

Weeks went by and finally we seen two huge balloons in the sky, one was coming down, but the other was going on moving slow to the north. We rushed in an called the Army, and said quick there we have one down with another that went over us. Airplanes were up within minutes, and here come the army at least 10 to 15 trucks with a lot of soldiers. We had told them where it had gone down, and entire two mile area was cordoned off, it took several hours before they were done with this one. Very few people knew that Japan had actually bombed the United States, Those balloons were carrying bombs from Japan. Then to stop and think for a minute how close they had actually came within just a few miles short of where the Rocky Mountain Arsenal United States Army was making Chemical weapons for the war. We later found a couple more small ones that they explained were collecting scientific data. The Army recovered the 2nd balloon that had went over, and it was confirmed from Japan carrying bombs. For many years this was the most exiting time of my life they treated me like a hero. I never once thought that one day the United States Army would be causing terrorism at home while protecting Shell Oil Company. It would seem they were competing in who would be able to create the most and the deadliest pollution. They were both told the 250 million gallon waste storage basin was leaking and if they continued its use, it would be on the premise it was leaking, and this continued for an additional 25 years.

Hanford emissions dissolve family's hopes; radio-nuclides that were released from the bomb factory would forever cling. Thinking back she often wondered "why the pilots that flew overhead dragging meters through the air would never wave back" to any of the kids on the ground. She knew intuitively over the years, she now knows it as fact. With tears rolling down her cheeks, she was trembling as she pulled out a blue pamphlet. Her, her husband and three of her five children all suffer from weakened immune systems and are on regular dosage of thyroid medication. This blue pamphlet is a 1976 report that was presented to the United Nations by Japanese survivors of the atomic bomb. She read "radiation-induced atrophy which leads to immune system breakdown and thryroid dysfunction.

Betty Perkes standing here with tears running down her cheeks, as she considers how her children's futures were stolen from them. "The Government lied from the beginning; They use us for guinea pigs." She glanced for a moment at her oldest daughters house thinking "She is such a worthwhile person. She was studying to be an educator and a counselor. Now society has lost the opportunity to have a real useful person who could really give, and it breaks my heart that she will never be able to accomplish all the things that she has wanted to do. It has literally ruined her life and the lives of my other girls."

The Government scientists insist that only those who received large doses of radiation over a short period of time are at high risk. But an independent scientist, believes that there are lethal effects from low dose or low level radiation exposures. The United States of America, even after the acute suffering of their family their government knowingly deceived them. It would not have been so bad but they didn't even warn us and now they refuse to help us. "They only ask questions such as do you drink your own milk, do you grow your own gardens, do you can your own fruit, they ask what our health was like but done nothing. They took their statistics back knowing about our health problems they done nothing nor said anything." HELP NEVER CAME.

Ohio residents that were injured by contamination from the Fernald nuclear weapons Plant. The Department of energy made arrangements to pay at least $73 million. The group had originally sought $300 million, the lesser amount was approved by Energy Secretary James Watkins, and approved by the U.S. District Court. It was surely known that radioactive contamination polluted their soil, air and water, causing them mental stress and reduced the value of their property, that will be forever polluted. The Energy Secretary blamed it all on the disarray in the weapons program, saying that key managers and supervisors lacked technical skills and that some lacked the nessary discipline for safe operation of nuclear reactors, or nuclear weapons plants.

In a complaint filed in District Court, health officials said improper disposal and handling of ZPP had contributed directly to three accidents at the plant in March and June of 1990. Arapahoe County employees still show up for work daily but OEA INC. has been virtually shut down and State Health Officials last week ordered the plant to stoop generating hazardous waste. This order was probably the most sweeping of its kind ever issued by the State Department of Health.

We as people have put in place the watchdogs that had been missing for many years, they either turned their head to wrongdoing, or we had those taking pay under the table allowing such toxic pollution to our people go rampit. The State Department of Health state “We fully understand the significance of what we have done,” Fred Dowsett of the Health Department said, “But the matter is so serious that we felt this rather drastic action was necessary.” Complying with the order, said OEA president Charles Kafadar, has stalled most manufacturing activities. One of the questions involved whether the hazardous waste generated at OEA are explosive.

The hazardous wastes generated by OEA, include zirconium, potassium, perchlorate (ZPP), making triggers for the vehicle air bags, ejection systems for aircraft and launching mechanisms for satellites. What was wanted from OEA was to stop polluting, and to draft an emergency-response plan that would provide approved safety training to all employees. To stop generating or producing explosive hazardous waste.

State Health Department Officials conducted a surprise inspection at OEA learning they had resumed production without approval from the State. OEA is interested in resuming production while everyone works on a solution to iron out the differences with the Health

Department. The Denver based company that has several other subsidiaries stretching out across the nation, had reported record earnings of $2.5 million for the last quarter, They had believed they was in complete compliance at the time.

It's good to see we have individual State Health Departments that have put their teeth in and are standing up to wrong doing by Individual Companies, Corporations, and our government across this nation for the protection of the people. The United States Environmental Protection Agency are now coming under the control of the State Health Departments. This is something that has been missing for years.

The DOE, Department of Energy is trying to solve the problem with nuclear waste, and where to store it. It's waste that no one wants and by storing it every where, with everyone. The national nuclear-waste policy is in rough water, and there desire to spread it through out the West raher than maintaining at one or two approved locations. There has always been at least three obvious candidates for temporary storage of low-level radioactive waste from Rocky Flats: Rocky Flats itself; The Idaho National Engineering Laboratory; then there is the one near Carlsbad New Mexico. The Waste Isolation Pilot Plant known as (WIPP) has been scheduled several times. However those options have been eliminated by a strong minded governor. Then Idaho's Governor has closed his state to any additional waste shipments. Then Colorado Roy Romer has threatened to shut down Rocky Flats if a new storage site is not found. DOE says alright if no one wants the stuff, we'll just spread it around then everyone can share. "This is their stupidity having a Governmental agency talking such trash as to how they are going to manage their busines. There is government Land in every state, there are military basis in every state, Your back yard is the back yard to something or possibly the school yards. The people of DOE have always been so damned irresponsible and that is why they find themselves in such a dilema now.

The pressure is building on WIPP, DOE accusing Congress of foot dragging, and the Governor of medling. Even under the most optimistic time table WIPP won't greatly relieve the build up of waste over the past several years. DOE says the governors can retreat "on tip-toe", of course from their it's my-way or the highway rhetoric. Frankly when Colorado drew the line in the sand earlier this year against additional storage at the Flats. The State took greater control over the plant as well as more respect and cooperation from the Department of Energy. Thanks to their bull-dog persistence progress has been made on all fronts. DOE says the real issue is we need storage for radioactive waste, and a decision must be made whether its genuinely wise to divvy up radioactive waste like so many barrels of beer.

Environmentalists convinced the U.S. Department of Energy to make a radioactive waste disposal site safe before shipping thousands of truck-loads of nuclear waste along Interstate 25. Through Colorado's populated areas. The Department of Energy hopes to soon open the disposal site known as WIPP the facility near Carlsbad, N.M., where they would bury radioactive waste ½ mile beneath the desert in salt deposits. While the agency is trying to bypass federal regulations that are designed to protect public health and the environment. Most of that waste will come from Rocky Flats Nuclear Weapons Plant in Jefferson County, CO. DOE officials claim the necessity of placing some waste at the facility in an experimental phase to validate all information on leaks, emissions, or any other potential problems. The research center and other environmental groups claim the agency hasn't even done the computer modeling that could predict any problems before hand. Larry says, "that would have been common 'sense" "Even if they acted like they Had Some 'sense it may be better that no sense at all. The Doe will probably not stick to the requirements, "The right approach would be to establish compliance with the federal environmental regulations up front DOE is trying to bypass those regulations."

The Center and the Environmental Defense Fund in Boulder have indicated they would sue DOE if environmental regulations are not met. Standards set by the U.S. Environmental Protection Agency "are the only means of assissing the repository's safety," We're asking that DOE show compliance before they put any nuclear waste under ground. DOE has come under fire and intense criticism the last few years regarding environmental and safety problems at all the forty eight nuclear facilities around the country including Rocky Flats. We've seen the results of the complete non compliance at the older DOE sites where they gave short shrift for all the destruction of the environmental and the safety disasters caused to our citizens, their children and their children................................because of the

Ultra Toxic, nuclear waste and wide scale pollution and contamination.

GENETIC MUTATIONS APPEAR IN ANIMALS AT URANIUM PLANT

CINCINNATI……. Genetic deforamties have shown up in tissue samples of small animals living around six waste pits and an incerneator at the Fernald uranium processing plant near here, a federal environmental official says. Additional biological and water tests that were ordered by the U.S. Environmental Protection Agency, were conducted at the site for a determination on whether toxic waste had in fact contaminated the water or soil and to find out if there is any reason to believe it could be causing animal mutations, and putting residents and plant workers at risk.

There were abnormalities shown in some animals tested. But the cause is unknown at this time. They did find changes in normal gene patterns in the animals. So far the changes that would affect animal life is in the area of reproduction. There is total uncertainty as the how this would affect humans, but it definitely raises concerns on our part, and has indicated further investigation is needed. The genetic mutations were found in animals, such as toads living near the plant's incinerator and waste pits.

Another positive sign something is wrong, when you see a complete change in certain bird population as it decreased any where near the facility. The six shallow waste pits contain about 475,000 tons of low-level radioactive uranium. Tests was conducted at the waste pit sites nearly three years ago by an environmental research company found that they present a great potential for ground water contamination. These biological tests were conducted by Miami University Zoology professors for Westinghouse.

SHOWERS HAVE BEEN LINKED TO LEGIONNAIRS' DISEASE

Legionnairs' traced to bacteria, studying an outbreak of Legionnairs' disease in a hospital have identified the showers as the missing link for transmitting the bacteria in these cases that causes Legionnairs' disease. Certain amoebas in any one water supply needs to be watched carefully, their presence indicates an abundance of Legionnairs' bacteria in the shower heads. It has been linked with certain amoebas found in the water supply.

These studies were conducted at the Federal Centers for Disease Control, and the Department of public health. Legionnaires' disease is a form of pneumonia, and is caused by strains of the legion Ella bacteria which is frequently found in water in small amounts without a known disease association. Research suggests the disease follows the inhalation of heavily contaminated water droplets such as those in a shower. With inhalation of heavily contaminated water droplets, being transmitted through the use of showers is only suspected by Scientists, but most of the outbreaks have been traced to air conditioning systems in large buildings. The disease is named after the first known outbreak in the United States at Philadelphia hotel in 1976, the death toll was 29 American Legion Members attending a convenction.

In the hospital outbreak, their were 26 cases of Legionnaires' disease developed over a four year period of time. Ten of the patients died. Showering was the only exposure risk that was identified in all cases, but most patients had underlying health risks that made them more susceptible, like smoking, steroid use or chronic lung disease.

The Scientists isolated Legionella Pneumophila bacteria from the shower heads in each of the patients rooms. The hospital eradicated the bacteria by heating the water and treating it with chlorine. It was claimed by an epidemiologist it would be an overreaction for patients to be wary of the showering. It is said researchers need to identify other risk factors, and explain the new founded link between amoebas and the bacteria.

"LOVE CANAL"

A grim reminder of after living there for nearly ten years, the government said the site poses a significant risk to human health, and is heavily polluted with carcinogenic chemical wastes. It is offering to pay and relocate over 50 families, many are wanting the government to pay for permanent moves. It echoed "If its that bad they have to temporarily relocate us why don't they just buy us out. A spokesman of the Untied States Environmental Protection Agency said the contamination is so pervasive every resident eventually will have to leave for good so remedial action could be performed....The landfill this trailer park was constructed one was located just north of a landfill operated by CECOS International Inc......A hazardous wastes disposal company. For a family to think the kids played here there were barbeques, family get togethers, and now we hear the soil and water is contaminated with Industrial compounds, such things as Anilin, Phenothiazine, and Benzothiazole all of which are very toxic, and "Carcinogenic", by industrial by products of chemical production. Long time residents knew this area was once used to bury chemical wastes. This all takes time these people don't really have and the government has no emergency plan for something like this. Before residents can be relocated, the site must be placed on the EPA's Superfund National Priorities List. EPA will ask the government agency for Toxic Aubstances and Disease registry to issue a public health advisory, is just the first step toward inclusion on the list. The agency released a preliminary assessment that found "Significant health risks in the park. Years prior to this just a few miles from Love Canal Forest Glen neighborhood was declared a national environmental disaster in 1978. State officials originally evacuated 237 families where homes were built adjacent to Love Canal, then the State later bought out homes in a 10 block area around the former Occidental Chemical Corporations Dump. Residents were very upset, scared and confused when they were only offered temporary relocation.

Love Canal at the time was the country's worst and most incredible toxic waste site. It was the worst only for a short period of time, then there were others that were even worse. It got the grassroot activists pushing government officials, politicians, and scientits stirred to find out what went wrong. As for the Industrial Leaders standing alone trying to take no blame for what happened but. "They knew right from wrong and chose to do it wrong" "All for the almighty dollar". Now there are parents and children that will pay with their lives and the lives in future history, because these chemicals scar the Chromosomal, DNA and RNA that will be past from generation to Generation into eternity. This is the part none of them will discuss, very few Dr's will discuss for instance a family with all five children suffering from learning disabilities, or why does heart disease or cancer run in the family history. Something in life could have affected the DNA they are discovering that chemicals and biological warfare or germ warfare practice have played the larger part in the original event or injury in a persons life.

The model city that never happened: Occidental purchased the site of Love Canal in 1942. For the purpose of disposing of toxic waste. 22,000 tons of mixed chemicals were dumped into Love Canal. Shortly thereafter Hooker Chemicals and Plastics Corporation, which is now Occidental Chemical, sold this property to the Niagara Falls School Board for a price of $1.00 with out telling any one what had been dumped on that property. An elementary school was built on that property and for over 20 years children attended that school, hundreds of families moved in took up residence. It was the unusually heavy rain and snow falls in 1976, and 1977, causing a high water table 55-gallon drums popped up to the surface, ponds and any other surface water showed contamination, cinderblock walls in basements started to ooze an oily residual noxious odors filled the air permiating everything.

The Agency for Toxic Substances and Disease Registry Would reveal a long list over 418 chemical records for air, water, and soil samples in and around the Love Canal area. The New York Department of Public Health ordered tha 99thst Elementary school closed evacuation of pregnant women and children. By the time Ring#1 was evacuated, an immediate evacuation of Ring #2 was ordered, it soon grew to the entire Love Canal area. The main concern was the extent of chemical contamination, and the immediate and future impact of the chemicals on the Love Canal residents. There was a total lack of readily available information to explain any scientific, or health impacts.

A CORPORATE, GOVERNMENTAL, AND FATAL ERROR

The blame has been placed on the local government, who despite continued warnings from Hooker Corp. The Local government went against all warnings trying to profit from the contaminated Land. The government knew of the chemicals but chose to rupture the clay containers which held the Toxic Waste. Not only for the sake of fill dirt, but to put sewer lines in place. Hooker Corp. When realizing it was going to be sold for development, they asked for attention to be brought to the site The governments irresponsible handing of the waste land, and their concern of money over the concern for the residents of Love Canal is exactly what caused the ultimate Love Canal Disaster.

Hooker stated they took every precaution, even compared to the standards of today in digging into impermeable clay soil adding the chemicals in this clay tomb, then placed a clay cover for that tomb. The government claims Hooker had to know when they sold the land to them, that it would be developed. "It sounds as if the government to which the Land had been given to for $1.00 failed to hear the warnings of Hooker Corp.

As the Author of this book thinks back at what happened at this place in time at the Rocky Mountain Arsenal, and Rocky Flats near Denver Colorado, The Sheer Terror he and his family felt at the time, and then the Terror he had to face for a life time all because he became what is known as a "WHISTLE-BLOWER". Larry finally took the IRS to court charging them with Harassment, Fraud, and Deceit, for over the past 16 years, It was a myriad of harassment every year since his bill had cleared the full House of Representatives in 1988. Within four months time he was called in for the first of many audits, It happened within 60 days of his first deposition by the Justice Department. The end results will be coming up in chapter 20, He is still waiting the decision of the trial that ended over twelve weeks ago in the Federal Courts.

WILL IT EVER BE CLEAN

This I s a question we all ask, as for the Rocky Mountain Arsenal, they started on a program that was going to be done within four years, then it became ten years, ten years became twenty years, now twenty years has become possibly by the year 2060. The last I heard it can never be cleaned, 18,000 acres of land sitting just north of the city of Denver, where once they had an eye on for the new airport. The government is trying very hard to at least clean the surface so they can call ("The most polluted piece of land on earth") Wild Life refuge, and the Bald Eagle Reserve. There still having serious problems keeping the animals looking alive or staying alive, they are on the watch daily of anything becoming sickly, They will quickly get rid of them or any dead animals, the Refuge so they call it, if one looks very closely you will see an eight foot fence circling it, and they have led people to Believe the animals passing through, most if not all these animals are trapped and brought into this refuge. Just like the pairie dogs the army watches them closely because they are the food for the eagles that fly over. The army had to cyanide all the prairie dogs on the arsenal because the hawks were eating them, and because they were so contaminated it killed the hawks. The the Army hired the Airforce at Buckly air field to trap the dogs out there where they were not contaminated and move them over to the Arsenal, as food for the Eagles.

As for Rocky Flats near Denver Colorado, again they are doing a clean up job at the same time they are manufacturing Plutonium. They have given up on all the plutonium that was in the air vents, They weren't even missing any Plutonium until they discovered in in the air vents, then they suddenly remember "OH" that's where it went. They think of it as trivial. "Just enough Plutonium to make at least six atomic bombs is all. Larry says, "If there is even a place left when they're done there are places that can never be cleaned, toxically polluted forever."

As for Love Canal, at least it woke up America, They have figured the decomposition rate for the chemicals at a minimum of 20,000 years. Genetic mutations will survive from now to eternity. The legend and stigma will live in our minds, and will go down in history as THE LOVE CANAL TOXIC WASTE DUMP SITE THAT OPENED THE EYES OF AMERICA. ARE THE ROCKY MOUNTAIN ARSENAL, ROCKY FLATS, AND LOVE CANAL.

CHEMICAL TIME BOMBS

Yes they are time bombs, and only a few around the Country, The life time of these materials are 10's of thousands of years. Especially when these materials are placed in tombs I believe its only a matter of time with the weather changes, records being lost, or in some cases there was no records kept at all. They hope these material encased in tombs, will over time bio-degrade, or de-compose and that is as far from the truth as one would want it to be, it will only preserve the life times. Larry says this reminded him of the Hill Billie's "When Pa says to Ma look at this just oozing out of the ground.", and they thought it was oil.

THE MAN MADE ENVIRONMENTAL TIME BOMBS

Scientific findings have now strengthened a controversial theory labeling Alsheimer's disease and other brain disorders affecting millions of Americans may well be caused by the environmental toxins, that humans have been forced to breath from the time they were a child and until the day they die. These toxins slowly destroy the nerve cells. Two of the Worlds worst have been banned from the United States, They are Mercury, and Lead. These materials because of their toxicity have been used in chemical warfare agents, as well as the Paint that everything was painted with. These materials continue to leach out deadly toxins many years after they have been used. The damage these two items alone have done to us, our children, and our childrens children thru eternity. Mercury is so toxic it uses the nerve route to the brain, which gives it immediate access to, and immediately begins to destroy the Neuro-transmitters in the Central Nervous System, and can within minutes of showing the signs and symptoms will have caused traumatic injury to the brain. The liver, Heart, Kidneys, Pancreas are the next immediate targets.

It has taken more than 25 years of research on the extraordinarily high rate of brain disease among the Chamorro people on the island of Guam in the Marianas chain of Micronesia. Parkinson's, Alzheimer's much like de-mentia and amyotrophic lateral sclerosis were at least 50 to 100 times higher than in the United States. Researchers from National Institutes of Health seem to be implicating genetic factors and so-called slow viruses, that will produce a disease many years or decades after the initial exposure to the toxins in our environment. Now an international group of scientists have established these diseases are caused by a "SLOW TOXIN". A toxic chemical that similarly produces disease years after ingestion, breathing, or contact with the skin.

Dr. Peter Spencer has now predicted the findings will lead to the discovery of a whole new class of environmental chemicals that do trigger the death of brain cells. Spencer, who is with the Institute of Neurotoxicology at Albert Einstein College of Medicaine in New York, headed the study group. Other participants are from Britain, West Germany and France. Results of the study are reported in Science, the journal of the Americans Association for the Advancement of Science.

Proponents of the slow –toxin theory of brain diseases believe that initial exposure to such substances causes damage to specific regions of the central nervous system. The damage remains hidden producing no obvious symptoms, for decades. During youth and middle age, the brain is able to compensate for the damage. But the slow toxin amplifies the effects of the gradual loss of brain cells that occurs with advancing age. The person exposed to the environmental toxin thus becomes especially vulnerable to the memory loss of Alzheimer's disease, de-mentia, for example, or the tremor and muscular rigidity of Parkinson's disease. "The Envrionmental Toxins" is the key to many brain diseases.

ARE ENVIRONMENTAL TOXINS THE CAUSE OF ILLNESSES AT THE CENTER OF THE STORM OF BRAIN DISEASE CAUSING PSYCHIATRIC ILLNESSES.

Those living in subsidized housing cannot be forced to have bugs in their homes exterminated with chemicals. The U. S. Department of Housing and Urban Development has ruled that because of people with environmental illnesses. While there are doubters who try blaming employees who have suffered environmental injuries on the job, calling those suffering malingering as the reason without looking at what may have caused the environmental toxins in their own workplace. Subconcious fears of workplace pressures or down right scare tactics, or threats to their families or their jobs and the frightening truth of the media reports about chemicals, biological chemicals, Anthrax, unsafe chemicals used at the work place, about employer induced environmental toxins, and if you are a whistle blower to the truth you will be living a threat for a life time from insurance companies, disgruntled employers.

Clinical ecologists, doctors who specialize in chemical sensitivity, meet the causal acts, with the enthusiasm necessary to find out, what has created the real problems. They have said the psychological explanation ignores the stunning similarity of the triggering exposures and the symptoms that have been reported and ignored. The question that concerns them mostly is when the sufferers are all coming up with the same symptoms? They are all so diverse yet they are coming up with the same thing, and that is chemical sensitivity. Whether it be at home, at your work place, or even out in general public Chemical Sensitivity is very real and to the sufferers it gets worse. It has been agreed by the Massachusetts Institute of Technology that more research is needed.

There have been dozens of proposals to both Federal and State governments, an almost overwhelming desire to set up an "environmental unit". This would be a contaminant-free chamber in which subjects would be placed and observed as chemicals were introduced. The scientists at this point have found four groups of people most prone to chemical sensitivity. Groups come from 1. Industrial workers, 2. Occupants of well insulated, and poorly ventilated offices, 3. Residents were communities are contaminated, 4. People exposed to or overexposed to household, or workplace air that is or has been contaminated.

It is very widespread in nature and is not limited to what some would describe as malingering workers, hysterical housewives, or workers experiencing mass Psychogenic illnesses. In addition to that and must be considered with Chemical Sensitivity, a person will suffer chronic fatigue, and appear to be stunned suffering total dementia, with difficulty breathing, the report says.

For those who feel and believe chemical sensitivity is a physical illness say the skeptics fear an ominous message: If people are getting sick from the chemical-laden world, we all know we live in, it bodes ill for Industry and huge Corporations everywhere. Insurance companies, and large corporate business have fought long and hard, spending billions of $ dollars trying to prove this world is full of psychogenic people, as of late it has not happened to many of them and their families, and they have become stronger supporters for better education in the use of chemicals, and use it as directed. A lot of injury could be avoided if only some of the users followed common sense and used directions, and if something went wrong the directions are their for a reason and that is safety of all. Don't ignore them the way my employer did. <u>There are Material Safety Data sheets available on all hazardous and toxic materials that by law are supposed to be posted for all employees in case something goes wrong or maybe directions for its use are not followed. What makes one ill may not effect others, or vice versa.</u>

CHAPTER 18

THIS IS A STORY ON HOW ONE MAN TRIED HIS BEST TO BE HONEST FOR HIS SAFETY AND THE SAFETY OF TWENTY FOUR OTHER EMPLOYEES, AND AT LEAST TWELVE OTHERS.

"ONE MAN WITH COURAGE IS A MAJORITY.

ANDREW JACKSON

COURAGE:

Courage is a special kind of knowledge; the knowledge of how to fear what ought to be feared, and how not to fear what ought not to be feared. From this knowledge comes an inner strength that subconsciously inspires us to push on in the face of great difficulty. What can seem impossible is often possible, with courage

This is one man's story of how he dealt with his employer who was knowingly using toxic, hazardous, and flammable materials in the work place. What they done was no accident it was done with purpose in mind. Calling it safe even though the employee's were being made very ill. They used unlabeled containers, no Material Safety Data Sheets that were required by law to be posted in the work areas, and it took over two weeks to get OSHA to investigate, and then OSHA issued Felony Citations to the employer, fined them nearly $2,000.00 and placed them on probation to the State of Minnesota.

I chose the path of courage of how not to fear what ought not to be feared. Oh how it inspired me to push on even in the face of great difficulty. What seemed impossible was possible with courage, but it cost me my way of life, along with many others. In 1985 the Washington County Minnesota Sheriff Department James Trudeau and the Board of County Commissioners as well continued to allow the facilities Department to intoxicate employees Using paint that was extremely toxic in fact required those using it to wear oxygenated air line respirators, and to have lead levels checked at least once a week. This was an industrial Bridge paint, one that had been banned by EPA in the United States in 1967 to paint the interior of the work area, with nearly no ventilation at all up to 12 hours per day for 21 days before

OSHA finally stopped it after getting the serial numbers from the cans of paint being used. They ordered Washington County to stop immediately warning them this was a bridge paint and will kill if used in an interior. They stopped, nothing else was ever done to inquire what effects this really had on employees. There were many seriously injured, scarred for life. Larry wants to relate his experience with this, it caused Severe Hypertension and identified from lead poisoning nearly a year after it happened. Larry now takes Atenolol for this, it relates to high blood pressure, that is an amount 800% higher than that which would be prescribed for a person who just had quadruple by pass surgery on a temporary basis. Larry must take this amount because of the lead poisoning on a permanent life time basis. Then he also suffers terrible joint, and muscle aches for which he is prescribed Motrin, and extra strength Tylenol. It took nearly four years to finally diagnose the reasons for all the horrific pain one Doctor's called psychogenic, this is something they tend to use when they are confused, but finally Larry was allowed to see specialists and discovered his pain was very real, Arthralgia, and Myalgia severe joint and muscle pain, that could have been caused by lead poisoning. Larry's lead level two months after the ordeal was still 400% above that of normal.

The employer assured everyone this would never again happen. But it did in 1990 they once again painting the work area and jail with a oil based paint making everyone very ill. The paint being used was Mautz-Line 89-00 Industrial enamel, Alkyd, Alkyd, oil based epoxy paint. This paint had also been banned for use in the United States because of it's toxic effects and long term illnesses. "WHAT THEY DONE WAS NO ACCIDENT, IT WAS INTENTIONAL." They kept calling it Latex Paint and continued its use, ten hours a day for thirty days. The Paint data sheets that were supposed to be posted were never posted. Larry took a label off one of the cans of paint and took it up to the County Health Department explaining this is what's being used. They said it can't be we have been given data sheets for latex paint not alkyd oil based paints. The Health Department then called Ed Kapler of Facilities Department, ordering him to immediately bring to them the Material Safety Data Sheets for the Paint being used. This is over two weeks after employees began complaining. The Data Sheets had been locked in a safe in Ed Kaplers office. Larry had been telling OSHA what was happening but they done nothing, until after he came up with the Safety Data Sheet. Then they had sent out an investigator who did not know what was happening or what they were using. They continued to paint not with the paint they showed OSHA they were using, by the time they got back to it the painting was over. But they did issue Felony Citations to the County and the Painters alike and to Captain Richard Becker who refused to listen to the employees, and spent very little time at his office during the paint incident. Letters were written to Sheriff James Trudeau by Doctor's, and employees alike stating the illnesses and they felt it was killing them. But Sheriff Trudeau failed to intervene. Don McGlothlin the office manager would not allow Larry to talk to any one in the front office as to how sick they were or how they felt, He was another who spent very little time at the office. Nearly Twenty five individuals again signed a legal document as to the overexposure to the paint, many will testify. It was talked about by the Investigators, and Deputies said Sheriff Trudeau has finally solved the over crowding of his jail, "He created his very own gas chamber, only one problem he is gassing the Jail Staff along with the inmates."

Larry got OSHA, NIOSH, and EPA documents that clearly showed the employees had breathed enough of one ingredient called Xylene to have killed them. They were showing every sign and symptom up to death. Then the paint also contained Mercury the one item Mautz failed to answer as to why they did not have it listed on the cans of paint as an ingredient. Mercury was banned in the United States altogether, and directive stating nothing painted with mercury can be used in an interior. Again the employees felt every sign and symptom up to death, per OSHA, NIOSH, and EPA.

Mercury was not discovered right a way, until one day about ten months after the incident occurred, one of Larry's Neurologist's states you have every sign and symptom of mercury poisoning, and that is one thing we have not yet tested for. Mercury tests were done right away, and the blood was found about 500% greater than normal, and well above safe standards for mercury ten months after the incident. How do DEATH FACTORY'S fit in here? The Death Factory Mautz Paint Company of Wisconsin failed to list as one of the paints ingredients on the cans nor did they give any Material Safety Data regarding Mercury, only after the Doctor had decided to have the mercury level of the blood checked did we know that mercury was a key ingredient in the Mautz Line 89-00 Alkyd Alkyd epoxy paint. For every Miligram of paint used was 170 times the safe limits for mercury. Per OSHA, EPA, NIOSH, for every Miligram used would increase each other milligram by 170 thousand times the safe limit for Mercury. This I was overexposed to on my job, it was no accident, and the fact the Death Factory Mautz Paint Company left mercury out of the ingredients was no accident, it was intentional of my employer, and it was intentional of the Death Factory“ ”Mautz Paint Co.) for using something that had been banned in the United States completely. During the 1990’s Larry had a very difficult time adjusting to what had happened to his life. He had taken this before the County Board asking for help, he was given none. The employer denied him any medical help whatsoever. Larry had to learn to talk all over again as he had suffered Aphasia, a devastating blow in communications, overwhelming difficulties talking, and, understanding, what other’s are saying. He also suffered a severe type of Motion sickness, motion such as laying down could cause him to vomit. He also suffered loss of balance, Ataxia, moving rapidly would cause him to fall over, looking up would cause him to fall over along with many other physical injuries from the toxic and hazardous paint. Mercury was used in Nerve agents because it was so potent, and it’s ability to enter the bodies’ nervous system through the nerves, it is a true nerve agent in itself and why it affects the Central Nervous System so rapidly……It’s ability to block the neurons in the brain from the part of the brain that transmits them. Causing learning disabilities, Short term Memory, and the inability to retrieve that what was learned making it almost impossible to learn any new things. Neuro ::::::/:::::/……/ Transmitter. Larry has done his best in trying to explain what happened to him, and help everyone from any similar pitfalls. Larry’s Work Comp. rating for injury to the Central Nervous System 40%.

For the first five years after this happened Larry was going through many changes in his life and loss of balance when looking up was one. His supervisor called him to the storage room and ask him to remove a box form an upper shelf. Seven feet with finger tips Larry could reach the bottom of the box, he ask how much does this weigh, he didn’t know, Larry tried explaining to him when looking up he loses his balance, the supervisor was insistent on getting the box down and ordered Larry to remove it from the shelf. <u>This was Absolutely intentional, it was no accident.</u> The box out from the shelf on his fingertips while standing on his top toes. The minute full weight hit the fingers went down to full hand and the tip toes to full feet but then Larry lost his balance and fell to the right against another set of shelves, with the full weight of the box on top of him, with assistance from the supervisor we were able to move the box forward to another shelf about four feet, but Land could no longer move, and was in terrible pain. The supervisor left and did not even come back to check on him, nor did he ask if he needed help or medical help.

Larry had got to his feet and moved very slowly trying to make the back work. He was able to get some trustee's out and using the laundry cart had two of them place it on the cart, taking it into a small room he showed the trustees what he needed crossed out on a certain page, and ask them to get a good count of the books. There were approx. 806 hand books, weighing in at 2.25 ounces per book, then Larry had the inmates take the box on the laundry cart to the nurses station where it was weighed on a platform scale 112 lbs. A very large box that was very difficult to handle by one person, so Larry used the two trustee's to carry the box back to storage and placed it on a lower shelf because of the weight. Larry was forced to go in for medical help, cat scan showed a serious herniated disc. Later follow up showed the herniated disc, with more that were bulging and damaged impinging nerve routes to both legs. The back injury was rated at 59.5% Workers comp. Larry was off work for a long period of time.

The New Law Enforcement Center was under construction, Larry was assigned to attend the new Correctional Academy, to run drills, to train, to edit and rewrite the Washington County Sheriff Departments Policy and Procedure. Most of the work being done in the new building that was still under construction. The circulation fans were running without filters, not being encased. As it was becoming more and more difficult to breathe, and coughing getting worse He looked around noticed people wearing respirators working in the tunnels with fiberglass where they were cutting and wrapping all the plumbing. Then thinking back to the desks, and tables and realizing this was all unfiltered and how everything was covered with a fine white dust. We cleaned it off and seen it was the same the following day. Larry turned this into his supervisor's who paid no attention to the fact everyone was coughing, eyes, nose and throat irritated and becoming more and more difficult breathing. "LARRY BEGAN SEEING RIGHT THROUGH THIS AND REALIZED IT WAS

INTENTIONAL AND WAS NOT AN ACCIDENT. The employees were at the hands of Captain Richard Becker the very same person who allowed the toxic Paint Over Exposure in 1990 at the old jail. With in the first three weeks Larry was seen on emergency basis, difficulty breathing, X-ray showed bronchitis with foreign infiltrates in the lungs. Even though he tried telling his employer there was something wrong they done nothing. Larry continued being seen on emergency basis with his condition going from bronchitis, to bronchitis and pneumonia with foreign infiltrates in both lower lobes of the lungs. Larry had ask they have the health department check the air, they refused. Larry had samples of the air taken and sealed in individual envelopes, and placed in a safe for future investigation. It went on until January 1993 when he was taken to the hospital by ambulance unable to breath.

The Doctors told him he should call the Health Department about the Construction dust as it was creating the problem. You now have Bronchitis, Pneumonia, Infiltrates in both lower lobes of the lungs and you now have asthma. Larry was reassigned out of the new jail back to the old jail shortly after being called on the carpet for calling the Health Department, and not going through them. Larry explained to them he had tried letting them know there was a problem and had reported it to them 24 times before calling the Health Department. The three Captains that did not listen to the warnings were Captain Richard Becker, Captain Donald MCGlothlin, and Captain Michael Johnson, Johnson was the safety manager, McGlothlin was Administrative Supervisor, and Becker was the Jail Administrator. They explained to Larry that they actually ran the Health Department out before they could do any testing, and if this ever happened again he would be fired. The only testing the Health Department managed to complete was for Carbon Dioxide, before they were ordered out without completing the testing on the dust. Larry knew there was reasoning behind the dust samples of the air he had taken. Larry soon learned the air was so dirty that the employees were breathing 12 to 15% fiberglass daily. Back now in the old jail Larry began to re-coop and he cleared up considerably with in days, but then the progressive Occupational asthma was getting worse by 1996 it was rated at 30 and 40% by employers doctors. As it got worse Larry contacted Legend Laboratory, who specialize in sick building air, such as asbestos. Their Microscopist examined, took photo's and made slides so it could be seen on a screen. Explaining this is devastating for every breath you were taking it was 12 to 15% fiber glass. He also explained fiber glass is not the same as asbestos, and probably even more dangerous and in many ways can be worse. The glass keeps moving around inside the lungs acting much like tiny knives cutting up the lungs into pieces, and leaving them open for secondary infections like Bronchitis, and Pneumonia, the only way your going to get rid of it is to cough it up and spit it out. Larry will list the signs and symptoms that were reported almost daily, and went unheard.

- PERSISTENT, DRY, HACKING BARKING COUGH.
- SORE THROAT, BLOOD TASTE IN THE THROAT OF BLOOD IN SPUTUM.
- BLOODY NOSE.
- PERSISTENT AND SEVERE SINUSITUS AND RHINITIS.
- PERSISTENT AND SEVERE RESPIRATORY INFECTIONS.
- HEADACHES, NAUSEA, DIZZINESS, INSOMNIA, IRRITABILITY DEPRESSION.
- ASTHMA LIKE BREATHING ATTACKS, CONSTANT WHEEZING.
- A REACTIVE AIRWAY DISEASE
- SWOLLEN, RED, WATERY INFECTED EYES.
- UNABLE TO WEAR CONTACTS.
- SKIN BUMPS, AND INFECTIONS.
- SHOWING AN EXTREME SENSITIVITY TO EVERYDAY AMOUNTS OF

- AMBIENT AIR POLLUTANTS, ESPECIALLY CIGARETTE SMOKE,
- CAR EXHAUST, PERFUMES, AND COLOGNS, CLEANING PRODUTS, PAINTS, OR
- VARNISHES, FRESHLY MADE FURNITURE, NEW CARPETING,
- CONSTRUCTION DUST.
- THEN I LEARNED THAT THE DEPARTMENT OF LABOR, AND INDUSTRY THE
- ENVIRONMENTAL PROTECTION AGENCY, NIOSH, OSHA, NATIONAL
- INSTITUTE FOR OCCUPATIONAL SAFETY AND HEALTH “”ALL WARN””
- FIBER GLASS MUTILATES DNA.
- ACCELERATED CELL GROWTH LIKELY PRECURSOR TO CANCER.
- MORPHOLOGICAL CHANGES COULD BE RELATED TO EXPOSURES TO
- FIBERGLASS.
- TESTS DONE INDICATE THAT GLASS FIBERS ARE CAPABLE OF
- TRANSFORMING MAMMALIAN (BALBL/c-3T3 CELLS IN VITRO AS A

- FUNCTION OF THEIR PHYSICAL PROPERTIES AND THAT GLASS-FIBER –

- INTRODUCED TRANSORMED CELLS POSSESS NEOPLASTIC

- CHARACTERISTICS.

- FIBERGLASS WILL CAUSE FRAGMENTATION OF THE CHROMOZOMES.

- IT ALLOWS DNA TO PASS ON INCORRECT GENETIC INFORMATION.

BOTH MANVILLE AND OWENS CORNING MICROFIBERS (AVERAGE .2 AND .18 MICRON RESPECTIVELLY CAUSED MICRONUCLEATED AND MULTINUCLEATED CELLS, THE CELLS GIVEN HIGHER DOSES-EXHIBITED RATES OF CHROMOSOMAL DAMAGE AS MUCH AS FIVE TIMES HIGHER THAN THOSE IN THE CONTROL GROUP.

Some people will develop severe intolerance to chemicals which they previously tolerated well. When people develop a rash upon exposure to a chemical, we usually say they have an allergy. When people develop wheezing upon exposures to a chemical, we usually say they have a form of asthma or allergy to that chemical When people develop headache or fatigue or confusion or other debilitating symptoms on exposure to a mixture of chemicals we say they have multiple chemical sensitivity syndrome. Surveys suggest that between 4 to 60 percent of the USA population has MCSS. About half the people with MCSS think they developed this condition after exposure to fumes from renovations at home or in the office. Some of the others developed MCSS after an accidental exposure to a high dose of pesticide. We suspect the remainder developed MCSS just from too much encounter with SBS.

Multi Chemical Sensitivity Syndrome, MCSS, tends to start with a mild exposure or reaction to a few chemicals. Then it develops into a stronger reactionary, and spreads in the sense that the more and more chemical mixtures will cause the symptoms to be more severe and widespread. (There are very few people who are exposed to pure chemical mixtures whereas almost all of us encounter mixtures of airborne chemicals all day every day.

ASTHMA IS A WORLD WIDE EPIDEMIC

Especially childhood asthma, about 15% of school age children are now considered to be asthmatic! There are known links between asthma and cockroaches, dust mites, and molds. Anderson Laboratories, say they cannot see how any combination of these factors can explain the current epidemic. They say that the rising levels of indoor air pollutants is probably going to be the explanation for the rising rates of childhood asthma. Various surveys show links between asthma and the use of wall to wall carpets, recent indoor painting, and the concentration of certain indoor air pollutants. Our Laboratory results show that the mixtures of chemicals released by carpets, air fresheners, fragrance products, mattress covers disposable diapers, marking pens, and certain toys can cause.

HEALTH EFFECTS OF INDOOR AIR POLLUTANTS, SUCH AS "SBS" SICK BUILDING SYNDORME. MCSS—MULTI-CHEMICAL-SENSITIVITY- SYNDROME, AND ASTHMA.

Many people are being poisoned or over exposed to toxic chemicals in the very air they breath at home, at work, or even at the zoo. There are many symptoms which can be caused by the toxic effects of indoor air pollutants: I will list just a few that I know you have all felt. (1) headaches, (2) dizziness, (3) difficulty with concentration, (4) loss of memory, (5) confusion, (6) severe fatigue, (7) burning eyes, (8) burning nose, (9) sinus irritation and infections, (10) soar throat, (11) hoarseness, (12) difficulty breathing, (13) wheezing, (14) cough, (15) palpitations, (16) nausea, (17) vomiting, (18) diarrhea, (19) weakness, (20) paralysis, (21) numbness, (22) twitching, (23) disruption of the menstrual cycle.

We are all so different in our sensitivity to sugar, salt, spices, medicines, and to air pollutant chemicals. "no one really knows why, we are all so different," but we are. Not everyone entering a polluted space and breathing that air will feel sick. Yet others entering that same space may have a severe reaction to fatigue, confusion or difficulty breathing, etc.

EXACTLY WHAT IS SBS "SICK BUILDING SYNDROME"

According to World Health Organizations, about one-third of all new buildings, world wide are not fit for human use, due to severe indoor air pollution. Your question is, does this include our homes? Yes it does, indoor air pollution arises from release of toxic chemicals from carpets, wall coverings, ceiling tiles, cleaning agents, copy machines, computers, marking pens, books, paints adhesives, mold growths at water damaged sites, and personal care products such as perfumes and after shave.

There may be "hot spots" within a building, one room you cannot go into because of its pollution. This is more likely in a home, where forced air ventilation is unusual. Schools and offices usually have circulating air systems, and the air pollutants move around, so all rooms become troublesome to some extent. People will tolerate problems less when it's hot, humid, or noisy, but don't get confused into thinking that these conditions cause SBS they only make us less tolerant of any particular level of indoor air pollution. Anderson Laboratories, "believe that SBS is primarily caused by the combined action of the many chemicals released into the air by a multitude of products currently found in homes, schools and offices. We have to remember it's not just one chemical, what about two chemicals maybe mixed will become very toxic. I've given you a list above of chemicals to start with. I have worked with my family in making our home almost chemical free, and feel much better about it.

Larry was given some information on fiberglass and the dangers, This was taken to the University of Minnesota to be examined by a Toxicologist who assisted in figuring out a method of therapy that would help rid the Lungs of the fiber glass. help to decide what to do about the progressiveness Larry was again tested for Occupational Asthma rated at 53% permanent and progressive. Dr. Katherine Lilley placed Larry on a therapy program of liquid and using the Pulmo Aide Nebulizer. Therapy was drink as much as possible each day, using the nebulizer up to at least six times a day for over a year. The idea was to rid the lungs of the glass. If it stayed in the lungs it would continue to do more damage and be more progressive. Larry was getting Bronchitis and pneumonia at least 10 to 12 times a year as it got worse by 1999 there was no more pneumonia, bronchitis only once and it stopped the progression of the Asthma. By year 2000 no pneumonia, and no bronchitis, asthma remains today 2002 absolutely stable 53%. Thanks to a Doctor who cared and done the right thing for me realizing the glass had to come out. Larry doesn't smoke, never before had Bronchitis, Pneumonia, or Asthma. What this employer done to Larry was beyond Negligence, and Malicious. Larry stayed with his job as long as he could, with Eval's significantly above average and superior. Larry was proud that he could still do his job well. But his Doctor's all agreed that he was disabled in several categories, and it was imperative to extend his life that Larry ask for Reasonable accommodations as he was an American with Disability's. With in two months after asking for reasonable accommodations they were denied, Larry was let go from his job and Washington County Sheriff Department has failed to pay for any of the injuries.

If it isn't Death Factory's we work in, or live next to, I guess what I'm trying to give you is a few clues as to what you might want to watch for in your everyday life, and the life of your children at your place of employment, or at your children's school. We must all be on the alert to keep ourselves and our families safe. Most toxic materials if handled as directed, and used with care can actually be safe, but if it's used carelessly, recklessly, and with negligence there will be a lot of sickness and death for no good reasons. There are those out there that don't care and will do things intentionally, negligently, or with malice in mind that could destroy our lives and/or the lives of our children if we allow it to happen. Everyone must do a small part to make sure they and others follow directions of the toxic materials for every ones safety. LARRY SAYS, "<u>ALWAYS REMEMBER TODAYS PREPARATION DETERMINES TOMORROW'S ACHIEVEMENT."</u> COULD YOU IMAGINE THE UNITED STATES EXPORTS IT'S TOXIC PROBLEMS "WE PRODUCE 10 TIMES MORE HAZARDOUS WASTE THAN ANY OTHER NATION. SO THE TRAFFIC IN TOXIC WASTE IS GROWING. THIRD WORLD NATIONS ARE IMPORTING TOXIC AND HAZARDOUS WASTE FOR CASH.

The toxic and hazardous waste problems that have plagued this country now for over 40 years...... There are barrels rusting through and oozing out pure toxic waste, Black, Orange, Green, and Yellow, puddles of unwanted chemicals on industry floors. There are barrels, and barrels of toxic waste on large storage lots, dumped in low spots or holes, huge pits dug, and in many cases by those who are not geologists, or hydrologists, digging theses huge pits and stacking the barrels ten to twenty high allowing them to leak into the under ground water. We have huge trucks hauling this stuff out into the country side dumping it in ditches along the roadways, dumping it into canals, or maybe low spots out in the middle of open fields. Barges laden full with garbage and toxic trash with no where to dock.

The legal, moral, and social problems caused by the United States dumping their toxic and hazardous wastes on any of many impoverished third world countries. Where some of it is packaged and sold on the Black Market. The trash third World Countries are importing is toxic trash experts in this Country have deemed toxic and dangerous. The problems with the reports of this, is suffering the same problems that have plagued environmental issues for many decades.......The inability to produce victims. Cancers and all the other health problems take decades to develop and often cannot be directly or definitively linked to any exposure to pollutants. Then Tort Claim Law protects the government, while at the same time protects Corporate America, and if they suspect any problems they have a way of making things disappear.

At least there were enough people to complain about odors, itches, and pain along with lower IQ's and other health maladies. It seemed to be enough to put 2 and 2 together allowing a complete detailing of the waste-exporting business. This issue is quite disturbing and along with more of a comprehensive view of something that should not be. Shipments have been traced to Africa, Mexico, Taiwan, and to the Philippines. This is plain and simple a business in Global waste dumping with no environmental controls. There is no explanation as to what some of the chemicals are and if they are so toxic the United States will not allow them into waste dumps. What it has done to lab-animals, it is doing to human health. Many of the health concerns seem to be anecdotal, but what about weeks, months and years, and the fact chemicals will mutate the human DNA. What is a cough or a rash today could be Cancer, Brain disease, vital organ disease, passing on mutations, learning disabilities, memory loss, many other diseases such as Parkinsons or Alzheimer disease, asthma or worse.

The waste exporters only claim is very simple, they are innovators well ahead of the times....... Its not recycling it is the surplus chemical business. The first indictment under U.S. environmental laws was to a company shipping toxic waste to Mexico. Colorado is being used as a corridor in the shipment of illegal wastes into Mexico, and it's not just a worry of the East or West Coast. An electric company from Colorado stores and ships one of the most dangerous chemicals in the world today PCB's which is a banned insulating material for electrical use, wrapped for resale as toxic waste. IBM, Hewlett-Packard are just two computer manufactures who have acknowledged shipping waste overseas to avoid the high cost of disposal in the United States. The United States produces more than 10 times more hazardous waste (and growing) than any other nation. Everyone today, tomorrow and into the future will be touched with chemical or nuclear waste, and must be prepared on how to handle your experience before its too late.

THE MIDNIGHT WASTE DUMPING IN AND NEAR OUR CITIES IS GROWING

The dumping of hazardous waste in cities across this nation is on the rises and most of them go undetected. The discoveries represent only a small portion of the total picture of hazardous waste dumplings. The problem will only get worse with the cost of legal waste disposal increasing. There has always been illegal dumping, and that is because it has gone unnoticed. Larry can remember, "working till midnight or going into work on the midnight shift can remember seeing neat little strips of something wet right down the center of the outside lane on highway, this was going on quite often on different highways. He finally caught up to one pulled up along side and motioned to pull over the person in the truck waived with one hand and only one finger sticking out. So he backed off and got the license number realizing he was letting go of about 10,000 gallons of toxic waste right down the highway. He turned it in along with a couple more he had caught but never heard anything back, and he knows this is the way millions of gallons toxic waste is disposed of right down our highways." So if you feel you smelled something bad coming from your car thinking maybe you hit something, It was probably one of those times you was driving down a highway of toxic waste. Take it to the car wash, don't wash it your self, your better off not taking a chance.

There are many ways, those getting rid of the toxic waste know it's wrong but they choose to do it for the money not really knowing of the dangers their dumping on their community. For another way of midnight dumping they are using tandem axle trucks to haul it out along a country road, or out in an open field. I've caught some driving down little plow liens, or field ends with no lights to get out to the center of a field then dump their load. Larry says he caught one trucker dumping his load in the irrigation ditches, and another good one is a small lake, they'll dump it any where they can get.

One in particular in Colorado, probably was one of the worst and most potent of toxic wastes was the Weaver Electric Company in Elbert County where they were accused of burying hundreds of pieces of PCB-contaminated electrical equipment, all to avoid paying the high costs of disposing of it legally.

A CRACKDOWN ON THE MIDNIGHT DUMPING OF TOXIC

WASTE. Here are just a few around Denver that were reported.

State patrol, and local police are working together targeting hazardous and toxic waste disposal methods across our nation. Midnight dumping is really hard to catch, we have those who have a pick-up truck, loads up a barrel or two, and maybe some cans, may go into an alley or a deserted road at that time, it only takes a couple of minutes to unload and be gone. Another one is a load of small containers and picking up on where there may be a huge garbage bin at some construction sight, they pull up dowse the lights and empty the toxic waste into these waste bins. A passer by spots some one dumping drums and cans along a bike path he called the police from his cell phone. When the officers arrived they found 40 to 60 barrels, and cans automotive paint and thinner. Specialists must be called in to remove the containers, and to remove all the contaminated soil and rid the area of any threat from Toxic waste.

Large trash receptacles are an easy target, seem to some being the most logical place to get rid of the toxic waste they were paid to dispose of, and real good spots to get rid of asbestos or fiber glass. These are just some of the small businesses getting rid of toxic wastes, when you realize how the big boys do it they just load 10,000 gallon tankers and dump it right up the highways or along the side of some deserted road, or out in the middle of a pasture or field.

There's another easy way to get rid of the hazardous wastes back in the 60's and 70's the producers of toxic waste just needed to know who the key man in the Army was at the Rocky Mountains Arsenal back then dollars under the table was big business. The same way the Army was being paid for millions of gallons of toxic waste from Shell

Oil Company. Under the table out there was big business until the Army was caught playing the game, and are now spending billions of dollars cleaning it up. There are some of the small businesses who are shipping their Toxic Waste out of state, to Idaho, Utah, Texas, North Carolina, and elsewhere. Tom Butts, a specialist with Tri County Health Department said, "The same regulations apply to smaller businesses such as dry cleaners, printers, auto body shops, furniture finishers, and strippers, they all create and generate hazardous waste, they don't know what to do with it and don't fully understand the same regulations apply to them as well.

If found guilty of dumping hazardous waste can be fined from $100.00 a can to $1,000.00 a barrel. Or maybe yet the offenders may be convicted of a felony and could receive a sentence from three to five years in prison. Over an eight year period of time EPA has prosecuted 60 cases in a six-state Region Eight office. 45 of those were in Colorado. It has slowed considerably but now there are an additional 60 active cases in the same area.

THE NUCLEAR POWER INDUSTRY SHOULD FACE REALITY, NUCLEAR WEAPONS AND NUCLEAR POWER PLANTS ARE DANGEROUSLY ANTIQUATED.

America is faced heavier than ever before with national security crisis that daily threatens the viability of its strategic nuclear deterrent. There are 48 federal facilities that produce materials needed for nuclear weapons they have been shut down for months on end because of safety concerns. These facilities were built in the early 1950's they are unsafe, antiquated and should be abandoned with new and more modern ones to take their place. If the government doesn't begin to construct new production facilities soon to meet the requirements of the millennium and beyond, this critical situation could cripple this nations nuclear deterrent, and beyond.

Events of the 90's show just how serious the problem is. The nuclear waste issues have been plagued with serious doubts everyplace they go. In the 90's the Department of Energy's, remaining production reactors at the Savannah River Plant in South Carolina was shut down indefinitely. "Huge Safety Upgrades". These reactors are the nations only source of <u>Tritium</u>, this element is an essential part of the warheads used in most of our nuclear weapons systems. There is a need to enhance the strength of the explosion and tritium is that enhancer used, and therefore allows warheads to be smaller and lighter than they would otherwise have to be. It has made it possible a smaller and more maneuverable Launch vehicle, as well as allowing multi-warhead weapons. Since tritium will decay rather rapidly. It keeps the government on its' toes in maintaining an ongoing program of tritium production. If supplies of tritium were totally exhausted, the United States would be forced to return about 1,200 warheads per year, effectively disarming us totally over a 20-year period. Even a temporary shortage could force the United states to reduce its existing arsenal by removing tritium from some warheads to stretch existing supplies. It may seem incredible when one thinks of what is at stake, the United States would eventually be forced to cannibalize some warheads to keep others functioning.

The United States would be severely crippled with this type of unilateral disarmament, and would severely jeoperdize our national security as we know it today. This nation cannot and I would repeat cannot stand by and watch something like this happen. There is a solution at hand. The total problem at Savannah River Nuclear plant are a direct and inevitable result of reactors that are approaching the end of their original design and lifetimes. We must face it the equipment is antiquated and is necessary to replace it with new equipment, and stop just putting bandaides on. Those reactors were built nearly forty years ago in the developing years for the United States, regarding nuclear technology. None the less one of those reactors at Savannah River just recently completed more than 35 years of production of tritium and without one lost day of work due to an accident. None of the safety incidents up to this time has ever resulted in a radiation injury to any employee, nor any member of the public.

Since time beginning for the Savannah River Reactors, our experience and expertise with commercial nuclear power plants has been completely revolutionized, by design, construction and operation of nuclear reactors. This nation needs new production reactors, so we as a people can take advantage of this wealth of experience, and expertise so we can avoid any temporary shutdowns at Savannah River.

Greater efforts were necessary to bring the weapons production facility into full and complete compliance with all this country's Health, Safety, and environmental protection standards in place, so it's not just a matter of building new and safer production reactors.

The high level and hazardous nuclear waste had been kept right here at Savannah River. Also being built right here is and all new and innovative $750-million facility for processing the radioactive wastes from the past and all future wastes that will be generated. The waste will be vitrified, or embedded in blocks of glass, then placed into steel canisters for permanent and a safe burial in an underground geologic repository. High priority must be given to the environmental clean up and disposal activities which is a necessary evil.

While the Department of Energy now proposing to build two new production reactors…one that will be an advanced version of the old heavy-water reactors that was remarkably successful as it met all the heavy production quotas of weapons materials over nearly 40 years of use at Savannah River. The new heavy-water reactor will contain state of the art computer and instrumental systems and all new passive safety features, that we have only been able to piece in as they came along. From this one new rector will be enough tritium to meet the demand and supply for the entire nuclear weapons program. The other reactor is a first of its kind. It is an advanced high-termperature gas-cooled reactor, and it will provide 50% more of the nations needed capacity, offering an alternate course for the vital material. Offering a much better maintenance program when necessary.

There are some that have proposed building what is called the linear accelerator for supplying tritium. However there is very little experience to draw upon than that of the heavy water technology. Two reactors setting up a dual approach seems unreasonable and gives us the opportunity to try even safer technology. The time is not only right but imperative in moving ahead. Any time old equipment can be replaced with state of the art equipment with a proven safety records we need to build on necessity. But doing it safer and wiser. It would take probably 10 years before one can be brought into service. Other wise the Equipment will not be ready when the nation needs it, and we will be disarming ourselves, unilaterally.

NUCLEAR WASTE STORAGE FACILITY'S SAFETY IS UNDER QUESTION

Congress continues to raise serious doubts about the ability of a nuclear waste repository in New Mexico to safely store radioactive waste from the Rocky Flats and other nuclear weapons facilities. GAO expresses sadness and concern about the waste drums scheduled to be taken from Rocky Flats to the New Mexico repository since it has opened. GAO's energy division, said the FBI and Environmental Protection Agency investigators probing plant operation supspect numerous violation in the way radioactive waste has been stored and labeled. In opening the New Mexico repository the DOE received internal Department of Energy documents that indicated DOE was rushing to meet its September deadline to open the facility, even while there were major management and safety concerns. The documents highlighted the problems at the repository, known as the Waste Isolation Pilot Project.

1. There was to be only one person at WIPP, and one at Bechtel Corp. who would know enough about the layout and design of the facility to give key briefings to inspection teams.
2. The officials that had set up the emergency training programs for New Mexico police, fire and rescue operations did not keep a complete list of trainees who would be called in on an emergency.
3. One particular and important waste shaft engineering data used to justify the design was killed from the computer. Stored haphazardly in boxes were all the important engineering and design records.
4. The DOE sitting back putting pressure on EPA to approve the environmental permits needed to open the facility, It had now been put off at least two more months.

A group of scientists in 1991 said the Rocky Flats Nuclear Weapons plant should not be reopened for at least five years. While DOE is expecting to restart plutonium operations next month, and is proceedings on its standard Cold War Trajectory. With the U. S. and Soviet relations having warmed considerably. Scientists with the Federation of American Scientists, and teamed with a Doctor for Social Responsibility and for the Bulletin of the Atomic Scientist, said "It may not be necessary to ever reopen Rocky Flats."

Mr. David Albright a staff scientist with the Federation of American Scientists at a press conference, said "Rocky Flats can remain shut for many years, if not forever, with these measures: more realistic nuclear weapons production schedules, less exploitation of START treaty loopholes, fewer nuclear weapons, safety upgrades of existing weapons and more reliance on reusing intact warheads and pits." Pits are plutonium-based detonators used in all nuclear weapons, and are made only at Rocky Flats. DOE warned of "SEVERE RAMIFICATIOINS" for the Navy's Trident Missile program if Rocky Flats were to remain closed. However they noted the Navy could get by with existing war-heads, It was noted a top Navy Official had testified to that effect. Lewis said nuclear weapons production had a "very high price" in human health and safety, "" and the people of Denver know this."

50,000 NEIGHBORS VERSUS ROCKY FLATS. WORKERS, NEIGHBORS BAND TOGETHER TO FILE LAWSUITS SEEKING MILLIONS IN ECONOMIC AND PROPERTY DAMAGES. Suits ask for independent study of health risks surrounding the plant.

Rocky Flats workers and nearby residents joined forces yesterday in filing two mammoth class-action lawsuits against the plant's former operators, Rockwell International and Dow Chemical Co. The Lawsuits seeking independent study of health risks posed by the nuclear weapons plant, $250 million in economic damages for devaluation of property and $300 million in other damages. For the former workers they seek unspecified amounts of money because of the stigma of their past employment is preventing them from finding new jobs or health insurance.

This law suit is not an attempt to close Rocky Flats, but to have it run properly," with safety for the employees and the neighbors. There are thousands living in fear and have the right to know whether or not Rocky Flats has caused adverse health effects to them or to their family's. They want to hear the truth by someone who is independent and not under the control of the government or the operators of Rocky Flats. The lawsuits represent 10,000 former workers, filed by several trade unions. and 300 current workers. The 50,000 residents all of Jefferson County. It was thought overtime the controversy over Rocky Flats safety was overblown. Until the FBI agents raided the plant as a part of a criminal investigation. The residents then believed they were actually a part of the unsafe practices at the plant, and were not saying there have been adverse health effects only We hope that there aren't.

DOW and Rockwell are accused of engaging in a continuous course of "negligent, Careless, and Reckless" type conduct, It was no accident "IT WAS INTENTIONAL." Conduct resulting in repeated releases of radioactive and other hazardous substances around the facility which they lied about and tried hiding it from the public. The Lawsuits cites fires in the 60's that released large amounts of radioactivity into the air. Leaking drums of radio active waste that caused 5 million pounds of contaminated soil that was shipped from the plant. Open pit burning of oil containing uranium chips and the release of radioactive waste into the Rocky Flats drainage system. Rockwell International and Dow made millions and should be held accountable and responsible to those they abandoned, the workers and the community. Dow ran the plant from its beginning until 1975 when it was taken over by Rockwell International up until they were fired in September 1989 after failing to comply with safety and health standards in operating the facility. It was at that time run by EG&G INC. up until 1995.

Rocky Flats won't come clean for years, and will likely continue well into the next century. Just to figure out what is going to be needed and what gets cleaned up first. Environmental Officials say will take at least five years. To fully clean up the pollution it will take well into the 21st century, both Federal and State Health Officials said last night. It will take at least five years to get site assessments according to the Environmental Protection agency. There is no date set to have the clean-up completed. That would be foolish since there are a lot of unknowns at Rocky Flats. It is good to have the time to look for the high priority sites, then look at the big picture. "Clean-up will go considerably into the 21st century.

A clean-up agreement was signed between the U.S. Department of Energy EPA, and the Colorado Department of Health. These agreements set deadlines and target dates for analyses & studies through the year 2000. Most deadlines involve characterizations of waste stored at the plant analyses of groundwater and soil pollution and short-term cleanup plans for the most contaminated areas. There were 178 sites pointed out at the plant that need to be immediately analyzed. Rocky Flats completed the construction of a groundwater treatment facility at an area known as the 881 hillside, where the chemicals threaten to seep into creeks that feed nearby community reservoirs. There are at least a half dozen other sites polluted by toxic chemicals or radioactive elements that are also targeted. The decisions are sometimes tough as to what comes first. The agreement targets priority areas and even though there is not a tremendous amount of contamination off site, it certainly is a priority concern of the communities and will be addressed loud and clear.

Rocky Flats has been illegally storing hazardous waste for at least a decade all the time claiming it to be recyclable. U.S. District Court left no question that if necessary it will force the nuclear weapons plant to meet state hazardous-waste storage regulations for dozens of areas where ash and residue left over from illegal incinerations are stored. This is a big step toward making clear what they can and cannot store and what is and is not hazardous waste. They had been basically incinerating materials under the guise of recycling when it was an illegal waste disposal technique that had probably been used for at least 20 years or more. The Plants incinerator has now been shut down for nearly a year. Rocky Flats historically claims they have burned waste and debris tainted with radioactive plutonium to recover the valuable metal from the ash.

Plant Officials continue to argue that the ash and residue were exempt from regulation by the state because the material was being stored to recycle radioactive plutonium from the ash. However records do show that Plutonium was actually recovered from less than 10% of the tons of residue being stored. DOE tried skirting the issue for a long time under state law. DOE Attorney's tried blowing a lot of smoke at Colorado not regulating the waste and would jeopardize negotiations under an agreement between DOE and the State. The States agreement contained only flexible deadlines and no final compliance deadline, in the Courts ruling for it "lends definition, direction, stability and impetus to the DOE. and Colorado Cooperative Effort." It lends credence to a ruling that has already been institutionalized.

In order for DOE to get a restart at the same time the clean up effort continues at Rocky Flats EG&G plans to violate many of its rules to restart plutonium production. The private contractor that runs Rocky Flats has asked for 69 temporary and permanent exemptions from safety regulations so it can restart plutonium production at the nuclear weapons plant. Rep. David Skaggs office revealed at least 15 of the exemptions involve nuclear "Criticality"...... These regulations are designed to prevent an uncontrolled nuclear reaction, according to documents released by Mr. Skaggs. Meanwhile the Department of Energy Officials in charge of the restarting plutonium work at Rocky Flats told an independent oversight panel that DOE has decided to restart production even though it will be out of compliance with at least 21 of its own safety environmental and other regulations.

The documents appear to be contradictory to statements from DOE Officials in public forums and Congressional hearings that Rocky Flats will not be restarted until safety problems have been resolved. The following statement was released from DOE. The department has said many times and we repeat again the resumption of the plutonium processing activities at Rocky Flats will not be initiated until all the necessary environmental, safety, health, safeguards and security requirements are met. Plutonium operations are still at a standstill because of safety problems, & potential safety violations. MAJOR OZONE DESTROYER, THE EMITTER IS ROCKY FLATS NUCLEAR WEAPONS PLANT, NORTHWEST OF DENVER COLORADO. OWNED AND CONTROLED BY THE UNITED STATES DEPARTMENT OF ENERGY. "DOE"

Rocky Flats Nuclear Weapons Plant at one time was the single emitter of one Ozone-destroying chemical. It emitted so much Colorado was placed ninth among the states in total emissions of the chemical. The chemical Carbon Tetrachloride throughout the United States and the World is a major contributor to the destruction of the stratospheric ozone layer. It has gone on "completely unregulated," and the general public knows nothing about it, according to the Natural Defense Council.

THE NRDC, is an environmental group which claims over 125,000 members, released the nation wide "who's who" of 3,000 major emitters of Carbon Tetrachloride.... A solvent used to clean metal parts……and there are two other known ozone destroyers. Deborah Sheiman is the co-author of the report with Doniger stated the groups purpose was to encourage and make the citizenry aware they could use the "Who's Who" data to demand that each company stop using these Ozone destroying chemicals.

The Stratospheric Ozone is the natural layer of the atmosphere anywhere from six to thirty miles above Earths surface. Its destruction allows inordinate amount of the sun's ultraviolet radiation to reach the planet, and that is a number one cause of skin cancer, cataracts, immunological disorders, crop damage and other problems.

Two other Ozone-Destroyers were reported by NRDC, was focused on Methyl Chloroform, and Clorofluorocarbon-113. Even though Colorado is one of the largest emitters, one that stands out is

Martin Marietta in Astronautics in Littleton. There are many more in every state of the United States that are the demons destroying the Ozone layer for the world.

Upon inspections some years ago plutonium accounted for as much as 20% of the dust and debris that was clogging some air ducts. That 20% is the same as six nuclear bombs sitting in those air ducts since they were unable to clean them out, put new filters in and started production again working with half clogged or some nearly clogged air ducts. The main objective to point out is in 1990 they had enough plutonium for six nuclear bombs. What this sounds like is an intentional accident waiting to happen, and how much plutonium over the last 12 years has amassed along side the other 62 pounds of plutonium that was already there.

Two former employees had said that the filters are subject to cracking, fires, and blowouts. The frames deteriorate rapidly when exposed to corrosive chemicals used in the plant, some of those frames needed replacement in building 771, and in building 881. The operations manager for EG&G, said the magazine story oversimplifies the filter system and overstates its potential shortcomings. They say the paper in the filters is thicker, and contains more fibers. It is said the system should detect high levels of plutonium escaping, but it doesn't say a word about low level of plutonium escaping. Exhaust from the plutonium processing passes through the filters and is cleaned at least for high levels of plutonium escaping. Larry says "that low levels of plutonium is quite dangerous yet here it sounds the detection devices are only set up for high levels."

They say the original metal containers have shown signs of age since their installation three decades ago. There should have been reports on those container on a 2 to 3 year maintenance check for leaks. What this sounds like is they used the containers and the filters for that 30 year period of time, and it is no wonder they had leaks. Those containers in such a critical operation should have had absolutely no disintegration. They claim that was much as 20% of the dust and debris that nearly clogged some air ducts was plutonium. The last I heard before they restarted production, the plutonium was still there and they were unable to remove it. Yet they claim the air around the plant is acceptable.

THE NUCLEAR TIME BOMB IS STILL TICKING FOR MILLIONS

Multi-thousands of cases of thyroid cancer were caused by the fallout. Experts now say the levels of exposure could justify special monitoring for some people....Especially the children of the 1950's and 1960's. The 100,000 page study by the Cancer Institute, that was ordered by Congress in 1982. The draft report was completed in 1994, and has been undergoing revisions and rewriting. A summary of the study was prepared for the Department of Energy. The report states that according to formulas in international use for calculating radiation damage, the doses were large enough to produce 25,000 to 50,000 cases of thyroid cancer around the country, of which 2,500 would be expected to be fatal. The accuracy of the formulas cannot be certain because of the available data on exposures. It could be a little lower then again could be much higher.

There is a new study that says the average dose received by the approximately 160 million was ordered by Congress in 1982. The draft paper report was completed in 1994, and has been undergoing revisions and rewriting. A summary of the study was prepared for the Department of Energy. The report states that according to formulas in international use for calculating radiation damage, the doses were large enough to produce 25,000 to 50,000 cases of thyroid cancer around the country, of which 2,500 would be expected to be fatal. The accuracy of the formulas cannot be certain because of the available data on exposures. It could be a little lower then again could be much higher. This warning is probably more of a warning for those born in the late 1950's and 60's, those at the age brackets of 40+ to 50+ years of age.

There is a new study that says the average dose received by the approximately 160, million people living in the United States during that time frame was 2 RAD's (radiation absorbed dose) this would refer to the amount of energy absorbed by the flesh. It would be people living in the western states to the North and East of the test site received doses averaging 5 to 16 rads. Children age 3 months to 5 years received doses 10 times higher than others the institute said. At that rate the baby could have very easily received 160 rads. Then you tend to say the child would be getting a double dose, because of receiving the same as others, then to again be overexposed to it by the milk in a babies diet. 160 rads for a baby could actually be seen as

320 rads in a dose could easily cause death, or Cancer, or a Myriad of injury to any other parts of the body such as brain cancer immediately or in the future. The main pathway for radioactive iodine exposure is through milk, which children consume in larger quantities that adults, especially in comparison to their body weight. Consuming contaminated milk, the human body will deliver the radioactive iodine to the thyroid gland, where it can cause the development of cancerous nodules. With children the thyroids are smaller, additionally and equal amount of the radioactive iodine to a smaller gland would deliver more rads per kilogram of tissue. Per kilogram of tissue delivered would increase several hundred percent. A miniscule amount to a child is to terribly dangerous, Two rads to an adult would simultaneously increase the childs chances to 10 – 16 rads. A child or a baby would be easily overwhelmed by the same dose that may not even be noticeable to the adult or older children.

Larry says, "Is there a connection between rads, since SIDS……Crib death". We do know there is a connection between rads, and Cancer. I hope someone who reads this listens and is able to find out "could there be a Connection"? Or perhaps point this out to someone who may have an answer, or be willing to search for an answer. AS I was trying to figure out the formula they gave for exposure to (Radio Active Damage)? "rads" 160 million people received doses of 2 rads in the 1950's and 1960's and yet what an adult will be exposed to a baby's dose will be up to 10 to 16 rads. This left me quite perplexed because the formula never really talked about the 10 to 16 rads the baby just received while saying it was dangerous when the adult in the same scenario only received 2 rads. There is something here that just don't add up. To think to day the babies of the 50's would be 52 to 54 years old. It would not surprise me if we see the cancer deaths in the next 50 years nearly double. Larry says, "I'm hoping this will inspire further study into radioactivity, and the negligence that cause so much pollution, and then it's the every day people that pay the ultimate price.

Larry says, "not trying to take the monkey off Martins back but possibly we have two types of pollution that caused the people at friendly hill such terrible illnesses with the children and so many that have cancer. Rocky Flats has been burning the radioactive materials at Rocky Flats since the 1950's and 60's. The wind does not always blow to the north along the front rage. It first blows to the south then turns around and blows every thing back to the north, northeast. If so Friendly hills lies in the area it could be contaminated by Rocky Flats also with Plutonium, and other radioactive materials. Then remember maybe it was just 2 rads not putting an adult in despair, the very formula they use considering 2 rads for and adult but using that same formula for a child 1 years to 5 years absorbing 5 to 10 rads. And that is seriously dangerous in possibly causing death and cancer. At the same time a baby 1 to 5 years old would be receiving 200 to 500% more because of weight and age, which would be 5 to 10 rads or more, 2 rads is thought to be cancer causing to an adult. This is something that should be checked out maybe by some environmentalist in that particular area. Wheatridge and Arvada sit just to the north of Lakewood, and it is them in that area that are having problems with being over exposed to radioactivity, and Friendly Hills is just South of Lakewood, right along the front range."

Public Health Service, studying thyroid exposures around a government nuclear bomb factory at Hanford, Washington, and has recommended medical monitoring for adults who as children absorbed 10 rads or more. In contrast to the 50 to 160 rads those children are believed to have received, federal rules for nuclear power plant accidents call for taking protective action when the dose to human thyroids is anticipated to reach 15 rads.

There's a reasonable association between radioactive Iodine exposure and cancer, said Dr. Robert Spengler, the assistant director for science of the agency that made the recommendation, the Agency for Toxic Substances and Disease Registry. He said the association was demonstrated by a growing body of literature, from people in the Marshall Islands where tests were also conducted, and elsewhere.

THE ARMIE'S WAY OF TESTING CHEMICAL, AND BIOLOGICAL AGENTS IN THE 50's AND 60'S, on it's own people.

Several Minnesota sites were chosen for the United States Army to test chemical and biological warfare agents of the time on school children, and in general people on the streets of south Minneapolis they sprayed the Clinton school and other large parts of Minneapolis. The initial report on the spraying was broadcast on KTCA, a public TV Station. Additional spraying was dispersed in most south Minneapolis, Rosemount and Chippewa forest.

The zinc- cadmium- sulfide, is a known Carcinogen, is known to cause liver, kidney and lung damage if over exposed. It has also been known to cause birth defects in animal testing according to researchers at the University of Minnesota. These sprayings were done without the consent or knowledge of the general public.

The United States Army's Biological Warfare Center conducted sprayings of Toxic And hazardous, chemicals and above all they sprayed live organisms Toxic and hazardous, with that. So what they sprayed Minneapolis, Rosemount, and Chippewa forest with what they called Bio/Chemical agents of mass destruction.

These are deadly Nerve agents, the United States Army secretly sprayed their own people. Before arriving in Minnesota then also hit Scotts Bluff, Omaha, Lincoln, and a few other towns further to the West, before arriving in Desmoine's Iowa, then up to siox City Iowa. Then across the farmlands of Minnesota and small towns in between then on into Rosemount then Minneapolis. But wait they didn't stop there. They had a supply of Zinc-Cadmium-Sulfide and For an Organism they had Heiminthosporium Oryzae. Morris, Waseca, LeSueur, Duluth, Crookston and Rosemount again. The fungus was considered harmless but the last information we received from an unidentified source close to the Army. It was thought to be Carcinogenic. It was suppose to kill Russian Wheat while leaving America's wheat crops untouched. Little was known because the Army had kept all testing and manufacturing secret and classified. Information didn't begin leaking out until the late 1970's and early 80's during a lengthy investigation. It Was made to cripple the Russian economy or food chain by depriving them of flour.

I'm very much surprised at the fact Senator Wellstone has just now heard of this incident. I have had the information since the 70's and 80's and it was reported to Congress in Washington D.C. It was reported to Senator's and Congressmen, and Senator's alike in Minnesota And Colorado, when it came out in my lengthy legal investigation. I will not mention all the Congressmen or Senator's at this time, However Senator Hart of Colorado was one who's attention we tried to get. And gave his office that information.

I can see we have got a Senator who listens to the whoa's of his people, and this is my chance to assist Senator Wellstone of getting the real truth and not just given the Run around by the Army. Larry says the Army's motto is, "Tell them some damn lie, or classify it."

U.S. Senator Paul Wellstone, meanwhile In a News Conference in St. Paul, to express his outrage over the sprayings, which were done without the general public's Knowledge. Wellstone is going to demand a full-scale disclosure of the sprayings of the Bio-Chemicals of Mass Destruction in Minnesota and other U.S. sites. Larry says, "going through Army Memos, over the years says he remembers seeing a memo concerning the wheat rust fungus as being carcinogenic." In that same memo was discussion as to where the fungus material was spread over 160 acres and cover up at the Rocky Mountain Arsenal In Denver Colorado. By the way EPA listed the Rocky Mountain Arsenal as the most contaminated place on earth. That is also in an EPA brochure.

The United States Army in their usual arrogant, "We did nothing wrong attitude" is going to put together a panel of scientists to expand their review of Cold War-era biological warfare tests so the study will allay public concerns of public effects. Larry says, "The statement by the Army above to allay the public is nothing more than a cover up. These items are toxic, and hazardous materials, and for the army to spray the general public and school kids with no warning is totally inexcusable, and Malicious Criminal acts of monstrous proportions to bringing great bodily harm toward others. This whole damn sham is just reprehensible.

THIS NATIONS GOVERNMENT, THE DEPARTMENT OF DEFENSE, THE UNITED STATES ARMY, AND THE PENTAGON, ALONG WITH INDUSTRY GIANTS, HAVE WAGED WAR ON IT'S OWN PEOPLE. THE UNITED STATES HAS BEEN DOING THE SAME THING SINCE THE 1940'S ON ITS' OWN PEOPLE WHAT THE UNITED STATES ARE CRITICISING SADDAM HUSSEIN FOR DOING TO HIS PEOPLE IN THE 1980'S AND 90'S WHEN HE SPRAYED THEM WITH WEAPONS OF MASS DESTRUCTION. "Is there a difference as to who sprayed the chemicals".

The United states has been using their own people as guinea pigs since the 1950's If you read this entire series you will see what they have done and what they are still doing today, "It was no Accident, There has been no Accident, It was intentional".

These materials the Army sprayed unknowingly on the general public, and at schools specifically, on school kids of the time are Bio-Chemical Warfare agents considered for mass destruction. First, they had absolutely no business doing such a horrific thing to their own people. Or to any one's people. Remember at the beginning of the book and what it says. TERRORISM AT HOME OUR ENVIRONMENT, DEATH FACTORY'S, IN OUR MIDST."

BIO-CHEMICAL WARFARE THAT THIS NATIONS GOVERNMENT, THE DEPARTMENT OF DEFENSE, THE PENTAGON, AND INDUSTRY GIANTS HAVE WAGED ON IT'S OWN PEOPLE.

I STAND IN AWE, with human terror that Saddam Hussein in Iraq used chemical Weapons of mass destruction, on his own people, the Kurds in the North, and the Kurds in the South. He used them chemical weapons on the Iranians, and the Kurdish Iranians in the North. Massively spraying entire city's and town's. For this the whole world looks down on him for using all the chemical weapons the United states had sent to him while knowing full well his intentions at the time. The first shipments to him were Lewisite, and Mustard agents. Then they started sending him the heavier materials, Nerve agents, sickening agents, and then we get into Bio-Terrorism agents, and it was the United States that had sent him everything. In fact during the Gulf war our own soldiers were blowing up his arsenal and most of it was United States ordinance. For the Bio-Germ warfare in Iraq, get chapter 10 Death Factory U.S.A. Part I, It will give you the names and addresses these materials that had been sent and addressed to A person and address in Baghdad Iraq.

This happened in 1980, and 1990 just before the Gulf War. Does Saddam Hussein have Nuclear power capabilities? "YES".

Can we now stand here and be just mad as hell at Saddam Hussein for what he did spraying his people with chemical warfare agents of mass destruction, many of them died, lived horribly injured, and eventually didn't make it.

WE HAVE THE VERY SAME THING IN THE UNITED STATES, PEOPLE WHO COMPLAIN OF BAZAAR HAPPENING FROM THE TIME THEY WERE A CHILD, OR THE CANCER WE FEAR SO MUCH TODAY, OR MAYBE SOMETHING BAZAAR THAT IS HAPPENING TODAY. We have people who are living and dying in horrible pain. If you have any idea what it may have been the Army has it very neatly "*CLASSIFIED". TELLS IT AS A LIE, THEN MAKES IT A STATISTIC.*

LIE"S COVER-UP, CLASSIFIED

I'm going to relate a story of the Rocky Mountain Arsenal, United States Army, RMA here that I had earlier in the book mentioned about their Cover up Techniques. Why is Lie's cover-up and Classified the same? The Government and especially the Pentagon if they don't want something known by the public, even if it is wrong doing, and even if it would mean the safety and well being of thousands. (1) They first tell a lie, If that don't work they lie and contrive, and if that don't work they get out the rubber stamp, to stamp it CLASSIFIED. It will then remain classified until everyone is dead, and gone, and the Officials at the time are gone and retired and know nothing about. "Larry says, "Its kind of funny how these type actions can cause short term memory loss.", but it does. Then just maybe it can be declassified but only if you get your Senator or Congressman to stand by you, and to get any help is very rare.

The story I wanted to tell you to exact methods used by the Army. At the RMA. In Denver Colorado, The Army had an exposed basin constructed in 1954, This basin was lined with an asphalt membrane that was to make it safe for storing up to 240 Million gallons of Chemical Warfare Agent waste, where with the arid condition of Colorado would evaporate the water from the hard salts. Within the first two years of filling the basin it was noted that part of the basins 3/8 inch thick asphalt lining had washed away or was eroded away by chemicals. This was supposedly repaired and they continued the use of it, by 1965 the United States Department of Health and Welfare informed the Army that Basin-F was leaking, and again part of the membrane was absent. They repaired that and by 1969 the Army Hygiene Department and the U.S. Department of Health had advised the Army with two Generals from the Pentagon present that basin F was leaking and had saturated over 25,000 acres off the arsenal property with Chemical Agents considered weapons of Mass Destruction. They were warned at that time any further use of Basin F the Army would be doing so with the premise it was leaking. The one General on paper stated "WE GAN'T LET THE PUBLIC FIND ABOUT THIS, HAVE IT CLASSIFIED". The Army then continued using this basin full well knowing it was leaking into the Country side

Chemicals that could destroy life. The Army continued it's use from then 1965 to the time they were permanently closed by the Congress of the United States 1983. The Army was allowing Shell Oil Company to dump chemicals they were manufacturing that had been banned in the United States. They were allowed many other chemical manufacturing Corps. Out of Denver to haul in tanker loads of unknown chemicals and Train cars of Chemicals that may have been shipped to them as hazardous nuclear waste from Rocky Flats. Train cars that were full of chemical materials was being shipped in to the United States Army who's production days were over. Yet the train loads of materials kept coming in. Through the grapevine in the clean up of Basin F, there were some radioactive elements confirmed & what they wouldn't allow Rocky Flats to Burn The Rocky Mountain Arsenal was more than glad to help. Some of the tanker cars that were going into the arsenal are believed to came from Rocky Flats. We feel that a lot of money passed under the table with all these trucks and Train loads of unregistered shipments for disposal at the Army.

BE SURE TO READ BOTH BOOKS, PART 1 AND PART II DEATH FACTORY U.S.A.

THERE ARE SOME THINGS I KNOW YOU THOUGHT WERE IMPOSSIBLE?

What happened in Minneapolis was more widespread than just at the Clinton school as they are trying to make it look, We understand there were many places in Minneapolis and elsewhere that were sprayed with Bio/Chemical warfare agents during the 50's and 60's and where else did the Army do Bio/Chemical testing. We've got a list of cities where tests were done. San-Francisco, Los Angeles, New York, Chicago, Denver, Duluth, some already mentioned and there's a lot more down around Atlanta, Georgia.

There are investigative reports that have been able to link the Army to the Legionnaires' disease that struck at the Best Western International Hotel in July of 1996, in which 34 died of Legionnaires' disease. Legionnaires' disease is a lot like pneumonia, and takes it's name from the outbreak at the Pennsylvania American Legion Convention in Philadelphia hotel in July of 1976 at which 34 died of complications. For some unknown reason it and a factor of the environment required. There also must be a congregation of people who would be at risk.... The elderly.....Those with compromised immune systems.....Cancer.....Or those with Chronic heart of lung conditions.....

We have inside information, that person cannot be identified. Legionnaires' disease was supposedly developed by the United States Army. With what we know from the past. They needed a very large group of people, and what was a better group than a Congregation of Old timers meeting in a hotel in Philadelphia Pennsylvania. Since that kind of information was easy for them to attain, they had exactly what they needed, and since they work on the basis of people are expendable. With what I seen happen in the past, was immediately able to feel certain of what happened here.

What felt we had heard the truth, and putting that together with the past behavior of the Army we felt certain. The bacteria the Army had developed did not have a particular name until after it had done its damage. as a sickening agent.

Up until they tested it and discovered it was technically a weapon of mass destruction, they could not let the public know even though they worked on the medical vaccination. We understand they classified all material and reports regarding the BIO/CHEM bacteria, they at that time let it run it's course as Legionnaire's disease.

The disease, that never struck before July 1976 now strikes 25,000 people annually, and without warning. It is treatable with antibiotics, about 15% of the cases are fatal.

There were 16 cases of pneumonia reported after the Lamar School reunion on September 22-24 at the Lamar Hotel where they stayed, and again where a large congregation of people had met for the reunion of those who graduated in 1939. There was three who died at the time who had been at the Legionnaire's Convention in July. 1 out of 16 who met was confirmed as Legionnaire's disease at that time the other 15 were listed as pneumonia, at least until tests come back, Up to this point we were never Informed of how many others were confirmed, with Legionnaire's disease or plain old pneumonia. and died.

WHAT IS THE FEDERAL DOCTRINE OF "SOVEREIGN IMMUNITY?

Unless sovereign immunity is waived, the Government cannot be held liable for its plans and policies. Through egregious government malfeasance, used miners as guinea pigs to study the effects of radiation, and did not warn them of the danger. Although the Courts often agreed the government was responsible, most decided in favor of the government under the Federal Doctrine of Sovereign Immunity.

In 1985 the 9th U.S. Circuit Court of Appeals said, "This is the type of case that cries out for redress, but the Courts are not able to give it. Congress is the appropriate source. Owens, who recently lost his brother in law exclaimed "I can remember, as a child growing up in Southern Utah, watching the sky light up with various colors from the nuclear blasts at Jack Ass Flats proving grounds in Nevada. Another man explained the rumbling feeling as in an earthquake. Southern Utah was hit real hard with and epidemic of Leukemia's and other cancers. The same as the Navajo uranium miners, there were no warnings of radiation exposure keeping it secret, and failed to warn is wrong doing, and is not any part of the doctrine of sovereign immunity.

UNITED STATES, ADMITS NUCLEAR PLANT EMISSIONS POSED DEADLY RISK AND YET THEY <u>FAILED TO WARN OF THE AFFECTS OF RADIATION!</u>

The reports on the Richland, Wash., plant indicated the implications are very serious. The Nuclear Weapons Plant in the Pacific Northwest had released enough radiation in the 1940's that would pose serious health hazards to nearby residents. Known as the Hanford Nuclear Reservation released radiation high enough to have caused illnesses, which includes cancer, and leukemia. The reports did contain estimates of potentially lethal, or large doses of radiation from the Hanford plant from 1944 to 1947. The doses were significantly high.....3,000 doses is technically a lethal dose. Rads are a measure of radiation to human tissue. The Nuclear Regulatory Commission requires that facilities it licenses limit yearly radiation exposure from airborne emissions to 15 thousandths of a rad. In one particular episode in 1945 350,000 to 400,000 curies of radioactive material was released. This study is the first major research into accidental or Intentional releases of massive amounts of radiation that drifted from Hanford into at least 10 nearby Washington and Oregon Counties.

The reports encompassed emissions but not the health effects of the radiation. The DOE Department of Energy is paying for this investigation. The Government has reached no conclusions yet about how many actual cases of illness may have been caused. That information is the subject of a separate study by the Federal Center of

Disease Control in Atlanta, which has been conducting a broad survey of Thyroid diseases.

Soviet studies show residents of the Ukraine received far less radiation to their thyroids from Iodine after the Chernobyl nuclear accident in 1986 than the Energy Department now says was absorbed by residents near Hanford.

What happened to the civilian victims of that insidious World War II, and then the long Cold war with the Soviet Union. People were unknowingly conscripted, and then gave the ultimate for our national security. If death didn't happen immediately many suffered for years with no relief in sight. Most are just now finding out what happened to them 50 to 60 years ago, or are now dying of some form of cancer or leukemia. Ionizing radiation as the atomic equivalent of a bull in a china shop. When it penetrates the living tissue, it wreaks havoc on the atoms and molecules in its path, setting off a chain of events that can destroy living cells or make them function abnormally. That is why large doses of ionizing radiation can kill quickly or inflict severe damage, and why non lethal doses can initiate cancers throughout the body that may prove deadly years later. Studies of people who have received significant doses (such as atom bomb survivors, Uranium miners, and radium watch dial painters) show that damage depends on how they were exposed, the dosage and the type of radiation. Scientists know that human organs can repair some radiation damage, although many questions persist. How harmful is the background radiation we live in? Why are some organs more vulnerable than others? And why do some individuals seem more resistant to radiations harmful effects?

EG&G Rocky Flats Inc. took over leadership in 1989 from Rockwell International under the U.S. Department of Energy, when Rocky Flats was closed for numerous safety and Health violations. They strived to have it reopened by late 1989 or early 1990 at the latest. At the same time the Union contract between the plant operator and the union expires October 7, 1990 Local 8031 has about 2,600 members at the plant.

Washington very concerned about the deployment of the Trident Submarines, and of course the modernization of the nations nuclear weapons seem to be an unstoppable push to restart the plutonium production at Rocky Flats. Congress is making sure the Department of Energy doesn't bow to military pressure in the restart of operations. A premature start would cause safety concerns that could mean another shut down. All operations at Rocky Flats were discontinued because of flagrant safety violations that no one could answer. DOE at that time terminated Rockwell International who ran the Flats until it became an uncontrolled mess. The Department of Energy had halted all operations vowing that operations would not resume until the plant met all Federal, and State safety standards. Any premature resumption at Rocky Flats will not be driven by the Military meeting the target date for the Trident Subs. Pentagon officials discussed their concerns at a closed meeting of the Defense Nuclear Facilities Panel of the House Armed Services Committee. Congressman Skaggs, and Congresswoman Schroeder from Colorado were there. They said after the meeting was over they didn't want to suggest a cavalier attitude about resumption. Skaggs said "but you could sort of see the argumentative clouds forming, of course driven by the Trident Submarine Launch. DOE officials said, "There will not be a restart until it is safe to do so, at this point probably June at the earliest………

Rear Admiral Jon Barr Deputy Assistant Secretary for military application stated, "We're not going to be pushed by anybody to do anything that's improper". At that time the first two Trident Submarines, each armed with 24 eight-warhead ballistic missiles, and are to set sail by the end of next summer. The Navy has enough nuclear weapons in its arsenal to arm both submarines with the newer Trident 2, Dl-5 missiles.

But now Navy Officials are contending that the deployment of the third Submarine, The West Virginia, might be delayed if Rocky Flats doesn't resume full operations by October. But then Admiral David Jeremiah, Vice Chairman of the Joint Chiefs of Staff did acknowledge that the West Virginia could still go to sea on time. If Plutonium production at Rocky Flats were delayed until next fall. Jeremiah told the panel the Sub would have to be equipped with older missiles that are not accurate or powerful as the new Trident 2 missiles.

Schroeder says she has no qualms about arming the sub with older missiles until energy officials can guarantee that Rocky Flats will operate safely….."It 's not like it's stopping the whole shebang, and it's very close to what they wanted anyway" she said. This is what a person might see as he drives by Rocky Flats, He will see rolling hills, that are so contaminated the ground looks scorched; not from the

Colorado heat, but from chemical heat. Not a good place to go skiing. They are using large 10 to 15 tower self-propelled Irrigation type sprinklers. The United States Environmental Protection Agency call this irrigation. It's not crops of corn, wheat or millet they are watering. Its called radioactive waste water Land application. There is nothing growing there nor could there be, 600 acres, or one square mile of restricted wasteland.

I often wonder the lives that have been affected, and are being affected by a questionable and most controversial hazardous waste management program at Rocky Flats. There are holding ponds that release up to 30 million gallons of contaminated water a month into Great Western Reservoir and Stanley Lake. That brown cloud that sits on Denver a good part of every day is probably radioactive, and has been for years, but they intend now to triple and quadruple that. When renovations of building 371 are completed, there will be a step up in the recovery operations. In other words the radioactive waste will be incinerated every day. This will allow more radioactive emissions from Rocky Flats then will inevitably find it's way into Denver's ever famous brown cloud. There mission is to recover plutonium from the older nuclear weapons.

Just to renovate building 371 Congress is already to appropriate millions, or even Billions if necessary. Congress before these considerations should at least insist on a total clean up first. Let your Congressman or Senator know exactly how you feel. The Rocky Flats Nuclear Weapons Plant should not be allowed to operate in such proximity to a large Metropolitan area.

Department of Energy has been working on proposals from private contractors, at the Trinidad Colorado site, hoping it could become permanent. The Budget hasn't been set yet for the facility, or the determination as to the amount of waste the facility would be expected to handle. The DOE hopes to have it in operation real soon.

The Department of Energy is looking high and low for private firms to handle radioactive waste temporary, or even permanent. especially for Rocky Flats who has radioactive waste ready to go. Rocky Flats is expected to run out of storage space for its midlevel radioactive wastes, such as plutonium. They say they are in need of Temporary storage facilities for several years and beyond, until and after the permanent storage is made available. This is how Trinidad Colorado came into the picture when a Seattle-based Pacific Nuclear

Systems INC. proposed to the Department of Energy a $16 million dollar temporary storage facility near Trinidad. Some 500 residents of the South Eastern town protested, and along with the Governor on their side, the Company withdrew the plan.

COMPACTION IS SAFE AND ROCKY FLATS ARE CONSIDERING IT

Rocky Flats now want to make “pucks” at the Nuclear Weapons Plant, by smashing barrels of nuclear waste into radioactive “pucks”. DOE states “Them little pucks won’t Endanger the environment and easy to store at the Flats. Officials at the Jefferson County Plant have planned for more than two years to use a large compactor to reduce the volume of midlevel radioactive debris that’s been piling up at the plant.

There was no significant impact on the environment per the DOE’s Environmental Assessment. The report had been made public awaiting comments for 30 days. The compacted waste will be stored at Rocky Flats until a permanent facility is completed. Of course there are worries and reports that to make pucks out of barrels would only compact and concentrate. It is true you would have 1000 concentrated puck barrels for every place that would store 1 barrel. Will this lessen the problem or will it magnify the problem? If the material is explosive, hazardous and toxic, and you take one barrel, & it would to make a large explosion. What kind of explosion would 1000 pucks make stored in the same a size area? Remember were talking about uranium, Plutonium.

Others think if it were so compacted and concentrated and it were going to leak radioactive gasses. With two 5 x 5 rooms where room #1, 10 barrels would fit, and being compacted room #2. you would be able to store 10,000 Pucks in the same size room. When the gases begin to seep through cracks or holes and escapes into the air. Which would give off the most gas? When you put you’re mind to work using the formula explained above, puck’en up the Nuclear waste already at a very toxic state one would cause a compaction of the amount 1 v A X 10,000 v B. A X B would give us a concentration of 10,000 times the normal contamination level. “Compaction, Contaminants, Concentration, would leave an unrealistic Concentration of Nuclear waste“. “CCC”. A Concentration of is type could lead to a Catastrophic Nuclear Accident. Unless addressed realistically and Completely. They would be putting 10,000 Puck

Barrels in the same size storage area that was designed, to handle only 10 barrels safely.

Then you stop and think a moment of what they said, they could then easily store the toxic pucks at Rocky Flats. But that could certainly concentrate the problem they already have with storing the barrels that are leaking radioactive gases into the environment. If DOE had an opportunity to now store 10,000 times that. They would! We know now they are already puck'en up the Nuclear Waste, using storage area more conservatively, but much more concentrated, every place that would hold 1000 barrels is now holding one MILLION puck barrels leaking nuclear, and radioactive pollution 1000 times before.

ROCKY FLATS STEALING COLORADO SKIES, THIS IS SKY-WAY ROBBERY

The Department of Energy is sky-jacking all air, banning all aircraft flying through a large area near the nuclear weapons plant northwest of Denver. Jefferson County Airport which is just a jets rumble away from the proposed area named by DOE to be banned. Denver's new airport will suffer. When there has been bad weather, the Jeff-Co airfield steps in and provides a landing site for corporate jets when the new airport is unable to accept additional traffic.

Jeff-Co would have to shut down, when the bad weather conditions force instrument landings, if DOE gets its way. This would worsen air traffic problems over Denver, and add to airline delays. I believe it's important to know Jeffco Airport pumps an additional $18 million a year into the Metro area economy. This would cripple any future Ideals of airport expansion, while denying surrounding communities an important tool for creating new jobs.

The ban would force small airplane pilots to fly over the near by foothills, back East they would call them mountains, where they would encounter dangerous wind sheer, putting the lives of the pilots and their passengers at high risk. This would also over crowd Denver Internationals already over crowded air space. The only thing this would create is a lot of safety problems, surely would not eliminate them.

Despite all these concerns, DOE has not even offered an acceptable explanation for this proposed ban on flying. There is no line on an aviation map that spells out a banned area that is going to stop an attack by an intent terrorists. This whole idea just seems a little far fetched to claim that airplane traffic threatens national security. There are already restrictions placed on pilots, such as other emergencies, necessary safety precautions. Pilots have already been advised to stay at least 1,200 feet above Rocky Flats, when flying directly over the nuclear plant.

To ban all flights over Rocky Flats is nothing more than a plan from DOE, to show their greed by grabbing unnecessarily a limited public resource…..safe sky passages. The Federal Aviation Administration will make the final decision on the DOE's plan. When it comes time to make that call, the FAA should ground this idea…..Permanently.

A very large gathering of pilots, and local governmental officials did testify that prohibiting flyovers of Rocky Flats is a bad idea, that would put a lot of lives at risk. What is Rocky Flats doing here in the first place, out just Northwest of Denver Colorado, with the North, and Western areas already polluted with radiation. To interject their power of doing what it is they want is just totally wrong, and is seen as bulldogging. DOE Officials responded by reporting the proposed prohibiting of any further flyovers of any of the nine other nuclear weapons facilities in their control. DOE takes its protective force operations extremely serious. Prompting them to say, "It is true our protective force operations do have use of deadly force authority. There are no provisions to shoot down any aircraft that violates air space. Director of security stated "The comments received at hearings such as this around the country will have great bearing whether the DOE pursues the proposed restrictions. Any aircraft willing to breach security will do so, and aircraft in trouble will take the shortest route to safety, no matter what restrictions are in place.

DOE did not allow the clean up affect the planned phased resumption of plutonium operations, and we will continue to evaluate all safety data and the necessary clean up operation. Although the cleanup nor the safety involved will hamper the resumption of operations. All safety improvements such as the ductwork will be removed and replaced, "but that is not saying that we will remove it all". The ductwork where the 62 pounds of plutonium will remain in place. Steps will be taken to better monitor the plutonium, and routinely cleaned from the ducts, in an effort to eliminate future build-up. He said the material poses no risk to the health of the public or employees. Although the danger is that if too much plutonium is allowed to gather in one place. It could cause an uncontrollable nuclear reaction that would spread radiation over a large area. According to President Bush, and Doe whatever the Department of energy decides to do, the Colorado Department of Health, nor Environmental Protection Agency will have any say in it unless large amounts of the plutonium are released into the air. “”SUCH AS THE POSSIBILITY OF AN UNCONROLLED NUCLEAR REACTION””. Where there is a danger if too much plutonium gathers in one place. “”DOE just stuck their radioactive foot in the publics mouth. There is already enough plutonium in the ventilation duct work for many 6—uncontrolled nuclear reactions, they know it and yet they are preparing to continue operations with 62 pounds of Plutonium, stuck in the ventilation Duct work, and that is not the part they are replacing. There is enough plutonium they don’t want to talk about can make or cause “6” nuclear bomb reactions.

No clean up plan has been established yet DOE is ready to phase in resumption of plutonium operations. They say the material poses no health risk to the public or employees. The Colorado Department of Health and the Environmental Protection Agency will have no say of low levels of plutonium, and can only begin saying something if high levels or large amounts of plutonium or beryllium, begin leaking into the air. When they go out and begin banging on the duct work with a hammer to clean any plutonium stuck in the duct work there will be times larges amounts will be released into the air. Colorado has no standards that regulates plutonium in the air & all federal standards are based on soil contamination or lifetime doses to people. The only interest DOE has is when are we going in restart operations they seem to block out all safety issues that still plagues Rocky Flats.

The Inspector general of the Office of Secretary of Health and Human Services, may take the problem from DOE of long-range epidemiological study’s. DOE was not willing to farm out their control over the Health studies, and resisted efforts of Congress but soon agreed that Research of this nature is better suited for an agency that regularly addresses such issues. The panel headed by the Secretary of Health recommended transferring the responsibility to the Department of Health and Human Services. 12 people working at Rocky Flats that were exposed to beryllium have contracted incurable lung disease. The panel recommended to DOE that it spend more time studying health risks from worker exposure to non-nuclear hazards.

For years of complaints and demands for changes in safety programs at the Rocky Flats nuclear weapons plant have earned the facility's largest union a prestigious national award. They represent about 2,600 workers at Rocky Flats. The workers claim they have been able to sit down with company representatives, identify safety concerns and move forward with solutions. Rocky Flats has been blasted in recent years for it's problem plagued safety programs. The safety problems were fast becoming chronic. The new general manager EG&G INC for Rocky Flats stated, "Our common goal is the safest working environment possible, as they gave praise for the concern, thoroughness and professionalism of the safety committee members.

United States Congress is now ready to allocate, and additional $1/2 Billion dollars the Bush administration has ask for and environmental clean up of Rocky Flats and other nuclear weapons plants. On the request from DOE, since there 1991 budget had a serious short fall, and they needed much more, in order to live up to legal agreements with Colorado and various environmental problems that had not been dealt with in the past. Millions spent on something that will never be safe or clean.

DOE plans to restart plutonium operations at Rocky Flats nuclear weapons plant while cleaning up the environmental contamination and with 62 pounds of Plutonium still in the ventilation ducts. Enough for 6 "UNCONTROLLED NUCLEAR REACTIONS" 6 nuclear bombs just waiting for a time to happen. They know this for a fact and refuse to do anything to stop it. The state of Colorado and the Congress of the United States of America have ordered an independent review before operations at Rocky Flats begins. DOE rings out in anger refusing to be pinned down as to when the review will be completed, before or after the restart. Doe told the Senator from Colorado he don't have that information to make that judgment. The Senate also challenged the DOE as to the $600 million they had been given to rebuild building 371 plutonium processing center, It has been built but has never worked right or with any measure of safety, and this is coming at the same time the department is already reducing Rocky Flats' environmental clean up budget, of course calling into question DOE's priorities.

DOE spokesman insisted building 371 will also be needed 20 years from now when Rocky Flats is shut down, and the old facility they are now using where workers are being over exposed to unacceptable high levels of radiation. Larry says this don't make good sense "After being overexposed for 20 additional years they can then use building 371, and DOE can use the money for a new facility for the environmental clean up somewhere else and we can give building 371 back to them in 20 years." And the Senator from Colorado was right DOE is "Cloaking" building 371 in environmental and safety terms to disguise its true mission, for weapons production.

The Senator from Colorado requested that the energy department complete a chart for the public showing the safety problems that have been found at Rocky Flats, and what measures are necessary to fix them, and when the repairs will be completed. The energy Department refused to produce any charts or reports of public showing, and told the Senator, the information you are looking for is in numerous report that can be found at DOE's public reading room. He also said you can call energy officials or members of the department's scientific advisory panels.

It is recommended by an advisory group the Department of Energy arrange for a comprehensive readiness review, before restarting plutonium operations at Rocky Flats. Because of the extended down time of Rocky Flats Usual practice calls for an "operational readiness review". This review and recommendation by the Defense Nuclear Facility Board released by the Senator from Colorado, then presented to DOE on Friday. Now it will be up to DOE to assemble a team that can actually conduct the review. EE&G Corp., who runs the plant for DOE says they expect to have two plutonium-processing buildings ready for restart by mid year and others shortly thereafter. They don't believe the review will have any affect on the restart.

Considering all options DOE has at their fingertips, It is simply not conceivable they can give sufficient attention to the fundamental issues raised by this boards recommendation, and it previous recommendation concerning design, construction and operation standards at the plant and still talk of an early restart. There is no comparison of Rocky Flats to private nuclear power plants that is the type, the Nuclear Regulatory Commission oversees which said operational readiness review frequently take only a couple of weeks. DOE realizes no operations would be restarted without public meetings or without the approval of two federal safety panels.

Congressman from Colorado is relatively pleased with EG&G's INC. approach, and performance displayed at Rocky Flats Nuclear Facility. However he is not happy with the Department of Energy's approach to a restart without having their ducks in order, and with no national security need to do so. Congress says that DOE will not resume operations until it is safe to do so. DOE is again showing that the resumption of production is the high priority on their mind instead of safety being the goal, and safety will come first.

Congressman from Colorado is very clear, there will be no resumption at Rocky Flats until two independent reports certify the safety of the plant, short term, or long term, and this alone will give everyone a better idea of what the time line will be. The timelines that are thrown in by DOE are not worthy of consideration. DOE Officials agreed the buildings would not be reopened until a restart plan is complete & reviewed by the two safety panels.

The Defense Nuclear Facilities Safety board established by Congress has recommended immediately, an extensive upgrade of all buildings, and the equipment at Rocky Flats to properly ensure safety. Rockwell International the previous manager, allowed a casual attitude toward safety, and the New Manager EG&G will not tolerate plant rules, violations, or a casual uncaring regard to safety and safety rules. Rockwell International's contract with the Department of Energy was ended over safety violations.

EG&G fired four employee's for disregarding safety procedures, and 53 others were disciplined for other rule and safety violations, and this they have done before restarting operations during a time they were trying to set up the safety program. Discipline for 53 ranged from written reprimands in their files to suspensions, with time off work without pay. Insubordination, poor performance, and failing to comply with security, were some of the other charges brought to bear by employees. EG&G also agreed and confirmed with Colorado Senator, that it was premature for the plant manager to have presumed the plant would have a restart before safety reviews were completed. The Defense Nuclear Facilities Safety Board, has recommended that a thorough review of all equipment, and buildings, once again with safety in mind for employee's and the general public, to ensure it would withstand earthquakes, high winds and fire. Would tornado's be a possibility?

The first release of medical records for some 600,000 workers at nuclear plants in Oak Ridge Tenn., Richland Wash., Los Alamos N.M., Mount Ohio, Rocky Flats, Co. going back to 1942, and (Manhattan Engineer District) Code name of the Army unit that built the first Atomic bombs.

ROCKY FLATS DUCTS

HOLD A QUIET KILLER

FEARED NUCLEAR ACCIDENT

WOULD OCCUR IN SILENCE

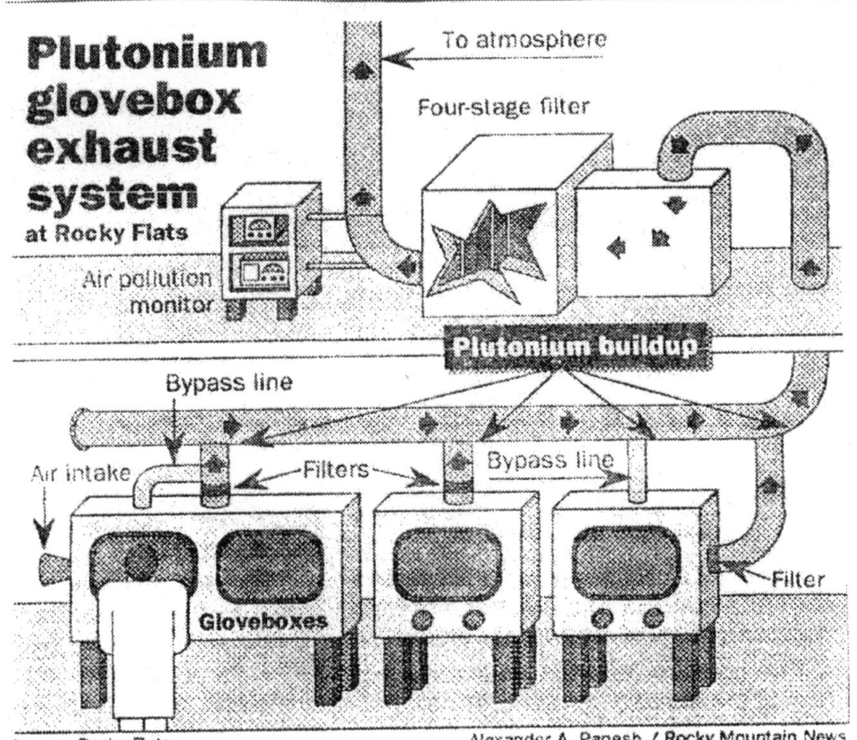

Source: Rocky Flats Alexander A. Papesh / Rocky Mountain News

DEATH FACTORY U.S.A. PART I, PART II, PART III WITH DEATH FACTORY'S IN OUR MIDST. DISCOVERING & UNCOVERING ENVIRONMENTAL DISASTERS. THINGS WE CAN DO TO HELP OURSELVE'S, OUR CHILDREN, & OUR FUTURE. WHAT WE CAN DO TO PROTECT OUR FUTURE, & OUR CHILDRENS FUTURE.

SOURCES OF INFORMATION

- New tips, Interested & knowledgeable citizens, Our sources from around the world,
- England, France, Germany, Belgium, Japan, Iraq, South Korea, & other Asian
- countries. Television News, Radio News, News papers, and any other news.
- Accountings here and afar, Life experiences, Investigation, and Discovery, of cold hard
- Facts about DEATH FACTORY"S, their products in our everyday lives.

New York Times	Public Information
L A Times	Public Information Bo Pal India
Denver Post	Court and legal publications
Rocky Mountain News	Bulletins, Reports, & public documents
Associated Press	Depositions of Military Officials
Stars and Stripes News	Libraries, and National

	geographics
Silent Spring	Environmental books of knowledge
Poisoned America	Engineering and Science
Army Officials	NBC news tips
Government Documents, from around the world	ABC new tips
Senate and Congressional Reports & Documents	CNN News
Environmental Protection Agency/EPA	TNN News
Colorado Department of Health an Environment	Personal Investigations

CHAPTER 19

HAS THE UNITED STATES ARMY, THE DEPARTMENT OF DEFENSE, GONE MAD, ONCE AGAIN WAGED WAR UPON IT'S OWN PEOPLE, OR COULD IT BE SOME TYPE OF TERRORIST PLAN USING VIRO-CHEMICAL WAR-FARE. THE BIO/CHEM. WAR-FARE HAS NOW POSSIBLY TURNED TO VIRUS, OR VIRO-BIOLOGY, OR MAYBE VIRO-CHEMICAL WAR-FARE. THEY'VE GOT IT ALL AT THEIR FINGERTIPS.

THIS IS A TRUE STORY OF TERROR, ONE OF HORROR, AND A GAME OF SUSPENSE. "WHO DONE IT?" WAS IT THE ARMY OR TERRORISTS.

"WE THINK THE UNITED STATES ARMY DONE IT AGAIN", OR POSSIBLY SOME TERRORIST GROUP ON HOW BEST TO CRIPPLE THE PEOPLE OF THE UNITED STATES. WE ARE ASKING OUR SENATOR MR. NORM COLEMAN TO ASK FOR A SENATE OR A CONGRESSIONAL INVESTIGATION ON THE "WEST NILE VIRUS".

This could be done at the same time the Congressional investigation into what happened in the 1960's in Minneapolis, Duluth, and across the American Wheat belt, Colorado, Wyoming, Nebraska, Kansas, Iowa, N. Dakota, S. Dakota, Minnesota, the Midwest, that has already been proven a time when the United States army, terrorized the nation with germ warfare, & chemical warfare agents, were used in schools on school kids, The same as what happened in San-Francisco, Los Angeles, Denver, New York and Chicago subway's as well as other large cities, & the Hotels where large groups of people will meet or congregate for Conventions.

We are asking Senator COLEMAN to please investigate the "WEST NILE VIRUS" WE BELIEVE The United States Army has already done what we would think impossible by using the people of the United States as Guinea pigs for war fare agents, nuclear agents, Sickening or debilitating agents, such as the Legionnaires disease. – Di-iso-propyl-methyl-phos-pho-nate, -Iso-proply-methyl-phos-pho-nate. Heavy metals, Mercury specifically.

Which we believe was again the army with germ warfare agents, now we believe the army has went one step further to Virus warfare agents. Using Virus, VIRO/Bio, or VIRO/CHEM, or Bio/CHEM warfare agents. Using them as teams, the Chemical, and Virus do the injurious, and debilitates the immune system then the biological germs take over to complete the job.

"WHAT THEY DONE IS UNEXCUSEABLE, IT WAS INTENTIONAL, IT WAS NO ACCIDENT". The U.S. Army has spread, sickness disabilities, and cancer as they spread fear and terrorism across the United States and the people were not aware of what had been done to Them, and their children, many still are unaware. Those millions in the Northwest affected by the Hanford nuclear weapons base, the millions that were affected by the nuclear testing done at Jack-as Flats Nevada are just now finding out what really happened in the 1950's and 60's, and how it affected them or will affect them now and in the future. What has been done has been proven "The U. S. Army holds the Intentional banner, and was caught testing warfare agents on the people of the United States of America, and targeting the young at schools, while targeting the old where they live. "The United States Army has lost all credibility". Would they be capable of testing the West Nile Virus on their own people? "YES' been proven they would be capable of testing Virus this time, and so would a terrorist if they had that capability. We talked in chapter 18 of what the army perpetrated, across the United States testing Ultra-Hazardous Bio-Chemical Warfare on their own people. This happened back in the 1950's, 60's, 70's, 80's, 90's, and now right into the millennium with the "WEST NILE VIRUS". There are 100's of thousands believed having been affected in Denver area living Any where near the Rocky Flats Nuclear Weapons plant who were all handed their experience by the Department of Energy when they violated every safety rule on record, so their products, radioactive plutonium, radioactive beryllium, could be produced at what ever the cost. For those who live around this nuclear plant or any other nuclear plant can take a lesson from those who lived around the Hanford Nuclear plant. The experience left them with pain, suffering, bizarre health maladies, shortened life spans and then they were handed what was dealt with over the years 1950's to the present where many are now dying from cancer. The young died of cancer at a very early age, after years of suffering.

ROCKY MOUNTAIN ARSENAL DEATH FACTORY USA

The Sister plant Rocky Mountain Arsenal who gave the Lands and 100's of thousands of Others their experience with years of pain, suffering, Bizarre health maladies, shortened life spans, and then the cancer's. Any person living near or around the United States Army, Rocky Mountain Arsenal Chemical and Biological Warfare Manufacturing Plant for many miles suffered for a life time. Many died of various forms of cancer, again the young were affected early on. Per Capita those around the U.S. Army Rocky Mountain Arsenal had the highest rate of Cancer in the World, even CHIR-NOBEL. The U.S. Army went through the Court system admitting they had contaminated people around the ARSENAL both by air and water, with chemical warfare agents, but accepted no responsibility because they acted under the plans of the United States, and the United States Court of Claims excused them for what they had done under a law called Discretionary Function, very few people like myself even knew this Existed.

Even though they had been found liable of "Extreme Malicious and Reckless uncaring Negligence and Wrong Doing. 1. Ultra-hazardous activities. 2. Duty to impound. 3. Negligence under both State and Federal law. 4. Airborne Trespass, 5. Fraud and deceit, with full knowledge of hazardous and toxic operations. 6. Willful suppressing of evidence they were using storage for hazardous waste that was leaking, in 1969 ordered by the pentagon to classify all information, and then continued it's operation for another fifteen years. Polluting more than 150 square miles, it was called by EPA the most polluted piece of Land on Earth, surpassing Russia's CHIR-NOBEL. In the Land's case against the U.S. Army, Judge Robinson was very clear in the record a case in the Court of Claims cannot by heard under the Tort Claims ACT, and this case in particular will be heard under Colorado Law, and not Tort Claims. It was pointed out several time during the Trial the Army was guilty under Colorado Law. The Army was found liable for what happened by fraud, deceit, by willfully suppressing evidence, Trespass by both air and water with Ultra Hazardous, and highly toxic materials. They failed to impound or control a dangerous, ultra hazardous and highly toxic material.

Judge Robinson, at the latter part of the Trial threw out Colorado Law, Would not allow evidence that had shown the water and the air were highly toxic and unsafe, and under Colorado Law there was every belief the water and air is what killed all Larry's livestock, his parents, and a child. He then placed it under a Tort Claims Act, which although there were hundreds of chemicals in the Land Wells, He then forced the Land's Attorney to name one chemical produced at the Arsenal and prove beyond a shadow of doubt that one chemical is what caused the death and injury. DIMP being the one chemical the Army admitted finding in the Land wells, using Colorado Standard for safety of water was several hundred percent above the safe level, But the Judge would not allow Colorado's safe standard entered into evidence, and forced what EPA is calling their standard, but under Colorado Law EPA cannot set a standard, yet that is the only range the Judge would allow. He affectively took Colorado Law out of it, and then forced it into the Tort Claims Law, and Larry's Attorney based the Cased under Colorado Law and the Judge would not allow him to challenge the Discretionary Function.

"DISCRETIONARY FUNCTION"

Claims based on the exercise of a performance, or failure to exercise or perform a discretionary function or duty on the part of a Federal Agency of Government Employees' are no Cognizable under the FTCA [Federal Tort Claims Act]. The purpose is to protect the "Government" from Liability that would seriously handicap efficient Governmental operations. "whether or not the discretion is abused is immaterial. United States V. 331 f.2d 498 [10th Cir. 1964].

Even though the Court of Claims was destined to by the law of the United States, Judge Wilkes C. Robinson stated in fact, "Plaintiffs' argue that the Army's conduct on the Rocky Mountain Arsenal [RMA], Manufacturing and demilitarizing of chemical weapons and disposing of chemical wastes, created a substantial risk. Furthermore this Court Agrees".

Judge Wilkes C. Robinson stated of record "Negligence is not a prerequisite for liability in a groundwater pollution case. "NO TRIAL IS REQUIRED ON THE ISSUE OF LIABILITY".

Once the Court had rejected Colorado Law completely, and stopped any further action against the government by dismissing all rights to a [Legal claim, or an Equitable claim] against the United States. That really didn't bother the Land's in their quest to be made whole again. The Courts then awarded claims called a Gratuitous award, funds for the award to be appropriated by the Congress of the United States. As the Judges stated, "That award is to be based on the facts present in the record before this Court and the Review panel.

The reason it is turned back over to the Congress of the United States is very simply, [A Gratuitous Award] is "An Honorary Obligation, the Nation speaking broadly owes a debt, to a party when it's claim grows out of general principles of right and justice". "A Moral and Honorary obligation such as are binding on the Honor of an individual, or a Nation". This debt obtained recognition in this Court of Law, it was grounded on the conscience of the United States of America and it's Congress.

The Congressional Bill H.R. 816 was then filed with the clerk of the United States Congress in February 1997, for appropriate action on the awards, and appropriations made to the Lands By the full House of Representatives. This taking the Army of the hook for any legal Or equitable claims for their malicious negligence, that the Court must decide the penalty or amount of damages and the legal claims could go into Criminal Charges that would be handled in a criminal court.

This bill was turned back over to Congress, in February 1997, and the Land's Senator's, and Congressmen from Colorado were to handle the claims handed down from the courts using the evidence and admissions of the Army of fingerprinting toxic and hazardous war-fare chemicals from the Rocky Mountain Arsenal Army Manufacturing plants directly to the Land Wells. The Congressmen, and Senators of Colorado "Have SERIOUSLY FAILED the Lands". Each one was asked how much more do I have to provide what happened, It's in the Courts records Washington D.C. The Defense for the United States Army stated specifically as near as can be remembered. Directly to the Court, "WE admit that chemicals used to manufacture deadly nerve agents, Sarin Gas were found in the Land wells, and can have only one source, The [Rocky Mountain Arsenal].

The Congressman, and Senators of Colorado were expected to do their job according to the law. Senator Wayne Allard, Senator Ben Night-Horse Campbell, Congresswoman Diane DEGETTE, Congressman Mac McGinnis, Congressman Bob Schaffer,

Congressman David Skaggs of course the Congressman Hank Brown that introduced the original Congressional Bill, and was approved by the full House of Congress. It was then turned over for investigation by the United States Court of Claims, and to return their findings to the House of Congress, which they did. Reported a Gratuitous award to the Lands, based on the facts in the records of the United States Court of Claims.

Congressman Hank Brown now the President of Colorado State University in Greeley Colorado. The Land's wish many times over he were still in office, because they know he would not have let them down as they have been.

The United States Army is known to use their own people as guinea pigs or subjects that are at their disposal, right here at home and it has been pointed out much like Legionnaires disease it seems to be geared toward the Young and Elderly……,those with compromised immune systems……, Cancer…..or other chronic heart or lung conditions… They are, and have been the guinea pigs the same as they were during the tests made on the legionnaires disease. They have used Biology, and radiology before, now it's new called VIRO-BIOLOGY OR VIRO-Chemical warfare, and where is the best place to test out the product of course the Army knows how to do it to their own people and not get caught, and then if they get caught they first deny it then they just classify it, where not even the Congress of the United States could investigate.

VIROLOGY

The West Nile Virus" was first heard, and feared in 1999, and only in New York, it is believed they did not get the results they expected, it was kept very quiet as they reset what we believe was their traps. It only killed 7 people then. So they went back to the drawing table, the difference as it spread across the United States in 2002 has been overwhelming as it has affected thousands, and killed hundreds. sure they lost a couple of years but in 2002 they came back with the new "'West Nile Virus", which has proven to be very effective. In fact so effective. The United States Government has now paid upwards of $50, to $60, million dollars trying to find a cure, because it is a virus there is no cure yet, Antibiotics do not work on viruses, will only allow it to do it's destruction, then allowing also the germ part, such as pneumonia, Bronchitis, Lung disease, Heart disease, to step in and take them out.

I sit in AWE wondering why did they do this. We know now that the United States, France, Germany, and Russia has handed out all the secrets and how to manufacture germ warfare, chemical warfare, Bio/Chemical warfare, Iraq was given instructions, and equipment. To several 3rd World Countries including IRAQ. A radical and piranha nation, one the World cannot live with. The Bio/chemical weapons of mass destruction the United States of America handed out to Iraq, during the late 70's and thru the 80's warfare knowing full well how Saddam Hussein was going to use it, and the United States stood behind him, up until the dessert storm, when Saddam Hussein turned the weapons of mass destruction against the United States, and the United States had no idea how to treat the ill and injured in a war that became BIO/CHEMICAL that was then turned against their own people, with their own weapons of mass destruction used.

The soldiers in the Persian Gulf war found themselves wiring for demolition of weapons, and weapons of mass destruction the United States had sent to Iraq during the 1980's in fact some shipments were even made as late as 1990 around the time of the Persian Gulf War. The massive Iraqi ammo dumps to be destroyed were all clearly marked, United States. The weapons inspectors never discussed what type weapons they were making sure were destroyed, but word came out that many was American made weapons and some were weapons of mass destruction, also BIO/CHEM where the chemicals and germ warfare used simultaneously.

The United States were not ready for this, they hadn't thought of how lethal the two would be together. Russia already knew how to mix germ warfare and chemical warfare together to triple or quadruple the consequences, and the United States should have Known, but didn't and to top it all off they didn't know how to treat the Vets. With exposures to BIO/CHEM weapons, who were not only exposed, but brought it home unknowingly exposing their families, The United States turned their back on the sick, and injured veterans of the Gulf war, because they didn't know what had actually happened. 1990 PERSIAN GULF WAR, VETS EXPOSED TO SARIN NERVE AGENT WERE NOT TOLD, for 10 long years, THE PENTAGON HAD INFORMED THE WRONG VETS.

OCTOBER 28, 1000 Pentagon sending letters to Gulf War Vets amending their status. The Pentagon is Reversing itself on who was exposed to nerve gas during the Gulf war and who wasn't. About 30,000 Vets are to be notified in the coming weeks that they probably came in contact with low levels of SARIN Nerve agent after being told over a number of years they had escaped exposure, even though they had thought to have been exposed to at least chemical warfare agents [SARIN & GB, and probably Bio Warfare agents, weapons of mass destruction are to be notified by letter, of the reversal based on a review of weather and locations. Whether or not it may have been CYCLOSARIN, And 30,000 believed to have been exposed will get letters saying they probably weren't.

The Pentagon is claiming that the Gulf ware illnesses were probably very low level exposures, low enough that there is no expectation of a health risk. This has been a decade-long awaiting the answer to health questions of those who served. This will raise a question of credibility for the government, and the Pentagon.

The revision seems to be based on troop location and whether conditions, and the issue in question is what was the health effects, and how many were exposed. The Khamisiyah weapons depot demolished by U.S. Forces destroying munitions and rockets containing the highly toxic sarin, and the even more lethal cyclosarin that paralyze the nerves, completely shutting down the lungs and other organs, both Ultra Toxic weapons of mass destruction.

At one time after the war was over they were treating some that were not involved in being exposed, and weren't really sick while leaving those who really needed the help without treatment, treating them as Psychogenic. Until they found out they were treating the wrong group, and paying no attention to those who needed it for nearly eight years.

It's now mid September & the West Nile Virus is still growing with many deaths left in it's wake. If you need a blood transfusion, "Don't do it" The FDA, suggests you consider alternatives, such as storing your own blood, or delaying elective surgery, to help avoid the risk of West Nile Virus.

The West Nile Virus has been found to be present in at least some of the nations blood supply. The sometimes deadly mosquito carried virus has health Officials doing everything possible to discover a test that could save hundreds of thousands of unit's of the donated blood bank.

The West Nile Virus has now climbed to 16 confirmed cases in Minnesota, 14 confirmed in Wisconsin. 2 have died in Wisconsin, and none so far in Minnesota. Encephalitis is a potentially fatal Viral infection of the lining of the brain that will cause the brain to swell. In Grand Forks North Dakota reported an elderly man died Officials say the virus was probably the killer. As far North as Grand Forks it becomes real scary, and there has been cases reported across the border with Canada.

The positive way this disease has spread across the United States, has all the potential signs of Terrorism. To be a Mosquito borne disease as they claim it is, means the mosquito laying the eggs was infected passing it onto the young. You can almost see the fingerprints across the States as this disease was spread by adult mosquito's from a Laboratory that were infected with West Nile Virus. They became the carriers that passed it on State by State clear across the United States. That is why I'm asking our Senator Norm Coleman who is already familiar how the Army was testing bacteria, Chemicals, and a deadly fungus across the Midwest on unsuspecting human guinea pigs, even tested on school kids.

This is 15, day of September 2002, the disease West Nile Virus has now blanketed 42 states, Massachusetts, and Pennsylvania have now reported their first cases. Illinois is carrying the highest figure for death's is now 16. With 1438 cases now reported across the nation. Protect yourself and your children, avoid going outside at dawn or dusk and use mosquito repellent. Don't stand in water around your home, they say the risk of getting sick is very low. West Nile Virus, is a poison, akin to venom and is the causative nature of an infectious disease, such as injury by chemicals, that open the gate to bacteria antibiotics do not work on the virus yet do wonders on bacterial infections. It takes the away the ability of the body's immune system to produce antibodies that tends to ward off the virus, however this sometimes depletes the immune system, allowing secondary bacterial Infections to take over. This is when medical intervention is very much necessary to rid the body of the infectious bacteria. Remember they always said you can't cure a cold if it turns into pneumonia it can be treated, and cured. Pneumonia is a secondary bacterial infection, can be fatal if not treated.

September 23, 2002 the death count across the United States from West Nile Virus rising, the cases reported at this time is nearing two thousand and growing rapidly. This has also caused the death of thousands of horses across the United States from encephalitis. Your thinking what good does it do to know it now? well this same situation could spread across the Country again next year, and you can be better prepared. A strategy for defense against smallpox.

The Federal Health Officials are issuing detailed guidelines today for vaccinating the entire U.S. Population against Smallpox, and all within 5 days should there be an outbreak of the dreadful disease. This is much like painting a picture, and is intended as a blueprint for all state, city an county health officials. This move is unprecedented and reveals an Over whelming belief amongst the Bush Administration, and even one case of smallpox any where in this Western Hemisphere would signify another terrorist assault.

Two of the experts involved in planning the mass vaccination response, to any terrorist attack, stated it would trigger a far more massive response than officials had previously planned for.

The work book being sent to all health commissioners in 50 states and the District of Columbia offers an explanation on how a mass vaccination clinic would have to operate. from the logistical issues of parking to the medical challenges of vaccinating thousands, and then treating the side effects that can be very severe and sometime fatal.

IT APPEARS THERE MAY BE FINGERPRINTS IN THE SAND

Does this mean Saddam Hussein has the Small Pox Biological agent, A Weapons of mass destruction. Just maybe it is in the hands of Al Quita terrorists. There has to be a reason the United States is gearing up for mass vaccination, do they in fact know that Saddam Hussein has that in his arsenal. Is it now being released to terrorists who threaten its release on Israel, and the United States.

Materials of mass destruction that were known to have been supplied by the United States of America to Saddam Hussein Iraq, to be used against the Kurds of the North and South Iraq, and was to be used on the Iranians during the 1970's and 1980's, Then in 1990 Saddam Hussein turned those weapons of mass destruction on the United States Military during he Persian Gulf War.

Several shipments of Escherichia

Coli (E. Coli) and genetic materials

[human and bacterial DNA] were

Sent to the Iraq Atomic Energy

Commission

Shipped: August 31, 1987, April 26 1988

January17,1989January 31, 1989.

- Other materials listed were shipped directly to Iraq
- SACCHAROMYCES CEREVESIAE
- BACILLUS SUBTILLUS
- STAPHYLOCOCCUS EPIDERMIS
- SALMONELLA CHOLERASUIS
- KLEBSIELLA PNEUMONIAE
- BACILLUS PUMULUS
- ACILLUS MEGATERIUM
- CLOSTRIDIUM TETANI
- BRACILLUS ANTHRACIS

See chapter 10 for more details on what was shipped

Had the United States sent Small Pox, or maybe I'm just not reading that in the scientific names given to the materials that were sent. At the same time he was getting Bio/Chems from the United

States, our source states he was getting all the technical support setting up his manufacturing plants from Russia, France, and Germany and also from the United States. I believe he does have in his arsenal Small Pox, and has made some sort of deal with terrorists.

THIS PLAN WOULD BE ACTIVATED ONLY IN CASE OF OUTBREAK

Smallpox one of the worlds most feared diseases had been eradicated in the United States now for several decades, and World wide since 1980. As a Biological weapon of mass destruction, it is contagious and often incurable leaving death and devastation in it's wake. There is probably little to no immunity left from vaccinations in the past since routine vaccinations were halted in 1971 here in the United States, and very little expertise in dealing with either the disease or the vaccine.

There are only two Countries known to possess any stockpiles of the virus, but there is some concern as to the possibility that the piranha nation of Iraq has acquired some and may have set themselves up in the manufacturing, and producing their own stock piles. The real concern here is Saddam Hussein and his capabilities with biological weapons, it is being predicted at some point Iraq may attempt or use smallpox, anthrax, amongst a large array of biological weapons. His supplies were already large during the time of the Gulf War. The list of the biological weapons he was given and set up to manufacture was quite unique since most of the technology and beginning supplies were shipped directly to him by the United States.

That is why the Government from the top on down contend that the risk of attack might necessitate inoculation for every American. One of the real concerns regarding Saddam Hussein and his arsenal of Chemical and biological weapons may also now have VIRO capabilities also such as Smallpox. Smallpox is or may have already been handed over to the Al-Qaida Network. We have also heard through the grapevine that Saddam Hussein has got radio active material in his possession, enough to build radiological Bombs, and has supposedly offered them to Al-Qaida operatives who would in a minute use on the Untied States or it's Interests.

"Ring vaccination" is a blueprint that was released by the CDC last winter for containing a smallpox outbreak. This was the same strategy that was used during the campaign to eradicate the disease. The way this works is by starting with those closest to the exposure and then working outward in concentric circles called Ring Vaccination.

The Ring Vaccination would probably not work in a scenario with terrorists. They would have the ability to release the virus in several locations simultaneously. The ring idea grew out of a naturally occurring outbreak during a time air travel was rare. A five day time frame, is what is expected for a mass vaccination, Federal officials are requiring all states to develop that plan. The vaccine given in that time frame will provide immunity for the disease. Terrorists will strike in many places simultaneously, and for that reason everyone must be ready. There are plans to ship the vaccine to each individual state letting them take control and handle the inoculation. The National Pharmaceutical Stockpile, will be immediately shipped and ready every city and town within 12 hours.

Each state will become it's own intricate part of the plan, each state would decide the locations of the vaccination clinics, staff them, and a plan of action as to how hundreds of thousands of people can be processed quickly and calmly. The logistics will be handled by the experts, the Military, UPS, Federal Express, perhaps even the post office. Large shopping centers, or large sports arenas' would have parking space as needed, and some metropolitan areas may rely on buses and subways to bring people in. Bio-terrorism is a huge challenge for everyone, and requires immediate response and action.

Federal officials have released very little information as to whether there has been a potential threat on Bio-terrorism or not. The Government is certain they want the country to be ready for whatever may happen a possible smallpox attack. There is going to be a mass necessity and need for volunteers, to staff health clinics around the clock for at least a full week. Vaccinating 288 million Americans calmly and within 5 days. This is the largest inoculation program that has ever been undertaken.

SMALLPOX VACCINE MADE WITH LIVE VIRUS VULNERABILITY HAS INCREASED

There are serious side affects from smallpox vaccine, many more today are at risk. Especially, those with skin disorders such as eczema, pregnant women, children under 1 year old, those with any type of cancer, immune system disorders, those with lupus, even those that may have the eczema in childhood are at risk. Today your looking at millions and not thousands, and all must be done within 5 days. Those vulnerable would be endangered not just from being vaccinated but also from being in close contact with someone who has recently been vaccinated.

For those with serious reactions, it can be countered with vaccinia immune globulin, or VIG. The availability would be far in short demand, since the nation has only 600 to700 doses available. The government has done everything possible to speed up production of VIG says spokesperson at the terrorism preparedness and response to the Centers for Disease Control and prevention. 600 to 700 doses of VIG would certainly not be enough if there really is a Bio-terrorism attack……. The risk of dying from small pox…… About 30 percent…….greater than the risk of dying from a reaction to the vaccine.

Federal policy directs health workers to isolate infected patients first, then vaccinate people in close contact with them, actually forming a series of rings of immunization around the outbreak setting up barriers to its spread. The vaccine is one of a few immunizations that can work even if a person is already infected. The vaccine can give full protection for people if given within four days of exposure to the virus.

"CDC"s level of preparedness is very high, the plans, the policies, the people, and the product, practice will be necessary to make sure we are ready to respond. The new plan addresses only the most comprehensive response to an outbreak of small pox, It fails to address giving vaccinations before an attack or for instance an outbreak, it deals only with after the fact. A real potential problem is how you ensure that a vaccination process is done orderly and people don't panic.

WHAT ABOUT THE GOVERNMENT, GEORGE BUSH AND THE REPUBLICANS, OR TOM DASCHLE'S & TED KENNEDY AND THE DEMOCRATS OVER TERRORISM.

Hoping to build support for tough action against Saddam Hussein, and Iraq, President Bush reiterated possible ties between Iraq, and Al-Qaita, with contacts that were made between Al-Qaita and Iraq. Saddam Hussein has been accused of training, sheltering, and supplying Bio/Chemical weapons to Al-Qaida. Even though bin Laden's religious driven motivation, is against U.S. presence in the Arab World, he has an overwhelming determination to overthrow

Secular politicians.....likened and to Saddam Hussein.....favors creating a fundamentalist Islamic regime.

Some Al-Qaida operatives have been given refuge, training and possibly given Weapons of Mass Destruction. Larry says the West Nile Virus, and encephalitis has been fingerprinted and foot traced across the United States, and with warmer weather now coming back would give the terrorists a second chance at lowering the immunity of the United States, like I said it seems awful strange how this virus is striking the young and the old. The West Nile Virus is the work of terrorists, it not only left foot tracks across the United States, if left fingerprints of terrorists. Or the U.S. Army.

Now we may have a new and deadly disease that will be spread across the United States, ""SMALL POX". It is far more deadly and sinister that the west nile virus. The President of the United States has done everything possible to warn the people of this country about the sinister deeds and preparation of the Al-Qaida, and Saddam Hussein Regime. We are being warned and the Democrats on capital hill are not listening to realistic happenings. The President has made it very clear to Congress and the World we may be faced with Smallpox yet Congress have dragged their feet for months, not just days. If this is not stopped the world will suffer a 9-11 much more sinister and deadly than that of the twin towers.

If we have a Smallpox epidemic, and it will be epidemic strong, I personally blame the Democrats in Congress, Tom DASCHLE, Ted KENNEDY, & even Richard GEPHARDT and those who drag their feet, and fail to stop it before it happens White House Spokesperson ARI FLEISHCER said it right; was "time for everybody concerned to take a deep breath, to stop finger-pointing, and to work well together".

Defense Secretary RUMSFELD presented evidence to U.S. Allies in a closed meeting of NATO ministers in Poland that established a solid link between Iraq & Al-Qaida Defense Minister, stated he didn't see anything new in RUMSFELD'S briefing. Larry says, "And rightly so, because RUMSFELD has been forced to repeat, what he has laid out, and it seems that the Democratic Congress & foreign ministers don't want to believe the truth.

Larry says again, ""Even though Rice's disclosure was significant, he was able to mark What the White House has said many times over, AlQuaida is operating within the Saddam Hussein controlled part of the city of Baghdad."'' "Should I say Let it be said AlQaida and Saddam have been seen holding hands."

Members of both parties have agreed this eruption was major and could slow the drive for the Joint Congressional Resolution authorizing President to use force in Iraq, as well as the Homeland Security Department....Congress is now making some headway on drafting a U. N. resolution that would give Iraq two months on giving Iraq, or Saddam Husssein a chance to demonstrate his willingness to cooperate with weapons inspectors, that too has now been slowed by divisions Amongst Western Allies.

It appears that President Bush and his administration has been trying to prepare the United States for a possible attack from Al qaida, or Iraq, with Weapons Grade Smallpox. Saddam Hussein, Iraq have now had ten years to produce a mass quantity of all his Biological, VIROLOGICAL, and Chemical Weapons. He has also been handed expertise in combining each, to "VIRO/CHEM, BIO/CHEM, VIRO/BIO, OR VIRO/BIO/CHEM. The United States of America and the World has not even seen or heard of any more than the tip of a deadly and sinister iceberg. I realize I have named some above that the United States is not even aware of, and I know where the technology came from. And have notified the FBI several years ago now. It has been said by Terrorists the best place to hit America is where they live, and what they do, and the worst would go unnoticed.

The foot dragging by mostly our Democratic Congress is going to drag our nation through the gates of hell on earth. I am Independent but I have enough common sense and have heard enough to believe our President is doing the right thing, and if Democratic Congress keeps his hands tied and restrict his powers. May God have mercy on the Democrats and what may happen with Smallpox, or even worse. The Risk of attack might necessitate inoculation for every American, and putting millions at deaths door. It's already been predicted that Iraq and/or Al Qaida will use Smallpox on the United States, The same as the Anthrax attack was predicted. They will have the mobilization to release the Small pox virus in many locations simultaneously.... Upon every American. Smallpox is so sinister and so real that even the vaccine has some very serious side effects, and will kill many if we are forced to do the mass inoculation. Those with even any type skin rash, any type cancer, Immune system disorders, or eczema, even those who had childhood eczema, and not as an adult, are all at risk. The tiny babies would even be at risk, although the CDD has taken special precautions in their case, but still must have the vaccination. Those vulnerable would be in danger of being vaccinated, and in danger with close contact with someone who has had a recent vaccination.......... The risk of dying from smallpox is about 30 percent greater than the risk of dying from a reaction to the vaccine. A mass vaccination could put millions at risk, no matter how you look at it, from taking it,, and would grow 30 percent from not taking it.

My advice to everyone write your Congressman, and advise them to give the support to Our President George Bush, he will need, to possibly stop the invasion from Iraq, and Al Quiata before it happens. If it is not stopped in its tracks it will happen, then I ask you how will you feel about your Democratic Congress who drug their feet in this World emergency. So we may be able to avoid a mass inoculation, by stopping an attack with Smallpox on our nation. Then we can all say we helped prevent a World Catastrophe of deadly Smallpox.

There is no doubt left that Saddam Hussein, and the Al Qiata network now have in their possession the deadliest weapons of mass destruction. Saddam Hussein has fooled the World now for eleven years giving him the time to produce the deadliest weapons in history. He is leaving nothing to the imagination when he continues for 11 years to turn his back on the United Nations, and the World, After 11 years to abide by the U.N. and allow weapons inspectors to return, “SO HE SAYS”. Iraq’s foreign minister pledged to allow United Nations Weapons inspectors to return to his country, “without conditions” for the first time since U.N. experts left Iraq. “That is what would have been expected”. KEY FACTORS THAT WERE QUICKLY USED TO STALL STRONG ACTION FROM THE UNITED NATIONS. UNITED STATES WAS SKEPTICAL.

Since Iraq agreed to allow the, “UNCONDITIONAL” return of U. N. weapons inspectors. Iraq also called on the U. N. Security Council, that includes the United States to, Respect the sovereignty, territorial integrity and political independence of Iraq.”

The white House had dismissed the offer a tactical step by Iraq aimed specifically at dividing the Security Council. A senior State Department official said Iraq’s motive here was, “not a promise to disarm, not a promise to allow unfettered inspections, not a promise to disclose the state of its weapons program”. The United States demanded a council resolution, "that will actually deal with the threat Saddam Hussein poses to the Iraqi people, to the region and to the world."

They have had 11 years to make SARIN nerve agents, Probably VX nerve agents, both VIRO/Toxins, and Bio/Toxins, such as anthrax, and Smallpox, & yes he was working on a master plan of nuclear weapons, Supposedly had in his possession plutonium, and Beryllium, other radiological materials, along with radio-logical technology, technicians, Equipment, and Uranium for dirty bombs.

"CHRONIC WASTING DISEASE" WHAT IS IT? WHERE DID IT START?

As for dead deer found in certain areas, there was no evidence of any of them being shot, poisoning is the only other possible cause being explored, besides Wasting Disease. Some were found near water, deer suffering from certain health problems will go to water, and that may have caused the clustering of the dead animals. One would surely not expect to find 12 to 15 deer killed by chronic wasting disease to be found in a very small area.

Wisconsin DNR are preparing to test at least 500 deer from every county during the fall hunt, to check the extent of the disease. The DNR also are also planning a 25,000 deer hunt in a 3889 square mile area around mount HOREB in an all out effort to eradicate the disease.

The chronic wasting disease is terribly contagious, an entire herd 350 elk to be destroyed by midweek. The slaughtering of Elk in Colorado continues. Because one of the cows in this herd showed positive for Chronic wasting disease, at a ranch near Del NORTE the entire heard must be destroyed.

Both State and Federal agriculture officials hope by mid week they can be done with the Euthanizing of 350 elk that were exposed to the fatal disease. There have been crews working constantly since Friday of last week to kill the elk on the Rancho Anta Grande Elk ranch near DELNORTE after one cow elk that had been purchased from a ranch near Stoneham had tested positive for the fatal disease.

There are so many diseases going across the nation, and the wasting disease is just another that has left it's fingerprints of terror on America. This disease again has not spread as a contagious disease would where the old method for controlling a contagious disease is by the ring method. This is another that pops up in groups far away from the last as if some one has containers and goes to the exact spot they need to contaminate and through a contagious disease into one of epidemic proportions. There is no way in Gods Green Earth can a disease spread so out of control as did the West Nile Virus, and to be mosquito borne, because it was human manufacturing of the disease that caused it. It would be very simple to breed in a laboratory and then take the egg laying mosquito's to an area where water may stand in puddles, long enough to lay eggs and you have Mosquito borne West Nile Virus. This is being done by terrorists right here in America that know exactly how to spread the disease from shore to shore. If your in Miami trying to control an outbreak, using the ring method for control. It's working well then it pops out in New York, to Minneapolis St. Paul MN. Then Tulsa Oklahoma, then Dallas Texas. Sioux City Iowa, to Chicago ill, for me to see this as it happened it was done by Terrorists from the United States or Terrorists from abroad.

The Terrorists doing this know exactly what they are doing, and this will continue until they are stopped. The smallpox virus if it does begin, can only be started by terrorism, and if it spins out of control and moves rapidly throughout the United States, it will have been terrorism.

Update on West Nile Virus, is known to be transmitted by blood transfusion, or Organ donation, & is mosquito borne, but health officials are now saying it could be transmitted through nursing as well since it has been detected in breast milk. The Centers for disease control are saying it is possible for West Nile Virus to be acquired through breast milk. It seems statistics show it to be rare if it occurs at all, since only four out of the 2206 U.S. cases reported have been in children under 1 year.

The possibility of how this 40 year old woman that had delivered her baby on Sept. 2, showed positive with the west Nile virus her and the new baby went home on Sept. 4, and by Sept. 17, she was re-hospitalized with a confirmed case of west Nile virus. She had received a blood transfusion the first and second day from giving birth. The baby was breast fed for two weeks, and did not fall ill, and mom recovered.

The discovery of genetic material from the West Nile Virus were found in initial tests of her milk. There have been no signs here, or abroad where the West Nile Virus seems to be running rampant that it may be spread through breast feeding.

Most blood-borne infections could not be spread orally because they cannot survive the stomach's acid. However a tick-borne encephalitis, a cousin to West Nile Virus, can be spread through the milk of infected cow's or goats, CDC said it would be theoretically possible West Nile Virus could also be spread orally. A sample of blood from the donor who's blood had made up her transfusion tested positive for West Nile Virus.

Thanks to Washington post, L.A. Times, Associated press.

We have vaccinations now for pneumonia, and we have the shots for a myriad of flu's. The West Nile Virus is here to stay says top federal scientists. A vaccine to protect the elderly from West Nile Virus, could be available within 3 years. Methods for testing the blood supply against the infection might be in place next summer, Federal Scientist's advised Congress.

This mosquito-borne virus has infected so far this year 2000 people in 32 States, and killed 98. Particularly worrisome and scarey was the recent discovery that West Nile Virus can be spread through blood transfusions. By those donating blood shortly after becoming infected. West Nile can also cause occasionally a polio like paralysis.

Public Health Specialists are expressing cautious optimism, while West Nile virus is here to stay they expect infections to be dramatically lower in the coming years......Possibly as early as next year....as more will become immune, and with quick response from the communities across the nation to act quickly each spring and summer to destroy mosquito eggs and the breeding grounds. West Nile Virus spread so quickly, and so rapidly to such a large area, and was so sporadic, with the fingerprints of a terrorist operation perhaps set up in a well known laboratory here in the United States. The transporting of the egg laying disease bearing mosquito's was done undetected, and with speed across the United States.

ACAMBIS biotechnology company, is planning to begin tests with a few dozen people soon to see if the vaccine is safe, then later testing will prove it's effectiveness, it could very well be made available in three years, Dr. Anthony FAUCI of the National Institutes of Health told a Senate hearing. More immediately, the Food and Drug Administration hopes to have blood banks testing for

West Nile Virus in donations....even if this means using an experimental test....next summer, said the FDA's Dr. Jesse Goodman.

Below is news by Lou KILZER Rocky Mountain News on "CWD", & "KCJD".

Something more adding to the dilemma Mad Cow disease & Chronic wasting disease, yet could be a significant finding in a treatment or cure fore mad cow and wasting disease.

A team of scientists, and a Denver researcher headed by a Novel Laureate, discovered high levels of infectious prion proteins in mouse muscle tissue. This being a significant finding after the debate of "mad cow" and related diseases such as the Wasting disease.

This new finding could raise issues about the safety of eating bear, and venison if the pattern found in mice are repeated in infected cattle, Elk, or deer. Patrick Bosque, assistant professor of neurology at the University of Colorado Health Sciences Center, said "finding the infectious prions in high concentrations in muscle was surprising." Previously researchers said the protein was mostly relegated to the brain and spinal cord.

Bosque, first author on a paper appearing today in the proceedings of the National Academy of Sciences, joined with Nobel Prize Winner Stanley PRUSINER to call for immediate studies to see if the Prions also exist in the muscles of cattle afflicted with "mad cow" disease, as well as deer and elk suffering chronic wasting disease.

Colorado is not alone in this battle it has spread rapidly to the North West and to the Midwest Wisconsin, Iowa, Minnesota and into Canada. The disease found in Ranched elk in Colorado has led to the slaughter of 1,500 in an attempt to eliminate the disease in the domestic elk.

Mike Miller, Colorado's Division of Wildlife veterinarian, has had the opportunity to read Dr. Bosque's paper stated "He lives here in Denver so hopefully we can work together on his research, and how it applies to deer, elk and livestock.

Earlier studies had found prions in very low concentrations were found, but it was thought that it had merely migrated from nearby nerves. Researchers have suggested the meat of deer, elk, and cows that show no damage in the brain, would be safe to eat. However this view could be altered with the new findings.

Dr. Bosque's research will be geared toward finding whether certain prion diseases might actually originate in the muscle. Mad cow [bovine Spongiform Encephalopoathy] "BSE" has infected 180,000 cows yet only 123 humans, he warned about overstating the findings. The disease in humans called vCJD is unknown at this time how it will progress, but the number of new cases last year compared to the previous year were down.

New research by the National Institute of Health suggests the infectious prions can cross species without being detected, but to cross back to the original species in a more virulent form. Much more is unknown that is known about the disease, Prion research is still in it's infancy. Dr. Bosque said infectious prions were not detected in quantity, only in the hind limbs. He noted that was really an odd finding and cannot totally explain it.

Dr. Bosque also noted that certain degenerative diseases favor certain muscles over others and a similar process might affect prion distribution. Remember the concentration was high in the hind muscle…a million or more particles per gram…with only 10 billion per gram found in the brain. The finding of prions in muscle could help provide a way to detect The disease in other animals and humans.

"TSEs" transmissible spongiform encephalopathies, as a class.

A. "scrapie" in sheep. Is another form of CWD.
B. B "CWD" chronic wasting disease in deer and elk.
C. Creutzfewld-jakob disease in humans.
D. D "BSE" in cattle, a form of CWD.

Veterinarians from all over the country are more interested in studying the Dr. Bosque report on his study of prions in the muscle tissue. It wasn't until BSE jumped the species barrier and started killing humans then became a real concern about CWD in deer and elk which had been considered an oddity rather than a serious issue.

THE UNITED STATES DEPARTMENT OF DEFENSE & THE PENTAGON, HAVE USED HIGHLY TOXIC BIO/CHEM WEAPONS OF MASS DESTRUCTION ON THEIR OWN PEOPLE.

Documents show how the military used BIO-CHEM warfare agents of mass destruction, during military exercises on American soil, Canadian soil, and British soil. The reports that detail tests that were conducted between 1962, and 1971, will reveal for the first time that the chemical warfare agents were used during exercises on

American soil, Alaska, Hawaii, and Maryland. A mild biological agent was used in Florida, and that CS gas a riot-control agent, was used during tests in Utah.

The Department of defense finally acknowledged much wider testing toxic weapons on its own forces as well as citizens across the United States without their knowledge. They then classified all document keeping it secret over 40 years during the Cold war and for 10 years after, and finally their going to tell half truths. Larry says, as far as he is concerned the Defense Department, and the Pentagon have lost all credibility across the America's and the World.

It was just during the 1980's when the United States Supplied Saddam Hussein with VX, SARIN, amongst others, and also supplied him with deadly biological agents which includes Smallpox. The butcher of Baghdad then turned the weapons of mass destruction on his own people, Kurds in the North, Kurds in Northern Iran, and the Kurds from the South, and included part of his efforts on the Iranians.

For this we have been calling Saddam Hussein the butcher of Baghdad. Tell me what do we call the United States, Department of Defense, and the Pentagon, do we blame all our leaders, or which one. The newly declassified reports describe how Bio/CHEM exercises previously undisclosed, VX, and SARIN, both deadly weapons of mass destruction used to test the vulnerability of American Forces to an unconventional attack. There is about a dozen reports describing the more benign substances were used to mimic the spread of the poisons in other tests.

Pentagon Officials are now trying to say according to their investigations indicated that no lethal chemical agent dispersed into the general population, they claim some milder substances did escape into the atmosphere. There are documents showing a plant fungus in an area of Florida. A naturally occurring bacteria in Hawaii and a mild chemical irritant in a remote part of Alaska.

Even though the military personnel were given protection from the toxins, This was the best available at the time though they conceded that the equipment was primitive didn't work well, at all compared to what is available today. The Pentagon also revealed that ships and sailors were sprayed unknown to them with BIO/CHEM agents merely as a part of the Cold War-ERA testing. The Pentagon calls those tests different because they took place on the high seas.

The more than 5,500 people who were believed to have participated in the land and sea tests it is unclear even today as to whether any of the soldiers and sailors were aware of the subject of the exercises, nor the potential risks. The Defense Department is supposedly working with the Veterans Affairs to identify those involved, and the affairs of their health, and or the cause of death.

Congress has scheduled hearing as we speak, and will continue to examine the documents, and try to figure the governments responsibility to any veterans that may be suffering from ill effects, and their families. The House Veterans Affairs subcommittee on health will meet behind closed doors.

I'm not sure I understand why the United States is condemning Saddam Hussein, and call him a killer of his own people. His testing on his own people was probably less sophisticated, and he killed his people right away. The sophisticated methods used by the United States was at probably low levels, you go through a lot of pain and suffering, and you die well before your time. There is no excuse for What the United States Army, the Department of Defense, the Pentagon done to their own people Far worse than Saddam Hussein, is thought to have done. "Possibility maybe his own military forces have done this and condemned Saddam Hussein the President of IRAQ". "Which President and which Congress does the United States Condemn for the same crimes on their people" They used their own people as guinea pigs, Intentionally sprayed & overexposed them killing and injuring millions. Where ever the Military is you can count on it, it will be intentional. What happened at the Rocky Mountain Arsenal, Rocky Flats, the Hanford site, the Columbia river, Jack-ass flats Nevada, what happened there was no accident, it was done intentionally. They manufactured warfare items of mass destruction, they had the EPA, and other laboratory's available, and used it's own people as guinea pigs. If you have not been affected by something like this be thankful, there are millions suffering, and have suffered, and will suffer even more as their children will grow up disabled because of some disease or the horrible affects of the nervous systems or mutations of the DNA. I speak from my heart I have seen, I have lived it, and I am Totally disabled today because of intentional wrong doing by others with deadly chemicals. These books DEATH FACTORY U.S.A. part I and part II. Will help you see what's out their and what you can do about it. without their knowledge here in the United States, in their homes & in the schools too their children, with BIO/CHEM agents "WEAPONS OF MASS DESTRUCTION", that was not all, they also intentionally classified in secrecy, used sprayed & overexposed our own military, our young soldiers with weapons of Mass Destruction in pure form, SARIN, GB, VX, BZ, and Biological weapons also were tested keeping them in the dark before, during and after. Offering no medical assistance "THEY WERE FED A LOT OF BULL-S____, and KEPT IN THE DARK'. for over 50 years right through today 2003. For many of our Grand parents, Parents, and our children, will never hear the truth.

During that same period of time the Army used the people in the Midwestern United states as their personal guinea pigs. They Intentionally and knowingly allowed chemicals of mass destruction enter water supplies, and was spread to the air in tests. Chemicals disbursed were Diisopropylmethylphosphonate, [DIMP] Isopropylmethylphosphonate, [IMPA] Nemagon, Deildrin, Endrin, Sarin and BZ nerve agents. This was done in a cloak of darkness, and secrecy by the United States Army, Department of Defense, and the pentagon. They classified everything pointing at them. DIMP, and IMPA have low levels of Sarin, and BZ gas, low levels are just as deadly and probably a lot more painful because you die a slow death. They both staying totally stable in the water or in the air. Low levels are Every bit as harmful as full doses, Low levels affect the person unknowingly for a life time of pain and suffering, and cancer & early death which is related directly to those weapons of mass destruction used intentionally on their own people. Colorado has a standard of 8-ppb, and EPA who has no right to set any standard as set it at 600ppb to protect what the Army has done to the people. Colorado Standard at 8-ppb is considered the maximum safe limit. Yet the Federal Courts Through that standard out in a case that was tried under Colorado Law. They will not use that as a Standard and to protect the United States for the Crimes they have committed they refer to EPA standard to protect themselves and the Army. We felt the Colorado Attorney General would intervene but they ignored the trial completely even though in Colorado that was the Law.

Under the little known law that allows them to bail out. It is called the DISCRETIONARY FUNCTION "Claims based on the exercise of a performance, or failure to exercise or perform a discretionary function or duty on the part of a Federal Agency or Government

Employee are not recognizable under the FTCA. The purpose is to protect the Government from liability that would seriously handicap efficient Government operations. Whether or not the discretion is abused is immaterial, United States V. 331 f2d 498 [10th Cir. 1964].

THE COURT HAD PUT IN THE RECORD THAT THIS CASE WILL BE HEARD UNDER COLORADO Law, and would not be heard under the FTCA, Federal Tort Claim Act. THE JUDGE SAID THE FTCA HAS NO PLACE IN THIS TRIAL OR IN THE COURT OF CLAIMS. THEN AT THE END OF THE TRIAL JUDGE WILKES ROBINSON AFTER HEARING ALL THE EVIDENCE FROM PLAINTIFF'S EXPERTS PURPOSEFULLY DESTROYED ALL CREDBILITY OF THE EXPERT WITNESSES WITHOUT CAUSE, THEN WOULD NOT ACCEPT THE CASE UNDER COLORADO LAW, NOR WOULD HE USE THE COLORADO SAFE STANDARD AS LAW, THEN HE WITHOUT WARNING BROUGHT THE WHOLE CASE BACK UNDER THE FEDERAL TORT CLAIMS ACT, THEREFORE DESTROYING THE PLAINTIFF'S ENTIRE CASE SINCE IT WAS PLEAD UNDER COLORADO LAW. NOW I FULLY UNDERSTAND WHAT THE LAW MEANT "WHETHER OR NOT THE DISCRETION IS ABUSED IS IMMATERIAL UNITED STATES." HOWEVER IN 1997 THE COURT DID GIVE AN AWARD TO THE PLAINTIFFS FOR CONGRESS TO APPROPRIATE THE FUNDS "ON ALL FACTS PRESENT IN THE RECORDS". THIS WAS GIVEN IN FEBRUARY 1997, THE COLORADO CONGRESSMEN & SENATORS HAVE BEEN RELUCTANT AND NOT WILLING TO PRESENT THE BILL BACK BEFORE THE FULL HOUSE SINCE NEARLY 15 YEARS HAVE PASSED, AND IT WAS NOT THEIR BILL FOR APPROPRIATIONS H. R. HOUSE BILL 816.

THE BILL H. R. 816 BE IT ENACTED BY THE HOUSE OF REPRESENTATIVES AND THE SENATE OF THE UNITED STATES OF AMERICA IN CONGRESS ASSEMBLED THAT THE SECRETARY OF THE TREASURY IS AUTHORIZED AND DIRECTED TO PAY OUT OF ANY MONEY IN THE TREASURY NOT OTHERWISE APPROPRIATED, THE SUM OF $______________________________, DOLLARS TO LARRY D. AND MARIE A LAND ET AL.

IT'S BEEN THERE SINCE 1997, AND IT APPEARS THAT THE COLORADO SENATORS AND CONGRESSMEN HAVE

NOT BEEN WILLING TO DO THEIR PART. THIS BEING 2002 WE HAVE AN ELECTION, HOPEFULLY WE'LL BE ABLE TO GET SOME NEW BLOOD IN OFFICE THAT REALLY UNDERSTANDS WHAT THEIR JOB IS, AND THAT THEY ARE SERVING THE PEOPLE OF THEIR DISTRICTS IN THIS PARTICULAR CASE COLORADO

IT'S BEEN NEARLY 15 YEARS AT ROCKY FLATS SINCE THEY WERE RAIDED BY THE FBI, & ALL OPERATIONS WERE HALTED, AS THEY PREPARED TO CLEAN UP THEIR ACT, RESTART PRODUCTION, & TO GET RID OF THE HAZARDOUS WASTE.

Rocky Flats is still trying to clean up their act, production is going forward, Plundering and polluting, and yes they still have problems with the hazardous waste, no one wants it. They have tried the Carlsbad caverns New Mexico, they tried Yucca Mountain Nevada, They tried Southern Colorado, the radio active pucks to store in a smaller are at the Flats, Then Cannon City Colorado, and their now working on South Carolina, who don't want it don't need it, and won't take it, Governor South Carolina Jim Hodges said, "I also think I'm doing a favor to the nation in trying to reach some kind of final resolution about what we're going to do with plutonium, What people in Colorado think, I could care less."

CARELESS DISREGARD OF SAFETY STILL IS A PRIORITY ISSUE AT THE FLATS.

Inspectors who entered a radioactive area without wearing protective equipment because a warning sign was hidden behind barrels. As a result, inspectors were exposed to low levels of radiation. They found signs remained posted last month, however they were posed in a room that no longer required respirators. They don't feel that employee health was threatened by the latest sign problem. This is the same type of problem DOE had last year, and a year later this same type of lack-of-control radiological event still occurs, according to the inspector's reports.

A team of experts arrived last week at Rocky Flats trying to find the source of radioactive strontium, and cesium. Strontium or Cesium can be formed only by a nuclear reaction. The discovery of these highly radioactive materials in groundwater under the Rocky Flats has led to speculation that an uncontrolled nuclear reaction known as a "Criticality" would have had to take place at the plant. There was no evidence of a chain reaction had occurred at the plant. The Federal environmental Officials are of course doing the back pedaling for DOE, when they stated, the levels of strontium and cesium are so small, and barely detected, and they are saying tests that identified them in the groundwater may be wrong. Now they're saying the levels reported were very much lower than the levels of radiation found in the groundwater through out Denver area because of fallout from past above-ground nuclear testing. "It was never before mentioned that the water for entire Denver area was contaminated with radio active strontium, and radioactive Cesium, for the very first time we have Federal environmental people admitting that as a fact."

From Rocky Flats manufacturers of nuclear detonation devices. It does not use a nuclear reactor to split atoms and produce cesium or strontium. Mishandling of plutonium, such as storing to much in one place, could create a chain reaction that could create these elements. Both the FBI and the Environmental Protection Agency EPA; that was investigating suspected illegal hazardous waste disposal at the plant seized records that do refer to strontium, they are investigating the possibility of a nuclear reaction at the plant. An internal memo to plant employees criticized workers for repeated safety violations including on the average of 2.5 nuclear violations per month.

It was expressed by the steelworkers at the Flats they fear DOE's latest recommendation to shut down the plant, and it's plutonium operation by 2006 spelled doom for long-term careers. However the Governor believes we have a talented work force and a tremendously fine facility out there. He feels that "Non-nuclear work belongs at Rocky Flats. Environmentalists blasted DOE recommendations and plans because plutonium work has continued on for years even after the shut-downs, the safety violations still occurring, the clean-up of nuclear waste that is still going on, as they continue to manufacture plutonium.

The reconfiguration of the nations nuclear weapons complex has been long anticipated. It was said that even though world tensions are high, the bombs are smarter, and the U.S. may need only 15% of the nuclear weapons it has now. It may make sense to spend a few $billion to redesign a smaller and leaner nuclear weapons as sited in other states. They feel Rocky Flats is too contaminated, has been plagued with trouble over the years and is too close to Denver.

Ground was to have been broken in the mid 1990's but because of exhaustive environmental studies at the alternative sites, the whereabouts of these sites will not be announced at this time. Plutonium triggers for nuclear weapons are manufactured here at Rocky Flats, and is the only site to fashion those triggers. Rocky Flats is the main site in the nation for processing plutonium from scrap, old weapons, waste and residue turning it back into useable plutonium. This alone has plagued the plant for many years, and has caused much of the left over nuclear waste at Rocky Flats. There is talk of moving the waste processing process to a new facility in Savannah River, S.C. The trigger processing would not be moved to Savannah, but to another facility in Texas, or Tennessee to be finished 2004, maybe 2006. The word is out that Rocky Flats will go by the way side for manufacturing of Nuclear weapons.

It sounds as if they may go ahead and move the manufacturing of stainless steel components for bombs. The Governor remains optimistic the non nuclear work will remain the possibility of moving in other non nuclear work to the Rocky Flats. However there is talk of moving that also to existing plants in Missouri, Ohio, of Florida. Their hoping when DOE looks at the cost figures they would find it cheaper to relocate more of that type of work here. DOE feels the heat from Colorado members of Congress as well as the protesters, have fueled the recommendation to move Rocky Flats Plutonium operation. They feel they have places that want it. Plutonium is scheduled to be out of Colorado, by 2006 the Governor promises he will work with environmentalists to make that happen. Congress is hot on the heels of DOE for more job security, Health insurance guarantee's, and money for extensive environmental restoration.

In November 2000 Rocky Flats has asked for permits to burn vegetation at the Rocky Flats. Where the soil is polluted with plutonium particles. Air Quality Control Commission will hold a public meeting on prescribed burns. These prescribed burns could possibly send plutonium particles from the soil into the air. Next spring would be the beginning of regular burns over the next eleven years. DOE's rational is the vegetation poses a wildfire threat during the dry season. However, they said what they really asked is to meet with lay, and professional people to seek any alternatives to burning.

During a test burn on April 6, a smoke plume was seen wafting over Boulder and drifting up Boulder Canyon to the pristine wilderness area above. We would be enduring this possible contamination for the next 11 years unless, with a sudden wind shift, Broomfield and areas to the east, suddenly were the recipients of smoke. The public meeting was held Nov. 17, at 9 am. At the Department of Public Health and Environment, 4300 Cherry Creek Dr. South, in Denver. There is known contamination at Rocky Flats and we shouldn't have to breathe it in the name of weed control, something else DOE forgot.

Ten Flats workers contaminated with plutonium, the source is a mystery clean up is delayed. 10 workers involved in the clean up of the most contaminated building at Rocky Flats was halted when plutonium in tests showed they were physically contaminated with radioactive plutonium. Even though the levels found in each of the workers were well below federal safety standards. Rocky Flats Officials said this is very unusual the workers were not suppose to absorb any plutonium, and the source of the material is a complete mystery. The contamination had to have occurred in building 771 well known as the most contaminated and dangerous structure in the United States [As was noted in to federal report] workers in respirators are combing the building said Robert Card, president of Kaiser Hill Inc., the company that is coordinating the clean-up efforts at Rocky Flats.

We call it the mystery stuff, and this is one of the risks, part of the plutonium may be coming from many parts of the building that haven't been visited or decades attics, ceilings, tunnels, and pipes that carried the acid solution, plutonium was found near a broken valve, contamination is being shook loose as the pipes are being cut.

The stigma of Colorado's Superfund list will be nearly erased. Latest news from State Health Officials who are predicting all but one of the state's most contaminated sites will be taken off the federal clean up rolls by the end of 2010. Rocky Flats is one of many to be removed from the list. According to the Department of Health and Environment, after 2010 costs to finish cleaning and to maintain some of the sites, and dealing with long term water treatment could cost Colorado tens of $millions of dollars in the coming years.

Water treatment at old mining sites in Clear Creek and Rio Grande counties, for example could probably continue for years well beyond 2010, and costing Colorado nearly $50 million over 30 years. The Environmental Protection Agency superfund Unit for Colorado believes the state regulators may be overly optimistic. When your talking about groundwater clean ups, sometimes takes decades to reach the standards that have been set out. It may take several decades to get a site de-listed. Experts believe Colorado is almost assuredly going downhill on its worst massive pollution and contamination sites.

Ken Salazar, Colorado's Attorney General moved up lobbying efforts to make sure the Department of Energy cuts in the budget don't stall the nuclear clean-up at Rocky Flats. Energy Secretary Spencer Abraham, knows full well all about Rocky Flats, and seems very supportive of keeping Rocky Flats moving forward on an accelerated schedule in order to use it as a model for other facilities in the country.

Rocky Flats has annual Federal funding of $657 million, and the Bush administration was finalizing the budget for the DOE programs, and may cut $425 million from the budget which will include the clean up at Rocky Flats. If there is a cutback in funds for those accepting waste from the Flats, could push back the target closing date of 2006. The cuts at other sites that support Rocky Flats waste removal, is all an integrated effort that involves several sites, all playing a major roll in the time table for the Flats to close. If their budgets are cut it would surely have a domino affect on the Flats clean up and closure. Mandated clean ups must be done and cannot be delayed, long term costs and legal penalties would begin mounting. It would be a penny saved and a dollar foolishly lost to cut the funding for nuclear waste clean up.

Talk about a serious security violation, and not just another safety violation when one technician was left along with the biggest part of a nuclear bomb in a can, 4 ½ pounds of weapons grade plutonium, in complete violation of strict security procedures. The two person rule is the first barrier; of theft of radioactive material, the contents of the can were verified.

The plutonium incident was a very serious matter of security left up to technicians not trained in security of something powerful beyond imagination. It was a job that wasn't even taken as a serious breech of security by supervisors. The federal agency is putting this in the same category as a safety violation. "Don't they have any security if not they should."

URANIUM MINERS, SOME 275 PEOPLE WITH IOU'S IN THEIR POCKET SOME TOOK IT TO THEIR GRAVE. They all paid the ultimate price "sooner or later". These IOU's they are now holding and cannot collect. The Federal Government promised $100,000.00 compensation to each of those who gave their all. Under the Law and enacted by Congress in 1990, because of a funding oversight in Congress last year there is a long list still in 2002 unpaid government IOU's. The Uranium miners found out many years later, and many with one foot in the grave, the federal government told miners about the deadly health risks they HAD, FACED when blasting an digging the four corners region, breathing radioactive dust had taken its toll.

At a time when Richard Leavell who suffers from pulmonary fibrosis and silicosis of the lungs which he is left gasping for air and constantly tied to an expensive, ever present bottle of oxygen. He says "he can't do much of anything, this is no kind of life" from his home in Cortez Colorado. The government did send him a notice telling him he qualified for $100,000.00 in compensation. But "Regretfully" the letter said, there's no money to back it up. Richard may not live until tomorrow or he could live a few years, government Officials don't seem to be in any hurry.

They told the miners the government accepts responsibility, and this was suppose to be some kind of an apology, one moment they tell us we got something, the next thing they say we haven't got it. The Radiation Exposure Compensation Act is in crisis. Even a quick fix to the problem would be too late for many of about 275 former miners, nuclear test participants, and the down winders or their surviving spouse still holding onto the unpaid IOU's.

Commonly known as TECA, finally got $10.8 million but needs at a minimum $84 million Just to pay the expected claims coming, that was asked for in emergency appropriations. Congress voted to increase the compensation for each victim, however President Bush put it on hold while he works on a $1.6 trillion dollar tax cut. I can't believe we have people dying with government IOU's still in their pockets, and nothing for the family. Spokesperson for RECA Reform Coalition of citizens groups that are advocates for victims covered by the act. It has been an injustice to delay any further appropriations or regulations because the people that have IOU's are dying.

The IOU's are a downright embarrassment, in fact they are an injustice to delay any further appropriations or regulations, People are dying, with IOU's, promissory note in their pockets. As you drilled holes with a jackhammer, and balsted out. Then you loaded Either with a slusher or by hand and a scoop shovel. Dust filled the air, but workers never wore any respirators nor were they ever offered any. Some of the miners remember many days when the only fresh air they breathed was what leaked out of the air compressors that powered the jackhammers. The United States government should have known better and at least followed very simple rules of safety. Instead they sent hundreds into the mines intentionally knowing it was wrong for them to be breathing all the dust, the Uranium dust and they allowed that to happen right up until recent years. Then gave IOU's to the dying, and their families while they are unable to pay what they had promised.

Rocky Flats, one of Colorados' biggest challenges in the Superfund Cleanup program. Work is now done at 6 sites, and progress is being made at the rest. All the weapons grade plutonium could be gone from Rocky Flats by the end of of the year 2002. Kaiser Hill Co., took over the clean up at Rocky Flats in 2000. The activity to close the now defunct weapons plant will be moving out the plutonium during 2002 during a continuous clean-up at the plant to keep in step with the proposed closure December, 2006. Since 2002 will be one of the most difficult times positively dealing with really hard issues, they absolutely believe 2002 will be our greatest challenge. Most of the plutonium will be going go Savannah River S. C. plant subject to availability of storage space, by September 2002. The removal of the plutonium was initially stalled awaiting container approval for shipping the plutonium oxide, and that will be the heart of the nuclear weapons at Rocky Flats.

Here are some scarey things that happen, You think you've been safe and all of the sudden for instance a Denver site where heavey equipment crews are digging up hundreds of residential yards and removing contaminated soil. Larry's thought is "what about the house itself that was nestled down into the same toxic and polluted earth. What about the people who live there, and what about those that have moved and your children who played in those yards and slept in the house that was built into the contaminated soil". Other places they are covering the waste with limestone to assist in neutralizing the acids.

They're moving in clean soil and planting trees to revegitate the areas. Another Contaminated area in Denver where they are cleaning the soil by vacuuming volatile Organic Compounds from under the ground, then distilling them in a garage sized treatment plant. Here they are hoping to build new homes and businesses for the area when they're done. Still another Radium Superfund site which was contaminated with radium, thorium, and uranium, amongst other pollutants. Home Depot contributed to the cleanup with an agreement with EPA wouldn't sue the company if more pollution showed up.

Well what about a thousand weed eaters, in the form of billy goats, The old saying is "Billy will eat anything". There was a plan by Rocky Flats Officials to burn the weeds. But run into a lot of opposition from Denver area residents. During 40 years of plutonium production they fear weeds and the soil have been contaminated and a fire in the 6000 acre buffer Zone could send radiation pollution over the city of Denver and where else.

University of Colorado biology professor Harvey Nichols and students in his critical-thinking class really got serious coming up with an idea "goats as an option" They would make the weeds disappear with little to no pollution. Knapweed and thistle that offend the human, is lunch to a goat. Nicholas said. We really don't want polluted smoke filling the skies, destroying someones health. Larry says, "I cannot remember any time that I was advised that plutonium was found everywhere because of the above ground nuclear tests conducted by both sides during the Cold war. So all the Forest fires California, Colorado, which the smoke infiltrated every city and town across the United States We in Minnesota could smell and see the Forest smoke from Colorado. In essence everyone is being contaminated with plutonium, Beryllium, Selenium, Cesium, all radiation that we are now told is as normal every where, as it would be at the Rocky Flats, where they have been known polluters for over 50 years. I don't understand why no one was every told our soils are so contaminated with Radioactive pollution from the Bomb tests, and they are now comparing every ones' back yard to the most polluted place on earth, Rocky Flats.

If we end up with 1000's of goats that all of the sudden turn green and glow in the dark warnings would go out that the area is radioactive. we could probably call it just another way to automatically puck up the pollution, what goes in must come out, and someone may have to follow the goats around with pooper scoopers. Then the droppings could be sent to a Radio-active-waste burial site, and the goats could just keep eating cause they would be too contaminated to be moved. They said it would take more than a thousand goats to get thru the 6,000 affected acres, Larry says either get more goats, or maybe sheep, to Burn & send contamination through the air is just wrong. It was Mr. Paul Danish, Boulder County Commissioner's idea to find someone to stay close behind the goats around with Pooper Scoopers.

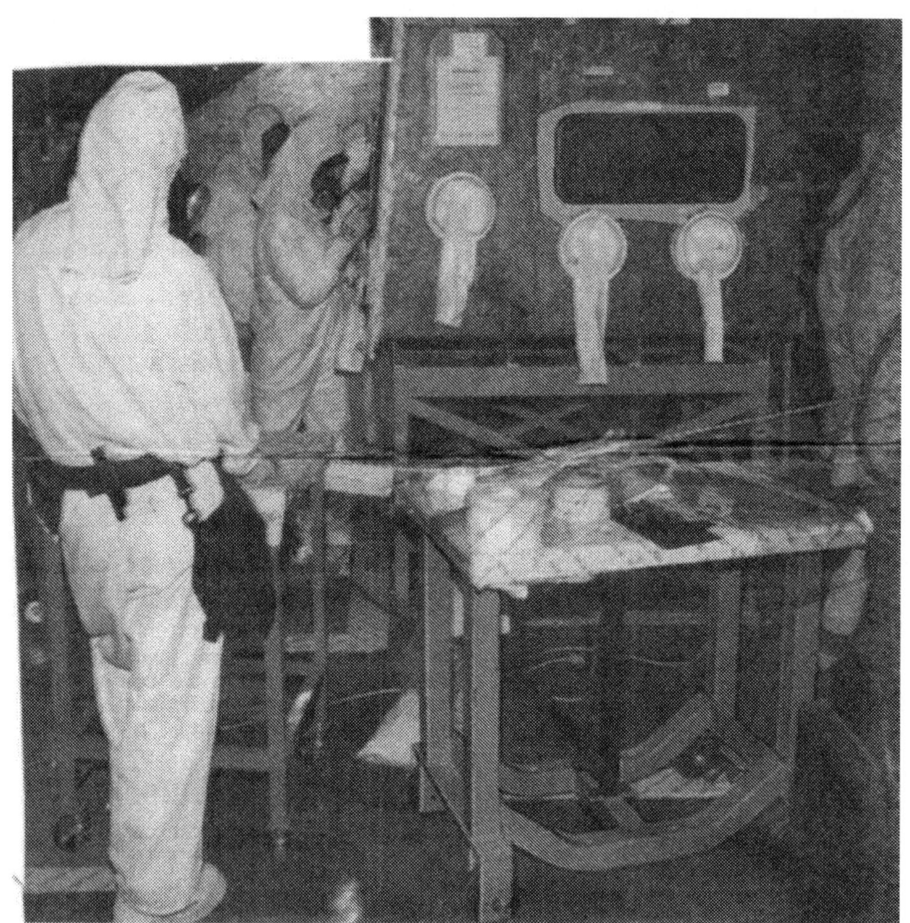

CLEANING BEFORE DEMOLITION.

Building 771 the oldest of the structures where radioactive materials were processed and numerous spills occurred. Rocky Flats workers will begin drillling through the floors and foundation to determine whether chemicals have seeped into the ground below. These drillings will show just how complete the clean up will be and whether it all can be completed by 2006. The main issue with the buildings being torn down by Kaiser Hill, is what's underneath and how far did it seep into the ground. The Communities real fear is that Kaiser Hill, and the Department of Energy will tear down the buildings and leave the contamination underneath.

There is talk now the Environmental Protection Agency must also help pay the costs of clean up, they claim the agency must live by the law it enforces. It has been estimated the soil will cost $6.7 million to clean up contaminated soils. This cost may rise to several $billion dollars if it is anything like all other starting in the millions and ending in the billions. Storage of chemicals is blamed for soil contamination that will require several more years of a costly clean up. Under federal law, so-called "generator's" of hazardous materials are held liable to pay a portion of clean-up costs, even if they were not directly involved in fouling the environment. Sometimes is sort of leaves the smaller companies holding the bag for the companies that are larger and more culpable of polluting.

South Carolina Governor says the Energy Secretary is breaking his commitment, not to leave South Carolina the final storage or disposal site for any nuclear waste. Sure we had an agreement to pack it and stuff it suitable fore waste storage, but it was suppose to be in and out, with no storage of nuclear waste. Energy Secretary is saying a strategy is in place of moving in out of South Carolina, and we remember Idaho Gov. Cecil Andrus calling out the State police to stop waste shipments from Rocky Flats, and we want them to know also we in the South have State Police that will do the same.

Governor Hodges' does not oppose the plutonium coming into South Carolina, but he wants assurances from the Energy Department that the highly radioactive material will also eventually be moved out. The controversy won't delay the startup of shipments of plutonium, were shipping the stuff, what savannah then does with it is a separate issue. But, it was shut down in the 1990's by New Mexico objecting to be it's radioactive dump. DON'T BLOW UP THE OLD BUILDINGS AT ROCKY FLATS. Neighbors don't want demolition using any explosives. Kaiser-Hill who's cleaning the now defunct nuclear weapons plant wants to blow up the reinforced cinder block guard tower to test methods to use on the larger of the structures. Profoundly dumb Says Boulder County Commissioner, Paul Danish. Kaiser Hill & Spokesperson said Words “Rocky Flats” in the same sentence as “explode” is not a good thing to say.

Rocky Flats Nuclear Weapons Plant, a very huge complex that sits just to the northwest of Denver Colorado in Jefferson County, and is known as one of the most contaminated places on earth. There are some areas that will never be brought back to the pristine state that it once was when taken by the government.

Rocky Flats in Colorado a former nuclear weapons plant, is in the midst of a $7 billion clean up, targeted for completion in 2006. The Bush administration has once again postponed a major part of the Energy Department's plan to dispose of left over plutonium citing budget constraints, part of that plutonium is at Rocky Flats. South Carolina Officials are saying the Bush action violates federal governments agreements with them to clean up nuclear waste they have still sitting in steel containers, instead of it being immobilized in glass as we agreed. Plans during the Clinton administration. Calls for the plutonium to either be turned into reactor fuel or enclosed in highly radioactive glass, & to a burial site.

Five years ago, the Energy Department set out to get rid of 52.5 tons of weapons grade plutonium as part of a deal with Russia, which had agreed to remove the same amount from its stockpile. The plan was to convert some of the plutonium from warhead form into fuel for civilian power reactors.

But then much of the plutonium in forms not suited for fuel, including a large quantity at Rocky Flats. This plutonium would be shipped to South Carolina where it would be baked into ceramic slabs, which would stabilize it, then it would be embed it in highly radioactive glass to protect it from being stolen.

The rest of the plutonium, at the same factory in South Carolina would seal the plutonium in storage containers. The Department of Energy with South Carolina, there would none sent into them unless they had a clear exit agreement. Governor of South Carolina says they have violated the agreement. The Government still has a plan of sending it to the Savannah River site where the Government will decide what to do with it and that is exactly what is happening.

Rocky Flats is preparing to package the remaining plutonium in 50 year containers. They will be transported to the Savannah River site for interim storage. Conversion to fuel requires two factories, they said; one to take apart the plutonium at the heart of the warheads and another to turn it from metal to oxide, and then to process it into the proper shape. To immobilize plutonium would require a third factory.

The federal government will open will open its first office to help workers win compensation for health problems to toxic materials, such as plutonium, and beryllium. Then similar offices will open near 7 other sites around the nation where nuclear weapons parts were and are being manufactured today, mainly to work on the Cold war erra. August 2001 the Labor Department was to begin distributing to the injured workers up to $150,000.00 and lifetime medical benefits. The Department of Energy's conservative figure or 3,000 people nationwide, and half of whom have contracted cancer. There are at least 115 at Rocky Flats, who became ill from inhaling beryllium dust, but then there is 100 others that are at risk of developing the disease. Beryllium is a metal that was machined into the bomb parts. While many others at Rocky Flats come in contact with plutonium a heavy metal linked to cancer, and is highly radioactive. Those with claims during the Cold War were brushed off their claims were ignored, they died with cancer, and could not even afford a place to live. Larry says, "He visited homes where anywhere from 10 to 20 of them were given a place to die, and it wasn't the government.

POST-COLD WAR PROPOSAL, THAT WILL SHRINK U. S. NUCLEAR ARSENAL

America's nuclear weapons will undergo radical changes including gradual elimination of all land-based intercontinental missiles and a sharp reduction in strategic bomber force. Described by experts as a revolutionary idea in nuclear thinking, since the end of the Cold War, The proposals were triggered by our President George Bush's repeated statements that the United States must move beyond the concept of, mutually assured destruction.

President Bush said America "can, and will, change the size, the composition and the character of our nuclear forces in a way that reflects the reality of nuclear strategy intended to answer many questions that will be asked. Underlying many proposals is the notion that the United States should pay more attention to China's small but growing nuclear forces and less attention to Russias huge but declining arsenal, and where is the nuclear material going or being sold to, even if its on the black market it is dangerous.

There is a law in place to stop the proliferation of nuclear material World Wide, however there are a few rogue nations that would use it and make and proliferate nuclear materials, for all the wrong reasons, such as Iran, Iraq, Syria, North Korea, Palestine, and Libya, and Axis of evil will never loose the stench of proliferation and terror until they learn to live in a worldwide community. I'm not picking on those countries named, and would like to see them all join the World and together there would be strength in what we do, and how we do it. In the world instead of War and terror. They continue to walk the edge between wrong and right. Like Libya has tried to turn it around but they have not done enough yet.

The thousands of nuclear weapons that are maintained in the United States, and around the World as a deterrent against attack, It is argued that the development of a missile defense system, and if successful the world could see the elimination of thousands of nuclear weapons.

About 50 Rocky Flats workers filed suit against a Cleveland Manufacturer conspired with the federal government to hide the dangerous effects of breathing even minute amounts of berylium 1-2ppb per cubic meter of air. Jury finds the manufacturer only 10% responsible, and the Judge released them from all damages. While the Department of Energy Cried Workers Compensation had been paid, and we owe no more.

Skull Valley Utah an Indian leader Leon Bear and a small band of Goshue Indians have wanted to host one of the courty's largest nuclear waste dumps, but Federal Officials are concerned that the group cannot even keep it's own drinking water clean. The groups leader Leon Bear has ignored and refuses to work with Federal Officials EPA, who are trying to clean up the reservations dangerously polluted drinking water. Skull valley is about 60 miles South West of Salt Lake City Utah. Over the years it has become one of the most volatile areas on earth. 1. Utah test and training range for testing F-16 fighters & cruise missile. 2. Dugway proving grounds test center for chemical and biological weapons. 3. Then we have Tooele Army Depot for the chemical and biological weapons storage. Bear who signed a contract with four years ago with private fuel storage, A Limited Liability Company representing a consortium of commercial nuclear power producers. The 840 acre site is designed to receive up to 40,000 metric tons of spent uranium. Yucca Mountain Nevada is still a planned permanent radioactive storage facility.

Rocky Flats is suppose to be cleaned the end of 2006 & ready to turn it into another government wild life refuge. But all of the sudden States Highway Chief shocks the local officials by letting the cat out of the bag. "State Transportation Director Tom Norton is considering a highway link between Golden and Broomfield smack dab across Rocky Flats. There will be a firestorm of criticism if the highway department recommends a highway thru Rocky Flats. Norton said there will be an opportunity to alternatives, and they must be on the table in how to connect the end of the [soon to be built] Northwest parkway in Broomfield with U.S. 6 in Golden. That's the deal. Highway may not be allowed if they turn Rocky Flats into a wildlife refuge.

Illegal dumping of dangerous radioactive wastes from Rocky Flats at Lowry air base near Denver Colorado. The new public communique, titled "Radionuclides and the Lowry Landfill Superfund Site," follows a rebuke from the EPA's inspector general last year. A program that will cost about $995 million at least 6400 people will receive the $150,000 by September 2002 as I write this book. The program started July 31, 2002 and will provide $150,000 to nuclear weapons industry workers who have suffered cancer, beryllium disease or silicosis form their jobs, plus medical expenses.

Yucca Mountain Nevada is near ready to accept nuclear waste for permanent storage while the Governor of South Carolina has waged a high-profile campaign to keep the plutonium away. DOE announced a delay, press secretary Courtney Owings called it a result of the governor's "showing of a much stronger resolve." All Governor Jim Hodges has ever ask the DOE, is to give him assurances the material won't be there forever, and they've never done that.

PLUTONIUM GAME PLAN STILL ON, FLATS WORK GOES AHEAD DESPITE S. C. DISPUTE.

Workers at Rocky Flats told to keep packin the plutonium for the planned shipment to the DOE's Savannah River site by mid October. Governor Hodges is willing to accept the plutonium from Rocky Flats to DOE's Savvanah River Site, and he is asking DOE for the plan to remove the plutonium. He threatened at one point to lie down in front of the truck if they don't come up with that plan. Rep. Mark Udall D. Colorado has been warning DOE Secretary Spencer Abraham that failure o fund a program to get the plutonium out of South Carolina would spark a political crisis.

DOE says an alternative hasn't been chosen for the pure plutonium, some of the diluted stocks could go to the burial site near Carlsbad N.M., where other is now going. Doe has nowhere to go with the pure plutonium, however the Rocky Flats workers are packing it in special containers for shipment. The plans call for the plutonium to be out of Rocky Flats by the end of 2002. DOE says studies are now on how putting the plutonium in storage will affect discussions with Russia about nuclear arms reduction. President Bush is willing to agree with Putin to reduce both the U.S. and Russian stockpiles to fewer than 2,000 a reduction of 2/3 of the current level of 6000 warheads apiece. The Russian leader Mr. Putin is flexible about Bush's plan for a defense against missile attack.

With Yucca Mountain Nevada's plan being so unpredictable, Doe states he will order shipments of plutonium to South Carolina Energy Chief overrules all objections. The governor of South Carolina says "Whoa who has the reins?" Abraham will announce this morning shipments of plutonium can start today, this is considered a milestone in cleaning the now defunct nuclear weapons plant. Governor Hodges' has not signed off on this plan and is not shy of lying down in front of any trucks carrying plutonium from Colorado Rocky Flats manufactured nuclear weapons from 1954 thru 1989 when it was closed amid a controversy over pollution and safety violations. There was a lot of preperation to continue in cleaning and constructing new buildings and equipment, and they tried start up time and time again but failed. With the Cold War now ended, it was never reopened. After nearly 10 years of cleaning, and redoing the equipment, in the late 1990's went defunct, never again to re-open it's operations, with a target date Dec. 15, 2006 to have all demolition done and cleaned up.

Scientists are saying no matter where the waste is put, it will be impossible to avoid unexpected problems during the next 10,000 years the material will be highly radioactive, and beyond. January, 2002 as the White House Prepares to possibly within weeks give the go-ahead for the Yucca Mountain waste project in Nevada. In a letter sent to Congress from the Scientists emphasized they were making judgment on whether Yucca Mountain, just 90 miles northwest of Las Vegas should be designated for longterm burial of 77,000 tons of nuclear waste. Even though the Scientific investigation order by Congress has taken 13 years, There are gaps in data, and even the basic understanding of how the volcanic rock and hydrology, as well as the man made barriers that would contain the waste.....will perform over the next 10,000 years. The policy makers will now have to determine "just how much scientific uncertainty is acceptable. The green light on this project is the responsibility of President Bush, even though Nevada Officials who have argued that the government is ignoring safety concerns.

Yucca Mountain Nevada still in turmoil, It sounds to me the Federal Government can't find a guarantee or anything to the people of Nevada and now trying to paint them unpatriotic. The governor who says their plan stinks, says Nevada will redouble its fights. He said no consideration at all has been given the thought of accidents, traveling through Cities, downtowns, and next to schools, and industrial areas, your neighborhoods, in 43 states. 3,000 to 4,000 truck and rail shipments per year for at least the next 38 years. This is not just a concern for Nevada but for everyone across our Nation.

Kaiser-Hill a company has been fined for violating safety rules. Their responsible for cleaning up Rocky Flats Environmental Technology site. They were fined $385,000.00 by the Department of Energy. This was done by DOE under the Price-Anderson Amendments Act of 1988, which requires the DOE to take enforcement actions against contractors for violating nuclear safety regulations. Listed violations in four areas.

☑ Repeated failure to assure quality of materials procured for nuclear safety, including an August 2000 purchase of 500 lids for 55 gallon drums that were later determined to be damaged and defective, and never used.

☑ Deficiencies in safety & work controls in resizing plutonium pieces for packaging, load of containers, & storage of waste containers after the plutonium had been measured.

☑ Non-compliance with the radiation safety program in building 771, in which 11 workers were exposed to plutonium in October 2000.

☑ Failure to correct previously identified problems in procurement, safety & authorization.

DOE remains confident that Kaiser-Hill is making real progress in the clean-up process to meet the closure deadline. While addressing the safety issues that have been identified. The majority of the items were self reported by Kaizer-Hill, with the ideals of safety in mind by doing it with the utmost of honesty.

Thanks to the Rocky Mountain News, and photographer Glenn Asakawa He took a very clear photograph, giving a closer look inside building 559 where plutonium is handled at the Rocky Flats nuclear weapons plant. 95 glove boxes allowing 95 pairs of protected hands will handle plutonium Iso-topes, made to make a weapon volatile before detonation. Lab 559 and five other plutonium buildings were shut down in November 1989 when the FBI, EPA, and the Colorado Department Of Health found numerous safety violations. They say the Lab is much safer now because of formalized procedures and training. As one Lab-technician is explaining his ten-years, and never believed the plant to be unsafe. But the Colorado Health Officials on tour noticed a dangling gasket from an unused gove box. They tried to clean it up, install new equipment, built new buildings but it never got off the ground, and now is just another remnant of the now defunct and heavily contaminated Rocky Flats "ONCE" nuclear weapons plant. If they are still on time clean-up alone will take until December 15, 2006 and beyond.

As of May 9, 2002 the shipments of plutonium scheduled for last year have again been postponed. Until it can be heard by a judge in resolve to a lawsuit brought by governor Hodges over the issue. Doe keeps pushing and yet cannot give their commitment to the State of South Carolina. Gov. Jim Hodges and for his State is demanding a firm guarantee from the Bush administration that the six tons, of plutonium from Rocky Flats, and the 28 tons from other sources will be transported out of S.C. after being processed. Doe cannot come up with a program of moving it out after processing and refuses to put anything in writing, and with that Hodges said "he will call out the State Police or lie on the road himself to stop the plutonium shipments". This could get messy before it is over, just how far can DOE push it's authority. It seems that is all they have ever done.

The Law now limit's the nuclear power industry's liability in the case of an accident. No other industry enjoys this federal protection. Price-Andersen requires that each plant carry only $200 million in insurance, with $9.1 billion for the entire industry. Unfortunately, Sandia National Labs has estimated the cost of one big reactor accident at more than $500 billion. Nuclear Reacter vessels are being checked after the discovery that acid in the cooling water had eaten nearly all the way through the 6-inch thick lid of a reactor at a plant in Ohio. There was only a stainless steel liner was less than ½ inch to hold in cooling water under more than 2,200 pounds of pressure. This could have been what is called a nuclear meltdown.

Gov. Jim Hodges ordered out State Troopers, and other authorities to S.C. borders. The transport of plutonium on S.C. roads and highways is prohibited, no person transporting plutonium shall even enter the State of S.C. June 15, 2002. A Federal Judge refused to block the shipments into S.C. Gov. Hodges has said he fears the government will end up leaving the plutonium permanently. D. C. making the state a tempting target for terrorists. DOE has a chain of broken promises, with no assurances that would quietly leave "S. C. the nation's plutonium dumping ground". S.C. has now been rejected by District Federal Court and are now in appeals June 18, 2002. Legally under Federal Law DOE cannot ship any plutonium if S.C. objections and complaints in the Courts are exhausted. A S.C. Federal court issued an injunction against Gov. Hodges allowing shipments into S.C., while it's in appeals, Hodges asked the Appeals Court to Issue an injunction halting all shipments of plutonium into S.C. while the environmental claims are being resolved. Like Gov. Hodges has said all along he doesn't object to plutonium coming into S.C. but he wants assurances as to when it will be moved. DOE has not been able to offer satisfaction.

JUNE 22, 2002, PLUTONIUM IS READY TO ROLL, COLORADO TO SOUTH CAROLINA.

Rocky Flats has been rolling since 1999, completing it's 500th shipment of plutonium waste to New Mexico. Hazardous waste shipments of all kinds dot the highways everyday. Trucks hauling the plutonium look much like the average 18 wheeler, unlike the more distinctive trucks they have a trio of squat cylindrical cases.

The Yucca Mountain site to be completed in 2010 went through an earthquake 4.4 on the U.S. Geological survey in Goldon. I hope this doesn't mean at sometime in the future we may have an atomic volcano, or maybe a plutonium-beryllium-cesium-strontium shower. ENERGY DEPARTMENT OFFICALS CONCEDE THAT TRANSPORTING WASTE NUCLEAR MATERIAL HAS NOT BEEN A PRIORITY. The Bush administration has refused to focus on the danger posed by hundreds of thousands of waste shipments, from Eastern third of the Nation. Yucca Mountain site will hold up to 77,000 tons. In 2010 when it is ready for storage, it would mean at least several dozen trucks would be on the highway's somewhere in the Country at least the next 24 years.

July 10, 2002, the Senate voted to entomb thousands of tons of radioactive waste inside Yucca Mountain in the Nevada Desert, rejecting the States protests and ending years of political debate over nuclear waste disposal. The vote to override Nevada's objections to the waste dump 90 miles northwest of Las Vegas cleared the way for President Bush to proceed with the project that has been studied and argued over for more than two dacades. This will open a lot of truck traffic through many western states carrying nuclear waste. Nuclear Waste is the monster we talked about earlier in part one DEATH FACTORY U. S. A., along with many terrible array of chemicals that will out live us all.

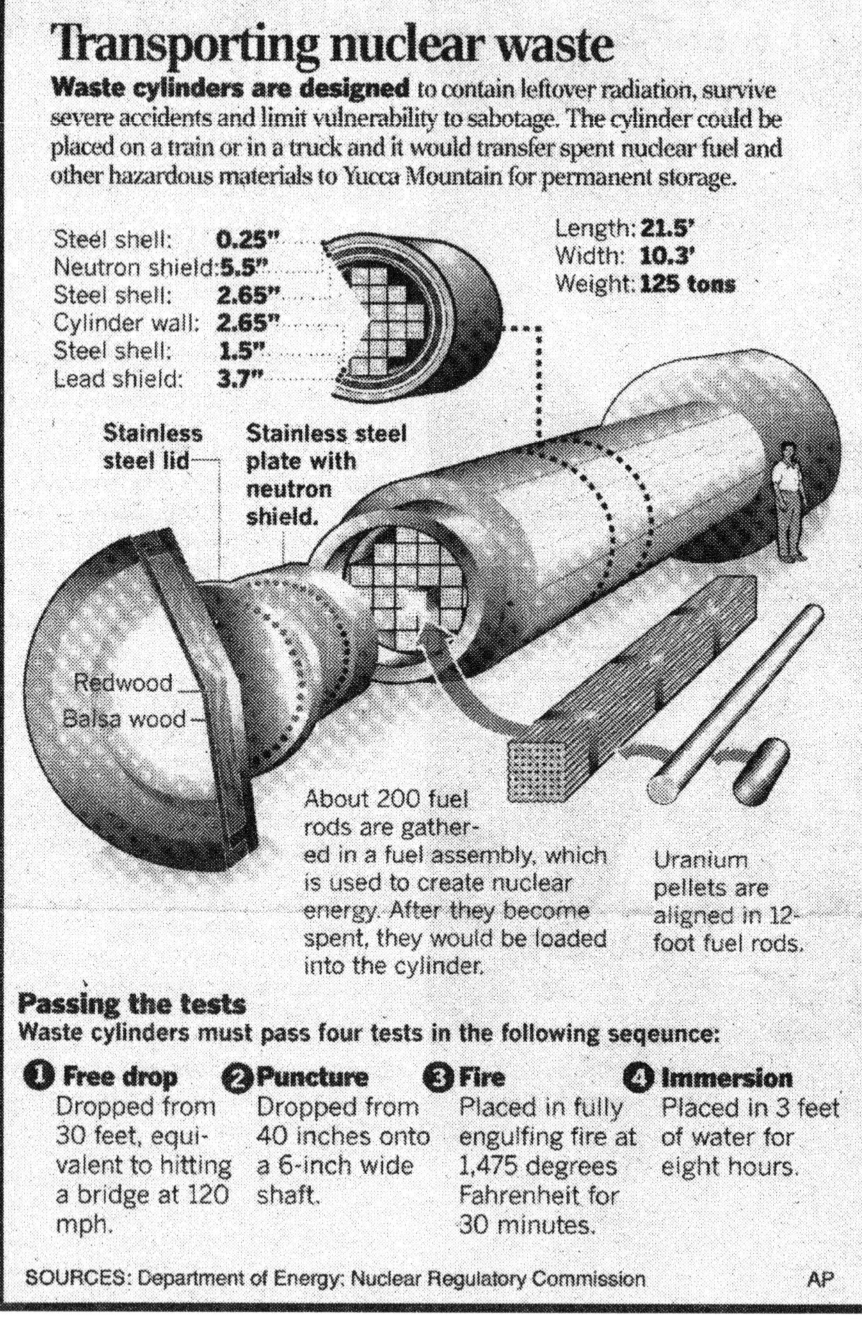

Transporting nuclear waste

Waste cylinders are designed to contain leftover radiation, survive severe accidents and limit vulnerability to sabotage. The cylinder could be placed on a train or in a truck and it would transfer spent nuclear fuel and other hazardous materials to Yucca Mountain for permanent storage.

Steel shell: **0.25"**
Neutron shield: **5.5"**
Steel shell: **2.65"**
Cylinder wall: **2.65"**
Steel shell: **1.5"**
Lead shield: **3.7"**

Length: **21.5'**
Width: **10.3'**
Weight: **125 tons**

Stainless steel lid

Stainless steel plate with neutron shield.

Redwood

Balsa wood

About 200 fuel rods are gathered in a fuel assembly, which is used to create nuclear energy. After they become spent, they would be loaded into the cylinder.

Uranium pellets are aligned in 12-foot fuel rods.

Passing the tests

Waste cylinders must pass four tests in the following seqeunce:

1 Free drop Dropped from 30 feet, equivalent to hitting a bridge at 120 mph.

2 Puncture Dropped from 40 inches onto a 6-inch wide shaft.

3 Fire Placed in fully engulfing fire at 1,475 degrees Fahrenheit for 30 minutes.

4 Immersion Placed in 3 feet of water for eight hours.

SOURCES: Department of Energy; Nuclear Regulatory Commission AP

Man has invented and mounted science, and it is now fast running away. I firmly believe that before many centuries more, science will be the master of mankind. The engines he will have invented, The towers reaching into the sky, super power of the sky's and outer space, superiority of the ocean. Nuclear power far beyond imagination, Chemicals that dare destruction and dare man to stop it, & germ warfare that could wipe mankind off the face of the earth. I'm afraid they have already gone beyond his strength to control. Some day Science shall have the existence of mankind in its power, and the human race may commit suicide, by blowing up the world and / or poisoning it, and all of us to death.

Larry D. Land 2002 / Henry Adams April 11, 1862

Science has radically changed the conditions of human life on earth. It has expanded our knowledge, and our power, but not our capacity to use them with wisdom.

J. William Fulbright, 1964

Chapter 20 HARASSMENT FROM UNITED STATES OF AMERICA JUSTICE DEPARTMENT, INTERNAL REVENUE SERVICE & TAX COURT. ROCKY MOUNTAIN ARSENAL MILLENNIUM CLEAN-UP. ROCKY FLATS MILLENNIUM CLEAN UP AND PRODUCTION

IMAGINATION

Rocky Mountain Arsenal U.S. Army, Rocky Flats Nuclear facility "DOE". The Hanford nuclear weapons plant, Jack-As Flats proving grounds in Nevada. Large Corporations much like Shell Oil Company, Dow Chemical, Martin Marietta, Rocky Flats/ Rockwell International Inc. also Seen the possibilities of what needed done, But they lacked the imagination of getting it done safely. Their own imagination stood out as their own personal laboratory, they could only envision dollar signs without any interest in people or the environment they stamped it all classified, the way they could not be investigated even by the Congress of the United States. The possibilities were rehearsed, they mapped out the plans, and only visualized what it would take to overcome obstacles, then took "SAFETY" out of the plan and put in it's place "TERRORISM". As they pushed forward safety was ignored, causing terrorism on the people of the United States. They seen disasters, in the environment, death and destruction of lives young and old, but they didn't care, they had a plan and it was deliberate, their chart was clear for those that would be expendable. Safety became no issue, it was taken out of their vocabulary as far as the environment. Their imaginations turned the possibilities into reality, the only cost that was not factored in was the environment, and millions of people who will have Terrorism today, no tomorrow they paid ultimate price.

Everyone in this nation can be suffering to a certain extent and not even suspect what may have caused it. Cancer, Mutations, Lung disease, Brain tumors, "SIDS" Sudden infant death syndrome, Learning disorders, Short term memory loss, Alzheimer's, Amnesia, Poor concentration, Dementia, Dizziness, Confusion, Irritability, Chronic fatigue, Stumbling, Tremors, Sleep disorders. Thyroid disease, or Thyroid cancer, Parkinson's disease, Hyper thalamus gland, Brain cancer disease, MS/Multiple sclerosis, MYALGIA, ARTHRAGIA, severe headaches, Nausea, Vomiting, Gastrointestinal problems, difficulty breathing, lung disease, asthma heart disease, Calcification of the ligament attachments to the bones, very painful, I got that by being exposed to low levels of SARIN nerve agent. E-COLI, Anthrax, Smallpox. MCSS/Multi Chemical Sensitivity Syndrome. This starts out much like an allergy and over time and exposures to various chemicals it gets worse, then much worse to cigarette smoke, carbon monoxide, perfumes, after shaves, deodorant, and cologne, Household sprays.

HARASSMENT BY THE IRS, COMBINED WITH THE JUSTICE DEPARTMENT.

The Harassment from the Internal Revenue Service began in 1988 when H.R. 816 was passed by the Congress of the United States of America. In 1988 Larry had been paying his taxes since 1958, 30 years and was never audited. Then depositions began by the Justice Department in summer of 1990 They were really trying to pry into my business trying to figure out how Larry was paying for all the investigations, contact with Congress, all the way to President Ronald Reagan. How in the world was he able to pay the $100's of thousands of dollars just in legal costs. Within 3 Months of the first deposition Larry got his warning from the IRS accusing him of owing over $28,000.00 in back taxes, plus enormous fines and penalties. Larry had his personal taxes, then they got into his company taxes, and then to his small S corp. Larry has an accountant and he himself is a trained accounting, and one very much up on all the laws on taxes and rules of the IRS.

The Auditors acted very head strong, and would not allow any of the business or personal deductions that were taken, even though Larry had used them for over 30 years and was never audited. and kept changing the meaning of IRS rules, then When Larry would confront them with the actual law's that governed the IRS rules, they continued to say no. So he would give the auditor what ever they needed such as receipts, that he had kept clipped together with each analysis sheet for a particular expense or group of expenses. The Auditor declared they would like to keep the box so they could go through it. Larry advised them he would not allow them to keep his evidence of all deductions. I sat in their office carefully taking the receipts apart so copies could be made. After each group they returned and was clipped back together while they made copies of another group. This took upwards of 6 to 10 hours. I truly felt like unclipping them throw them all in one box shake it well and give to them. But then I couldn't put myself on the same level of where they operate from. I would have return visits sometimes several, as each year went by. It always seemed the same auditor would disagree with nearly all deductions. Then I would ask to see someone in appeals, this would be set up and I would meet with the appeals officer. After the regular audits, it seemed to keep me well prepared for the appeals auditor, It was always presented very professionally, They seemed pretty much ready to agree with my presentation and stated they would contact me within a couple weeks. We only moved very close to Tax Court once. 1992.

1988, 89, 90, 91, were all cleared and accepted the very way they had been done I owed nothing they owed nothing. 1992 was somewhat different at the time of the Audit Larry found more expense receipts, and since IRS had opened 1992 for audit it then gave him the right to add the new deductions found, $23,000.00 dollars. Well that immediately blew the audit, they would not allow them to be brought in as expenses. This year went to appeals and Larry stayed firm and even the appeals officer wouldn't give. So it was set for trial, and two days before trial the appeals officer called and accepted my complete presentation, this gave me quite a deduction to carry forward to tax year 1995. 1994 was very easy the auditor wouldn't give and it was taken to the appeals and after one visit the appeals officer approved it as was. Well needless to say 1995 taxes was done by March, and Larry had to be prepared for Trial in Denver Colorado, attorney's, experts, against United States of America United States Army, Legal and Equitable claims against the United States. Trial was held in Denver where the disaster occurred in 1972.

The trial hadn't even gotten it's feet wet before Larry was again notified by the IRS that he owed several thousand in back taxes, plus fines, and penalties. Larry called Congressman Hank Brown's office explaining the harassment by the IRS. His spokesperson stated he would begin and investigation that this was not right. There were several letters written to the Commissioner of the Internal Revenue Service, and it all came down to the fact since they already began

1995 they would go ahead and do it, and if it came out with no problems it would end the audits.

Same auditor as was assigned in 1993, and 94, he would not allow anything even the deductions that had been approved by appeals. Each one of them was pulled out of 1988 thru 94 tax years as proof they had been allowed. Each though they had been allowed before, even though their rules stated them as good deductions, even though his accountant testified for him, he would allow absolutely no deductions at all, you then could almost see his arm was being twisted. So it went to appeals, and in fact special assistant to the commissioner of the Internal Revenue Service Blaine Holiday was assigned to work with the appeals officer assigned. Larry thought this is real strange, but he went in with everything very much in order. This was a standard deduction schedule the same type of filing Larry had always done before.

He was met by Blaine Holiday who set him up in a small conference room, the called in the Appeals Officer, she seemed ok as we discussed the deductions there was some that she said she could not do as cost deductions, and there were some she allowed because they were very clear in IRS Rules deduct-able items. It was Blaine Holiday who would not allow even the deductions the appeals officer had allowed, claiming that Larry should file an itemized deduction schedule, and take certain costs off as long term investment loss to the business, and not to forget to take off previous year state taxes. Well Larry said he agreed to that also since he knew he had a pretty good figure for schedule –A itemized deduction schedule, and was still going to use the expense items not allowed for long term capital gains investment, and loss to carry forward to 1999 tax year, and beyond $3,000.00 per year capital loss deduction. then Larry also found another $8,000.00 for long term capital gains (Loss), he had failed to take and with opening up 1995 now gave him the right to recapture it in 1995, and carry it forward to 1999 and beyond.

Larry prepared the itemized schedule, and as in fact able to remove all expense items that were in contention, on the standard deduction and replacing them with itemized deductions all allowed by IRS publications. In respect Larry amended his taxes in 1995 as the Assistant to the Commissioner had suggested. to a schedule A itemized, and did use the last year state tax as a deduction. Larry called and set up an appointment with the Appeals officer and again was shadowed by Blaine Holiday who was not about to allow any deductable items. The appeals officer had again allowed the expense items she had allowed before, and agreed with the Itemized deduction schedule. But Blaine Holiday was not at all happy about the new amended schedule, showing the IRS owed Larry and additional sum of taxes returned. He said go ahead and get it prepared and he would personally file it with the Court as an amended tax schedule, and if there were any problems it could be discussed in Court as that was the judges job to solve problems.

The trial was set for the 18, September 2000, Larry thinking Blaine Holiday had filed with the Court the Amended tax schedule, he was not worried of the outcome because the appeals officer had allowed certain deductions and had accepted the Schedule A, as proper. The amended tax filing was sent to Blaine Holiday on the 9, September 2000. Larry has an item proving it had arrived at his office on the 11th ample time to file with the Court. At the trial Judge PAJAK ask Larry if he was ready, with everything filed he said yes. Blaine Holiday give a short dissertation for the IRS, and filed documents with the Court, and naming those documents he did not say amended tax schedule for the Land's he had agreed to file. Judge PAJAK then asked if the taxpayer had anything to say, Larry immediately said he would like to file some documents of evidence with the Court. Judge PAJAK stated "NO" not allowing the petitioner to file any evidence with the Tax Court. Blaine Holiday had lied to the Court in the preparation of statutory notice of deficiency and evidence the IRS was allowed to file. He never once mentioned the taxes had been amended, nor did he mention the Schedule A. Blaine Holiday in his preparation lied to the Court claiming this was all that was done, and the IRS tried to be flexible and creative and were not able to settle this case.

Even though the Law is very clear that an amended schedule was proper Blaine Holiday clearly states to the Court that there was none. Judge PAJAK then states that in the transcript he Blaine Holiday is not accepting the amended tax form. Whether or not discretion was abused. After trying again to get Judge PAJAK to allow him to file evidence with the Court he refused, Then stated If at any point Mr. Holiday wants to make a stipulation about something he's perfectly able to do so. Judge PAJAK who is making the decision on what is deduct-able and what is not, along with Blaine Holiday, neither of them knew what is prototype expense, or what is, manufacturing expense, or what is expense for research and development, all are tax deduct-able under IRS Rules yet they didn't know what these deductions were. Mr. Blaine Holiday states very clearly that's correct your Honor, There is no Schedule A attached to his return, and there is no itemized deduction schedule. Larry states he has a document that clearly shows that Blaine Holiday received the Amended schedule A itemized deductions in the mail. Blaine Holiday even brought back to the Court items that the appeals officer had agreed to accept. To put it very plain and blunt when anything was presented to him by me at any of the meetings he would tap his pen or finger on the Gross Income and say this is the only figure I'm interested in.

Was there ever a chance for appeals, Was there ever a chance for an honest audit, No because someone had their thumb on them. As for Blaine Holiday is nothing more than a cheap white colored criminal who is the assistant to the Commissioner of Internal Revenue Service. He has now been charged with perjury in a Civil action in a Federal District Court, and the best part of all, it's in the transcript. He is also charged with Misrepresentation, Deceit, and Fraud. All Federal Criminal Charges. "TERROR AT HOME"

The best part of it all Judge PAJAK is charged with Constitutional violations in Federal District Court civil action, Article 7, amendment 14, for not allowing the Taxpayer Due Process of Law, and not allowing the Tax payer equal protection under the law. He also violated the Taxpayer's right under Amendment 1, Freedom of speech, and Freedom of the Press, when he refused to allow legal documents or evidence presented to the Court. When presenting his opening to the Court Larry said very clearly that Judge PAJAK had violated the Constitution of the United States Evidence presented to the Court is the Tax Courts Transcript of Trial of Larry Land. "The Courts Transcript is Judge PAJAK's 'own worst enemy, and one that will stand and will not waiver". Judge PAJAK was asked by Tax payer "eleven times" if he could file evidence of wrong doing with the Court and the Judge PAJAK refused each time. Even when the Taxpayer ask if he could show the Court that Blaine Holiday did in fact have the amended tax schedule in his hands with the itemized deductions The Judge even knowing that if the Evidence was allowed before the Court charges would have to immediately be brought against Blaine Holiday, so again he refused the Taxpayer to place any evidence before the Court, a clear violation of Article VII, Amendment XIV, "1868" and Amendment I, "1791" of the Constitutional of the United States. "YET THE IRS, AND TAX COURT TERRORISM CONTINUES".

If Larry wins his case in Federal District Court, it would end 15 consecutive years of harassment by the IRS. This all started in 1988 when Larry had H.R. 816 a Bill passed allowing him to sue the United States of America, Agency U. S. Army/. They have been relentless and seem almost bound and determined that he will not get passed the IRS/Justice Department, Larry is a remarkable accountant and has beat the IRS at their own game. even though Larry's Legal and Equitable rights against the Army were Froze in time, only to protect the Government operations, "Whether or not the discretion was abused is immaterial". United States V. 331 f2d 498 (10th Cir.1964). He was still given an award by the Court of Claims that is to be appropriated by the House of Representatives using all the evidence. The Conclusion of the Court of Claims Review Panel of Judges report: "Recommended that the Chief Judge advise the House of Representatives that Plaintiffs were given an award based on the facts present in the record, to be determined, and appropriated by the Congress of the United States".

Larry was an environmental whistle blower, Trying to warn everyone of what to watch for that may harm them, or the children because we owe them a clean environment, and not one that is so full of pollution it is no longer safe for anyone. The Government, the Military services, and large Corporations need to know we are serious about cleaning it up. He blew the whistle in 1972, "A whistle heard around the World" against the United States Army, Shell Oil Company and all other large ill meaning Corporations, or Company's The Department of Energy, and the Pentagon. His thirty (30) year investigation took him around the World & back too Washington D.C. the Congress of the United States, Congressional subcommittees, met President Ronald Reagan. Larry was visited by many who were interested in his environmental battle, The Chemical, Biological, and VIRO-logical, War waged on their own people, the Army, Pentagon, Department of Defense, the Department of Energy, and any large Corporations, or Companies that have polluted the Nation, and are now polluting with materials that are harmful to it's people without their knowledge. Larry says, "OUR POLLUTED

NATION WAS NO ACCIDENT, IT HAS BEEN DELIBERATE AND INTENTIONAL, NOT ONLY BY CORPORATE AMERICA BUT GOVERNMENT AMERICA AS WELL."

Larry thought that this thing with the IRS would have stopped by now, In 1996 when they first called for the audit of 1995 tax year, he had called Congressman Hank Brown's office told them what was and has been happening. He placed the IRS under investigation of harassment, an agreement was made they would like to complete the audit of 1995 and if it was right they would withdraw. "YET THE TERROR GOES ON". Like I said that was in 1996, this is the end of 2002, and they are still going hot trying to make sure Larry fails. Larry says "That will never happen, he gave them all the rope in the world they needed to hang themselves". They audited that one year now for seven years probably the longest ever against one single taxpayer. Even though the trial has already begun and by law should end Larry is anxious 15 years of harassment needs to end. This will be the first time in history the IRS has been brought before the United States District Court, This is the first time the Tax Court Judge has been brought before his own Court. This is the first time ever for the U. S. Tax Court to be brought before it's own U.S. District Court. It was brought before the U. S. District Court after waiving the IRS Sovereign Immunity automatically under the Tucker Act, and then an independent action under Fed R. Civil automatically under the Tucker Act, and then an independent action under Fed R. Civil Procedure Rule 60. (b) making fraud an express ground for relief. Rule (b) gives express relief from judgment by motion, (2)newly discovered evidence which could not have been discovered in time for a new trial, (3) fraud, whether heretofore denominated intrinsic or extrinsic, misrepresentation or other misconduct, or an adverse party, (4) The Judgment is void, the judgment has been satisfied, released, or discharged, or a prior judgment, on which it is based has been reversed or otherwise vacated or it is no longer equitable that the judgment should have prospective application. Larry says, "IN REALITY IT IS A MOOT POINT FOR THEM TO CONTINUE".

Rule 60. (b) A motion under this subdivision (b) does not affect the finality of a judgment or suspend its operation. This rule does not limit the power of a court to entertain an independent action to relieve a party from a judgment, order, or proceedings, or to grant relief to a defendant not actually or personally notified as provided in Title 28, U. S. C., & 1655, or to set aside a judgment of fraud upon the court.

THAT THE UNITED STATES AND LARGE CORPORATIONS AMERICA HAVE DONE. TOGETHER THEY HAVE CREATED A POLLUTED AMERICA, SOME WILL NEVER BE RETURNED TO ITS ORIGINAL PRISTINE STATE. THIS WAS NO ACCIDENT, IT AS INTENTIONAL WITH EXPEDIENCE AND GREED IN MIND.

LARGE CORPORATIONS, AND THE UNITED STATES GOVERNMENT WORKING HAND IN HAND WITH FIGURES THAT WILL SHOW JUST HOW MANY AMERICANS ARE EXPENDABLE. (MEN, WOMAN & CHILDREN). WHAT STARTED IN THE 1800'S EARLY 1900'S GREW RAPIDLY WITH GREED & EXPEDIENCE DRIVING IT FORWARD. THE UNITED STATES BECOME THE MOST POWERFUL INDUSTRIAL NATION IN THE WORLD. FROM THE 1940'S ON THE UNITED STATES INVENTED EVERYTHING THAT KEEPS MOVING THE WORLD. THE CHEMICALS, BIOLOGY, AND THE NUCLEAR AGE WERE BORN. BOTH INDUSTRY AND THE UNITED STATES INTENTIONALLY USED IT'S PEOPLE AS GUINEA PIGS, TO SEE WHAT EFFECTS RADIATION WOULD HAVE ON HUMANS AND WITHOUT THEIR KNOWLEDGE. CHEMICALS THE SAME WAY AND BIOLOGY WERE USED ON THE PEOPLE CLEAR ACROSS THE UNITED STATES WITHOUT THEIR KNOWLEDGE. MEN WOMEN, AND CHILDREN IN TEHIR SCHOOLS HAVE BEEN UNKNOWINGLY SPRAYED WITH OR EXPOSED TO BIO/CHEM FROM THE UNITED STATES ARMY. WITHOUT DUE CARE FOR THE ENVIRONMENT OR THE PEOPLE. THE UNITED STATES ARMY HAS SOMEWHERE CLASSIFIED MATERIAL THAT WILL TALK ABOUT HOW MANY PEOPLE WOULD BE CONSIDERED EXPENDABLE IN THESE TESTS.

WERE GOING TO TALK A LITTLE MORE ABOUT THE ROCKY MOUNTAIN ARSENAL, THE MOST POLLUED PLACE ON EARTH, ALONG WITH SHELL OIL COMPANY WHO HELPED MAKE IT THAT WAY, WHILE CLEARLY HIDING BEHIND THE SKIRTS OF THE FEDERAL GOVERNMENT.

SECRET TOXIC ARMS TESTS CONDUCTED ON GI'S IN U.S.

The defense Department acknowledged a much wider testing of toxic and biological weapons on its own forces during the Cold war in secret documents declassified and released. According to documents BIO/CHEM warfare agents during military exercises on American soil as well as in Canada, and Britain. Sixteen newly declassified reports describe how BIO/CHEM exercises remained undisclosed using deadly substances such as VX, GB, Sarin, to test the vulnerability of American Forces to an unconventional attack.

The reports, which detail tests conducted between 1962 and 1971 reveal for the first time that the chemical warfare agents were used during exercises on American soil in Alaska, Hawaii, and Maryland, and a mild biological agent was used in Florida, and CS gas a riot control agent, was used during tests in Utah.

The Pentagon Officials said during there investigation indicated no lethal chemical agent was dispersed into the general population. Some milder substances did escape into the atmosphere, the documents show, with a plant fungus dispersing in an area of Florida, a naturally occurring bacteria in Hawaii and a mild chemical irritant in a remote part of Alaska.... The Defense Department states, military personnel were given protection from the toxins available at the time, though he conceded that the equipment was primitive compared with what is available today.

In May it was revealed by the Pentagon that sailors and their ships were sprayed with chemical and biological agents as a part of the Cold War era. They claim the tests were different because they occurred on the high seas. The Department of Defense is working with the Department of Veterans Affairs. The Veterans affairs are taking this action, because of possible Health and any potential Health problems, They feel the soldiers and sailors were not fully aware of the risk or potential risks during these exercises.

ONE MILESTONE ON TOP OF ANOTHER, TURNING VISION INTO ACTION

One would wonder how do you remove 50 years of history from the terrain and call it progress?......The progress is stopping the United States Army and Shell Oil Company from any further Ultra Hazardous pollution in which during their 50 year history they raped and polluted more than 60,000 acres of land in Colorado they say will take at least to the year 2060 just to repair, with some raining polluted into eternity. They laid waste to the land, and tried to dump it back on

Colorado without cleaning it up. Thanks to the Colorado State Health Department and one man in particular Jeff Edson. The Rocky Mountain Arsenal that sits just north of Denver Colorado was pristine prairie and farms before the Army took over and spoiled the land forever. The 18,000 acre wildlife refuge is all behind 8 foot chain link fence, to make sure the animals cannot leave the area because they are so contaminated, as they die their picked up an buried, and new stock is brought to take their place.

THE ROCKY MOUNTAIN ARSENAL A 32 SQUARE MILE, ANIMAL REFUGE WHERE ANIMALS ARE BROUGHT IN, PLACED BEHIND AN 8 FT CHAIN LINK FENCE SO THEY CANNOT ESCAPE. ONLY TO LIVE & DIE WHILE BEING HELD CAPTIVE.

NORTH PLANTS AREA BUILDING 1501 NERVE AGENT GB/SARIN WAS MANUFACTURED DURING THE COLD WAR ERA. AS A PART OF THE NORTH PLANTS 50 ACRE SITE LOCATED IN THE NORTH CENTRAL PORTION OF THE ARSENAL. CLEAN UP TO BE COMPLETED BY FEBRUARY, 2003. AFTER THE PROJECT IS COMPLETED, MORE THAN 60% OF THE CHEMICAL WEAPONS PRODUCTION CAPABILITY IN THE U.S. WILL HAVE BEEN FOREVER ELIMINATED

The Irondale groundwater Treatment Plant was one Of the first in operation For more than two decades. Workers succeeded in it's demolition. Water flowing through RMA was piped and Treated at the plant, located on the RMA grounds near the intersection of Quebec street and highway 2. Over its lifetime, nearly nine billion gallons of water were treated by the Irondale System…….designed to accept and treat 1,400 gallons of water per minute.

Today November 2002 groundwater is being treated at a smaller, alternate plant located closer to the source of the contamination on the Arsenal. Because the size of the plume and amount of flow requiring treatment have been significantly reduced over the years. Use of the smaller treatment system is much more efficient. Approximately 100 to 150 gallons of water are treated per minute at the new smaller plant.

With Regulatory approval and oversight, the Irondale Treatment Plant was safely demolished in April of this year. Rocky Mountain Arsenal used five additional on-site water treatment plants during remediation work to ensure contaminated groundwater does not leave the site.

Another structure in the North Plants area, building 1606 another of the DEATH FACTORY'S U. S. A. along with it others at the North Plant "Known to be the deadliest DEATH FACTORY IN THE WORLD". is coming under the wrecking ball, as heavy equipment tears into the building 'beginning of the end' for another of the buildings that was involved in making several of the worlds worst chemicals of death.

It is understood they produced at the North Plants, VX nerve agent, BZ nerve agent, GB/Sarin nerve agent, Cyclo Sarin nerve agent, and Hydrazine, and Aerozine Rocket Fuels that are also known as the deadliest chemicals in the world. Most of these chemicals are stored at various Army bases such as one in Utah, known to be the most secure.

One of the Arsenals most extensive soil excavation projects will be completed by the end of this year. It was begun March 2002. The second phase of the South Plants Soils remediation project is designed to remove contaminated soils, foundations and sewer lines from an area that was used for many decades as a full-scale chemical operations area for several DEATH FACTORYS' that started back in the early 1940's.

By the time this project is completed, 120,000 cubic yards of soil, 220 building foundations and 18,000 linear feet of defunct chemical sewer line will be grouted in place or removed and placed into the on-site Hazardous Waste Landfill and Basin A. Though not anticipated to pose health risks to surrounding communities, odorous chemicals may be encountered during excavation work on the South Plants Soils project, where enhanced air and odor monitoring, and odor suppression methods are being used. Regulatory oversight and site-specific controls such as hourly odor measurements are ensuring that workers and neighbors are protected during all stages of this project.

MUSTARD AGENT DISPOSAL DECISSION HAS BEEN DELAYED AGAIN.

Once again the pentagon has delayed or postponed its decision on the best way of destroying 2,600 tons of Mustard Agents, they claim they didn't have enough information to make that decision. They said the pentagon will take at least another month Originally a decision on whether to incinerate the mustard agent, and more than 780,000 projectiles or to use an alternative technology was accepted already this year. There are four different methods for disposing of the mustard agent are now being considered. Two involve incineration. Two others would be to use water to neutralize the mustard agent and the either bacteria to consume remaining chemicals or heat and pressure to vaporize them. The Pueblo City council, and Pueblo County Commissioners say the community supports the neutralization, and biodegradation as their first option. How it works; the mustard agent is deactivated when mixed with water. Then bacteria similar to the kind found in municipal sewage plants are introduced. The bugs will then devour the material creating a non hazard substance. At least nine government and civil organizations in Pueblo are united behind this method.

Mustard agent causes blistering of the skin and the lungs. The weapons are held in reinforced steel and concrete igloos 15 miles east of Pueblo. The city has been waiting since 1988 for a final decision on how the military will dispose of the mustard agent weapons. Under a treaty, the United States has agreed to destroy nearly 30,000 tons of chemical weapons by April 29, 2007, although it is now said the Army will miss that deadline. The worst scenario played out would be if the Army takes their 2nd option of the treaty which would give them an additional 11 years to destroy all of the chemical weapons.

Four mortar rounds have just been discovered at the Rocky Mountain Arsenal during clean-up they have been secured at the Rocky Mountain Arsenal for examination. First examination, two of the mortars were not marked to indicate their contents. A third was marked indicating smoke, and the fourth was thought to be empty, and are thought not to contain chemical agents. The United States Army material review board have reported that The two unmarked were found to contain liquid smoke, a third was correctly marked for liquid smoke, and the fourth was in fact empty. The now Defunct Rocky Mountain Arsenal is known as the most contaminated place on earth, and if they are able to clean at least the surface for animals, it is being turned into animal wildlife refuge, and can never be any thing else. It is to remain forever inhabitable for humans.

NEW REMEDY TO BE CHOSEN AT THE ROCKY MOUNTAIN ARSENAL CALLED THE HEX PIT CLEAN-UP FOR THE SOUTH PLANTS AREA WHERE MOST OF THE ULTRA TOXIC CHEMICALS WERE USED AND PRODUCED.

The plan involved placing approximately 300 evenly-spaced heating wells below the ground to thermally destroy the waste. The gasses given off would be collected by a special vacuum system and filtered through an air treatment system that has met regulatory requirements before being released. The very first Hex Pit has been in place for forty years, Without any cap, and is has not been sealed from the underground water stratus however efforts must be made to seal it off in the ground, drilling holes around the toxic waste and sealing off the ground trying to stop it's movement, then by placing a forty foot cap on top that will have regular vegetation. Actually what we have here is a Tomb of toxic waste, and their will be several hundred of these toxic tombs on the now defunct Rocky Mountain Arsenal wild life reserve, with uncertain bottoms to protect the ground water.

A test explosion in the Armies vapor containment structure, That had been brought in as part of a plan. that had been fitted with a filter system to become a "vapor containment structure", as part of a plan to dissolve the bombs in a caustic solution. The doors were blown open, and if you look close you will see the roof is also bulged open. This was suppose to keep the Sarin gas from escaping. It was described as if it were as safe as a breeze in a calm day. It was tried and trued yet they claimed the test blast was not entirely incorrect because the blast was stronger than any explosion that could occur during actual destruction. Yet this structure was built to withstand exactly what they used nearly blowing up the structure. This is a shed that was suppose to keep the Sarin gas from escaping in case the bomb exploded. Each of these bomblets contain 1.3 pounds of Sarin, plus and explosive charge. These bomblets were manufactured at the Arsenal according to records until 1957, They were also stored at the Arsenal by the thousands. Destruction was never published or mentioned in any material that I've ever had. I would image it is still classified.

The nerve gas bombs were talking about is a part of a cluster bomb filled with deadly nerve agent such as Sarin, BZ, or possibly VX. And enough explosive to turn the gas into an even deadlier form of Sarin. The little grapefruit size bomb by itself is deadly to every thing in its path for about 900 feet each direction. But with adding the larger explosive charge, they get four times the coverage changing the gas into a vaporized gas covering nearly 1 ½ square miles.

These cluster bombs are exactly what it says, a cluster of about 260 to fit in a specially designed rocket called the Honest John rocket. Designed to at a certain destination and height over a city, the rocket would explode, dispersing the bomblets in every direction over the city, the bombs covering many miles the explosion and the vaporized gas from each bomblet saturating 3600 feet in every direction. This would be a rain of terror but death would come rapidly, within seconds, or minutes. A drop smaller than the end of a small needle is all it takes.

The Honest John rocket was some maniac's dream back in the 1950's and 60's on how to end all life in an entire city, without destroying a single building, The Neu-Tron bomb was the answer. It destroys the Neuro- Transmitters in the brain in a matter of seconds causing near instant death.

What they have been finding out at the Rocky Mountain Arsenal is these little bomblets, Made back in the 1950's for the Honest John Rocket, and they are loaded with both nerve agent and an explosive making them very deadly. Workers discovered the first bomb in a pile of scrap metal, then later as they were erecting the tuff shed around the first bomb, Two more bombs were spotted. They were later test for sarin gas and explosives.

ARMY MISLED PUBLIC, STATE CHARGES, BECAUSE IT HAD REPORTED ADAY EARLIER THAT A TEST SHOWED DEADLY NERVE GAS BOMBS COULD BE SAFELY DETONATED. Larry says, "The Army no matter what they have done it's always been trial and error. They don't examine and make realistic judgments on anything they've ever done, History is very clear on the Army's mis-deeds". "They promise you the world, and give you pieces of a polluted earth as they run like hell lying, and classifying everything in their destructive wake.

Test bomb detonation conducted in the tuff shed definitely showed a risk factor, and Showed just the opposite of what the Army had promised the public, according to Howard Roitman, he is from the Health Departments hazardous materials unit. If the explosion had been a real sarin bomb it could have released it's reminder "DEATH

FACTORY U. S. A." Into the atmosphere around the Rocky Mountain Arsenal, where there has already been three grapefruit-sized nerve gas bombs found recently at the defunct, Rocky Mountain Arsenal.

A drop of sarin as small as the end of a pin can kill in seconds. This is a drop so small you would not be able to see it, and you wouldn't smell it, you really wouldn't even know it. If it happened you may hear a small explosion, "then gasping for life". The Army's misrepresentations were misleading and confusing to the public. It was so Wrong, they then implied that the Colorado Department of public Health & Environment agreed with them. The Army's vapor containment structure after the bomb was detonated the outer walls were bulged, the roof was bulged, it blasted off two venting tubes. the doors blown open and it wouldn't hold air. The Army stated "though the structure was slightly affected by the blast, the Army is confident the tests were a positive indication that the current destruction plan under review is a safe option." Larry says, "what the Army is saying, though it didn't work, & it could have polluted a large area, were going to try it again.

All public use activities will be suspended until the Army safely disposes of the ammunition. There are parts of the Arsenal such as the old industrial core that will never be open to the public, and an area that will never be a part of the national wildlife refuge. The annual return of the wintering bald eagles has begun. This is actually a very scary incident, the finding of a 1.3 pound bomblet that contains sarin nerve agent was found in what they call the bone yard, the shell of the bomblet is made of aluminum. They were made between 1953 and 1957 for the Honest John Missiles that would carry up to 300 lethal bomblets making up what is called a cluster bomb. Each missile would be capable of killing every living thing in 5 square mile area. A drop as small as the end of a pin could kill a person. Then the explosive charge was added making what is believed to have been called Cyclo-sarin, a vaporized version of sarin gas that would cover an area Multiplied by 4.5, or 22.50 square mile area. Destroying all life, but leaving all structures in tact, this is believed to have been called the neu-tron bomb I can remember reading about it when I was just a kid. It destroys the neuro-transmitters in the brain, death would follow as the body would begin to tremble, the victim becomes very weak, his nose would run and he would begin drooling, the eyes watering, vision, smell, feeling, and taste gone with in seconds, then vomiting, loss of all motor control fall to the ground, and go into convulsions, losing all ambulatory skills, legs, arms, and eyes, convulsing in time until death. Death is not instantly, the victim of sarin gas could lay where they fell convulsing for hours or days.

I talk from experience you see, Sarin and BZ nerve agents found its way into our water supply.......Larry drilled two new wells on the new place he purchased, within days 635 head of dairy cattle lay dead or destroyed. Some survived the first disaster, but they failed to thrive, failed to grow any further, some had cancerous tumors, severe calcification of the ligament attachments to the bones, eyes were partially destroyed. He had about 65 heifers with calf, the abortion of all were almost instantaneous after they drank the water. Larrys wife Marie was also carrying their child, and the pregnancy was instantly aborted.

In less than two months they have now found six M-139 bomblets at the Rocky Mountain Arsenal, (RMA), they are all treated as if they were filled to capacity and explosively configured. The claim now is the M-139 bomblet were made to form a cluster bomb known as an Honest John Rocket, which would carry 300 bomlets capable of gassing an entire city like New York, or maybe Chicago, London or Paris. The secrets of Sarin nerve agent came out of Hitler's Germany in 1938.

Once cleanup is complete, the Arsenal will officially transition to the premier, urban wildlife refuge in the nation just north of Denver, The site now provides sanctuary in chosen areas for nearly 300 species of animals that most have been brought in, including deer, coyotes, the bald eagles and the white pelicans.

The Army brought in what's called the destruction chamber in February & successfully Destroyed the six M-139 bomblets that had been found last fall. The entire area where the bombs have been turning up was covered with an enormous shed to contain the nerve agent in case of an accident. The explosive destruction chamber had been set up inside the Steel shed. The shed remains in place as workers comb the area for more bombs. Four additional bomblets were found on Friday, and placed in ammunition canisters, and surrounded by sandbags. They are under guard 24 hours a day, The destruction chamber called "EDS" has been called back into the Arsenal to destroy the four recently found bomblets. EDS neutralizes the sarin, the process takes 12 to 14 hours including pauses for safety checks. Inside the chamber there is a small charge that cracks open the bomblet similar to a coconut releasing the sarin agent. The sarin gas is mixed with a caustic solution that is injected into the chamber by remote control, It will then be heated to 140 degrees and then agitated for four hours. At the end of the process the neutralized liquid will be drained into drums for shipment along with the bomblet to a toxic waste dump.

June 2001 a 27 inch long M155 mm projectile was discovered in the same dirt pile where the six grapefruit sized bomblets had been found last fall. Experts plan to X-ray the projectile today and then conduct a portable ion neutron spectroscopy test to determine what if anything or any of several nerve agents made at the Arsenal.

July 26, 2001 the Army has safely destroyed the third of the last four bomblets, the fourth is set to be destroyed immediately. EDS used to destroy these bomblets, and other warfare munitions when recovered. The EDS was placed inside the “LAMS” Large area maintenance shelter, which had been built over the same scrap pile where the first of the six bomblets were found. It’s the length of a football field, and ½ as wide. LAMS is equipped an air filtration system specially designed for containment and treatment of air inside the LAMS with still another containment structure within specifically designed to house the EDS and its air monitoring system.

The two Generals that were sent out to the Rocky Mountain Arsenal by the Pentagon to deal with the States concerns about the disposition of the bomblets, are undoubtedly very qualified to determine the best and the safest methods of disposal. On behalf of all the families living downwind of the arsenal, (RMA). The Governor or the State should not try to become overnight experts in disposing of bombs loaded with Sarin Nerve agent or other known or unknown nerve agents would be considered futile. What ever choice or whatever solution the Army, and their two generals, and their experts decide on should be fine with us. After all they are the experts.

Whether they melt them down, cut them up, or just blow them up, it’s all up to them; We’ll let the professionals decide. However there is just one condition and the Governor should ought to insist on it;......Let the Generals be present when the bombs are neutralized.... We would like to see them standing in ordinary dress, at the place and the very minute their professional decision is implemented.... If any thing is going to go wrong the decision makers should be first to know......The mistakes made by ordinary generals who have made critical decisions, other's have paid for it with their lives.... The nearly one million men women and children who have made their home near the Arsenal.... It would seem on this great occasion, the governor should insist the decision makers share any risk that might exist.... The mandatory imposition of this condition should immediately enhance everyone's confidence in how our military deals with all critical situations safely. This may help restore the Army's credibility that was lost over the last 60 years or so. It all started with the nuclear age in the 40's & the Bio/Chem warfare in the 50's & 60's.

SARIN BOMLETS HAVE BEEN DESTROYED SAFELY ON BASE

The Army has safely neutralized ten of the deadly canister's containing Sarin Nerve agent that were found at the Rocky Mountain Arsenal. The last of the ten sarin nerve agent bomblets has been destroyed this morning at the Arsenal, RMA. It has been very tense out here, as a bomb expert in a bomb suit picked up the bomblet with a 10-foot rod & placed it in the destruction chamber known as (EDS) that is sitting inside the (LAMS) Large Area Maintenance Shelter, as long as a foot ball field and half as wide, just in case something goes wrong with EDS.... It was very tense and very emotional. It seemed as if everyone was holding their breath. Then once the chamber was sealed with a series of bolts. Then an explosive charge cracked open the bomblet. "Like cracking a cocoanut". The Sarin agent was released into a caustic liquid that was pumped into the chamber, and then the 12 to 14 hour process Including several safety checks, began the complete process of neutralizing the sarin nerve agent.

By mid-afternoon, tests were showing the sarin agent had been neutralized, although there is a second dose of the caustic liquid was pumped into the chamber, this is just to be sure every last drop has been neutralized. Then all the liquid, plus what's left of the bomblet's all aluminum shell, will be sent to a hazardous waste dump. It was announced the Army feels they have, State of the ART ordnance destruction operations after successfully completing the destruction of the tenth cluster bomb. At this point the last of the grapefruit-sized bomblets has been successfully destroyed. Rocky Teter, a hazardous materials expert with the South ADAMS County Fire Department said, “I hate to say this but they outdid us“…. Of course the State, Local, and National officials were watching the entire operation on closed-circuit television. They were all very impressed with the Army’s extensive safety procedures. “They have tests to test the tests.”

THE ARMY HAS LOST ALL PUBLIC TRUST WHERE THE ROCKY MOUNTAIN ARSENAL AND THE PENTAGON ARE CONCERNED.

“Safety is a top priority in the Arsenal cleanup efforts” Gen. John G. Coburn Misstated the facts in his attempt to calm public concerns about the Rocky Mountain Arsenal. Instead he should have provided more information or even an explanation, for the surprising discoveries of 10 deadly sarin bomblets, Instead he makes a flag draped appeal for more blind trust from the surrounding community. Every official statement from the arsenal officials since the first bomblet was found on Oct. 16, has been designed to minimize the danger, sometimes to the point of absurdity. Public relations seems to be a greater priority than public safety for the Army, One lie then you have more damned lies, and all that is left then are statistics.

It was only hardball negotiation by the State of Colorado that prevented the Army from recklessly exploding the first bomblet, in the open air, within a few feet of other bombs that were later found. It was also the State that has provided the information on up-to-date technologies to deal with these bombs.

The Army and Shell Oil Company have continually refused to provide documentation to the citizen advisory board members, to support their claim that the arsenal is safe for visitors. Last year, when 10 board members requested that meetings be held off-site until that information was available, the Army changed its rules to require the public to meet on the superfund site.

Members of the public who had serious concerns about their safety were dismissed from the advisory board due to their absence. The accidental discovery of these bomblets proves these concerns were well-founded, but this important public input has been deliberately derailed and discarded. Because of their own actions, the Army no longer has our blind trust. Confidence may be restored if and when the full extent of the hazards at the arsenal is admitted.

From the 1940’s thru the 70’s if the Army or Shell Oil had something to dispose of they just buried it. Among the many items were the whole body suits worn by people who worked directly with dangerous and hazardous materials, Pallets materials came in on or were used and soaked with deadly chemicals the items were placed in trenches, then burned before being covered over. Unknown amounts of toxic and deadly chemicals were expelled out into the air. Weapons production ended in 1957 but the huge amount of storage remained. Shell Oil Company kept production going all the way thru 1982, the last four years in business and no restraints, they doubled, tripled, and quadrupled production, and filled basin F, over capacity and reached the over flow stage with 250 million gallons of the deadliest chemicals known to mankind. There have been lots of stray munitions found that may have failed to explode during testing. A napalm bomb was found, along with mortars, and some other large rounds. The sarin bombs began turning up in October 1999, and up to 2002 10 of these deadly bomblets have been found, and safely destroyed.

Acute has been redefined, even minute amounts of Sarin, VX, or BZ, can be fatal. Victims in most cases lose control of the muscles, they lose all senses, and will go into convulsions and die. Each of these tiny grapefruit sized bomblets made of aluminum, were designed to contain 1.2 pounds of Sarin or VX. The missile called the Honest John Rocket would carry a cluster up to 368 bomblets. It would disperse high over a target area such as Denver Colorado, or Los Angles, or New York, sending the bomblets in every direction, at a certain height above the target they would explode turning it into a shower of death. The explosion would vaporize the deadly Sarin gas turning it into Cyclo Sarin, 400 percent more potent than Sarin itself. Each bomblet coverage is over ½ a square mile. Leaving a shower of death over any particular city of 184 square miles. Death would be widespread with no structural damage. Larry says, "you can now see why the Rocky Mountain Arsenal was called DEATH FACTORY U. S. A." There are DEATH FACTORY'S IN OUR MIDST.

This system was never used in combat. It was the talk of the 50's and 60's "THE NEUTRON BOMB, it would kill all forms of life yet leave the structures intact and in place. It was meant to wipe out entire cities while the residents slept, or to wipe out entire army battalions. It could well have been used in Korean war, or the Vietnam war. It was so deadly, so cruel and unusual, the only thing I can imagine the Congress, or the Presidents stepped in and said no to the neutron bomb, and ordered its destruction.

When they began discovering the bomblets at the arsenal, it was no surprise to EPA. Environmental Protection Agency. There are two other sites besides the 27 acre site that would justify being searched. One army official states we're going to find it all in a safe manner, and take care of it. Discovery of the bombs was quite a mystery the workers were always careful with dangerous ordinance. Most of this took place during the Cold War period…. The munitions each had a serial number, so it could be individually tracked to the places they were tested, and it was carefully marked and recorded……Workers could not remove them from buildings without the proper paper work. The people were very responsible about where it was they placed dangerous objects…they had set disposal sites where these things were taken.

They had speculated the recently found bomblets were dummies used to train the workers who would be loading the Honest John Missiles. They say they can't understand why they were in a pile of scrap metal. Larry says, "probably just waiting for pickup by maybe terrorists". Because the Army cannot explain why they in fact were loaded with Sarin, and the blast to vaporize the contents of each bomblet. In their analysis it became a shocking reality when they discovered it was not just antifreeze that would have been loaded in the dummy bombs, It was in fact Sarin. Larry says "what it shows they weren't really very careful, nor were they carefully marked and individually tracked. Another spokesperson stated it was not of any concern back when they were made. Only now days people are more scared, and more observant as to what is going on.

There have been 10 of these bomblets found, and possibly more will be found. The search has been widened. They are all treated as loaded and ready, One drop of sarin can kill on contact. The bombs were built in the early to late 50's, the Honest John Missile would take a cluster of nearly 400 of these grapefruit size bomlets to the potential site, exploding and dispersing the bombs in every direction over its target. Then at a certain height the bomblets would exploded showering an entire city with vaporized Cyclo Sarin, That is capable of causing respiratory paralysis convulsions, and death. Each one capable of destroying every life form in up to a ½ square mile, and figuring 370 of them with a total destruction and devastation area of 184 square miles.

The Army locates still another munition at the Rocky Mountain Arsenal. The 4.2 inch mortar that was found weights in at 42 pounds. Initial tests show it does contain a liquid substance, and since these weapons were filled with a liquid chemical agent at the site. Officials saying it could be smoke, or a stimulant. Mustard agent was ruled out because winter weather conditions would have transformed it into a solid state.

THE OLD LOWRY AIR FORCE BASE AND TESTING RANGE IS UNDER FIRE.

Residents of a Denver neighborhood have sued the Air Force with complaints of pollution from the defunct Lowry Air Force Base has lowered their property values. The chemical at issue now is Trichloroethylene, found in the ground water, the chemical turns into vapors that can rise through the soil and enter homes through cracks in the foundations. The toxic fumes can cause health problems ranging from headaches, to dizziness to liver and kidney damage. Jeff Edson of the Colorado Department of Health's remediation and restoration manager stated that a mechanical ventilation system has been installed in one apartment house in the area where fumes were discovered. The suit claims property values have declined since the discovery of TCE in the area. Edson said, "the 15 homes are sitting on top of or just above water that contains more than 200 times the state standard for TCE". The level of the chemical decreases where the water fans out toward the former Stapleton Airport. There are many stories that surfaced, the Rocky Mountain Arsenal, and Rocky flats both used the Lowry Landfills to dispose of several Ultra hazardous chemicals and solvents they are now finding.

The clean up crew at the Lowry bombing range found an aluminum M-139 bomb, they immediately sealed off the area and summoned a team of Army Tech Escort Munitions experts, and the Aurora Fire Department hazardous materials crew to the site. If the bomb was already blown apart years before why did the clean up crew believe the area still to be a threat to human safety. Is there something their not telling the public? Like how long will the area remain hazardous to humans after an attack or accident. However the one found did not contain any nerve agent. The clean up of unexploded ordinance at the old Lowry Bombing and Gunnery range, and now Ultra hazardous waste disposal site is being conducted by the

Army Corps of Engineers in conjunction with a 1998 agreement with the Colorado State Health and Environment Department.

THE DISCOVERY OF 10 VIALS OF LIQUID RAISE FEARS ABOUT LOWRY.

There were 10 vials containing a suspected but unidentified chemical warfare agent has reinforced official's fear about what may be lurking in the soil of the former Lowry Bombing and gunnery Range East of Aurora, or even in the Waste disposal sites that have began leaking deadly TCE into Denver water.

This area was to contain only jeep demolition and supposedly only standard explosive weapons were used......Crews digging through this burial pit discovered vials of suspected and unidentified chemical warfare agents. Two of the bottles were labeled mustard agent. The contents of the other vials have not been identified, said Jeff Edson, manager of the Colorado Health Department's remediation and restoration unit. Edson's unit oversees cleanup of all Department of Defense sites in the State. The vials were moved to the Pueblo Chemical Depot, just east of Pueblo by the Army's chemical specialists. Edson agreed and the State decided it was far safer for the Army's Technical Support Team out of Aberdeen Proving Ground in Maryland, to determine what is in the vials in the security of the Pueblo facility. The Chemical Depot houses nearly 800,000 mustard – filled munitions.

"The whole issue is about what is in the bottles" Edson said, That's why we chose to put them in an empty and isolated igloo down in Pueblo". "It is probably one of the safest areas in the country to store this unknown stuff." Larry says, "had proper records been kept of this Ultra hazardous materials, they would have immediately known what was in the vials. Since the bombing range is a secure facility, and because of all the publicity as to where it was found, near both Aurora's Reservoir, and Arapahoe Park Racetrack. The Officials didn't want to cause any risk to the public, so they were move to Pueblo a secure facility. Jeff Edson said, "we don't really know how many more bottles are in that pit". The unearthing comes after numerous discoveries in recent months of other chemical and biological weapons were found at the bombing range Edson said. Edson said "I am a little anxious to go and get the bottles identified," "But the contents of the bottles is only a snapshot of what's out there.

Edson said, right now none of the chemicals pose a danger to the public, he said. They are sealed in containers and are buried. But the existence of the weapons was enough to make him call an Army Corps of Engineers team in from Huntsville, Ala., that specializes in chemical weapons to discuss how to proceed in cleaning up the jeep demolition area and other parts of the bombing range. Larry says, "There are no records, no serial numbers, and no information as to why these sealed containers of what is believed to be chemicals or biological weapons of mass destruction." Why would they just bury materials like this? There may be many more found like these, and now is the time to find them before they can hurt someone.

A new law in Colorado took effect immediately, and will cover the Army's pending decision on the method of disposal, which is expected soon. The Governor signed a bill giving the State authority to review a site permit and impose impact fees for the disposal of 2,600 tons of mustard agent now stored at the Army's Pueblo Chemical Depot. This bill S. B. 41, was sponsored by Sent. Bill Thiebaut and Rep. Joyce Lawrence, and signed by Governor Bill Owens. This will expand Colorado's Hazarous Waste Incinerator Act, and will cover the alternatives which will include water-based neutralization of chemicals inside 780,000 rounds of munitions that have been stockpiled by the Army just 14 miles east of Pueblo.

The Rocky Mountain Arsenal was just dealt a statewide Grand Jury on July 31 2002, among issues being dealt with. (1). Malfeasance of a governmental function. (2). There will be no confidentiality agreement violated. (3). A fair consideration of a criminal matter will not be prejudiced by filing a report as a public record. (4). Report released includes but not limited to those written responses which are requesting release as well.

Investigations into the handling of hazardous and ultra hazardous waste by the US Dept. of the Army and Foster-Wheeler, Inc. at the Rocky Mountain Arsenal was first initiated by the Colorado Attorney General under the authority of an investigatory referral from the Colorado Dept. of Public Health and Environment to the Attorney General under state law. Presiding Judge J. Stephen Phillips authorized the investigation, and conducted the State Grand Jury.

The State Grand Jury was based on their investigation…and it was further determined the following report and its release thereof are in the public interest…. The allegations of misuse of misapplication of public funds, allegations of abuse of authority by a public servant, or allegations of the commission of a class 1, class 2, or class 3 felony……and does not involve these allegations. Release of this report is in the best interest, because the matter does involve allegations of, and this report also supports the findings of, misfeasance and malfeasance in regard to a governmental function…as government functions are defined by CRS….

The excerpt that referenced and interview with Jim Green, a project manager with the Army, did indicate the basis of the deception was from concerns for how the Army would be perceived when they became engaged in the "bone-yard" search without having proper authority…It was related that the Army did not believe that it would look very improper to have found the M-74 before getting an approval from the DESB or CDPHE on the plan. But when asked how a misrepresentation as to when the ordnance was found, and was asked how the report could have been prepared with this misrepresentation….It was explained "it was not a big deal", that it was just a perception error. When the Army and Foster-wheeler entered the bone-yard, they violated the provisions of the ECO, (emergency compliance order) and its first amendment. It as found therefore, that entrance into the bone yard without prior approval was unlawful and therefore constitutes malfeasance by the Army and Foster-Wheeler.

It was placing a potentially explosive material back into an area containing a deadly nerve agent, so therefore it constitutes malfeasance by both the Army and Foster-Wheeler. They misreported the discovery, and covered up the original discovery, and therefore does constitute malfeasance again by both the Army and Foster-Wheeler.

Certain excerpts on the discussion of findings of the State Grand Jury….The only way to ensure proper clean up was for the Army to work cooperatively, honestly and actively with the State. Representations that had been made to the CDPHE that was concerning the date given in the report of the discovery of the M-74 round in the "bone-yard" were material and false as was contemplated by the CRS….They were made in order to avoid any inquiry into the unauthorized activities in April (2000). A particular individual or set of individuals was not identified by the State Grand Jury who made the decision or decisions to deceive the CDPHE. Several individuals within the top hierarchy of Foster-Wheeler & the Army apparently contributed to or participated either directly or indirectly In the decision of deceit.

But the State Grand Jury could not base criminal culpability on supposition. Therefore, we find that probable cause has not been established as to the identity of any particular culpable party and choose not to return an indictment, (with emphasis added). The following are crucial for the success of any environmental remediation project. **Compliance with all applicable requirements**, ** As well as full cooperation and a complete and open line of communication amongst all that are involved.**.

The U. S. Fish and Wildlife service (USFWS) & X-cel Energy corporation signed a "memorandum of understanding" that will help prevent electrocution of raptors (eagles, hawks, falcons & other migratory birds. This will not only help in preventing electrocution Of the raptors here at the Rocky Mountain Arsenal, National Wildlife Refuge, but through out 90,000 miles of electric distribution lines across Colorado, Arizona, Kansas, and 10 other states, and more as the safety mechanisms can be put in place.

In January 2001 environmentalists nationwide, waged a political jihad against Gail Norton, former attorney general who was the Interior secretary nominee. She's portrayed as an ideologue who lent her political muscle to conservative calls for protection of property rights, and less governmental regulations.

She has been portrayed as a polluters dream come true, as Colorado's top law enforcement Official, dropping or dribbling the ball when time came to carrying out Colorado's environmental and pollution laws. The center stage is already set when she begins her confirmation hearings before the Senate, Energy & Natural Resources Committee in Washington. The Sierra Club and other Environmental groups and interested parties are pointing to Norton promoting of Colorado Law allowing Companies to self-police for their environmental violations. She is also accused of her indifference in at least some of Colorado's most notorious pollution cases.

Norton a Conservative Republican who was to have carried the ball as attorney general in Colorado from 1991 through 1998. Environmentalist groups say she failed to take action in the Summitville mine cases in the early 1990's. She is accused of ignoring a Western Slope wood products plant that was intentionally spewing illegal levels of air pollution, and that she discounted Clean Air Act violations by a coal-fired power plant in northwestern Colorado. Larry says, "May the best person win, it's usually said 3 strikes and your out".

Her Ex-cohorts deny she was a patsy for polluters, and her former colleagues contend environmentalists have oversimplified situations or are wrong. Her resistance was given when state regulators sought her assistance, only time will tell in her nomination for Interior Secretary. Larry says, "Her promotion of State Law allowing polluters to police the polluters, sounds much like the Fox being appointed to guard the Hen house".

The Attorney General lacked the original authority, and cannot do anything the client doesn't ask them. The Attorney General's office does not have any authority to file it's own suits says Norton's deputy attorney general. The State Health Department was pretty much a mess in terms of enforcement. Legal cases could not be built against polluters with our referrals from State Health Regulators, and those referrals rarely come. However how well did she carry the ball, when it came to the way she enforced Colorado's pollution and natural resource laws. This could make or break her nomination.

Her colleagues say her politics had nothing to do with her environmental issues, as she assigned all case work to Bangert, a life long Democrat who oversaw & assigns caseloads to a staff of 65 attorney's she was given free rein to build and pursue pollution cases. Larry says, "Democrats Liberal, or a conservative Republican, and somebody drops a ball, or several balls, and we all know what happens under the table. Done without the necessary oversight (management), You quickly have something that does not work, and that sounds like what happened with the regulator's, they just got tired of trying to enforce something that was not enforceable even though the law was in place". Colorado is no different than any other state in America, the same thing we see here is happening elsewhere also.

It is stated the law had the support of then governor Roy Romer a Democrat, & the top administrator's at the Colorado Department of Health. Larry says, "That is not true that law is not something believed to have been accepted by those out in the field, the Health Department Regulators who are believed to have had very little support in their efforts". Larry says, "he seen this played out in real life too many times". Even the EPA, Environmental Protection

Agency agreed when they opposed the law and effectively killed it by penalizing the self-reporting companies under the Agency's federal authority. Then however the EPA had a change of mind, and supports the Law. That change of heart and mind came last year after the State agreed to give its regulators discretion to penalize the same self-reporting companies for egregious violations. Larry says, "The Company's the Corporations, and the government, could no longer use the laws as an excuse.

Most environmentalists still despise the law because it gives the polluters a free pass, and at the same time it does some good. "Now lets weigh the difference for or against". But State Health officials argue that more than 30 companies have come forward since 1994 with violations regulators might not have found otherwise. Larry says, "he believes the regulator's would have in-fact found those violations since 1994 had they been given a rein of their expertise in the field". It seems a bit like nonsense, if the polluter has been doing it over a long period of time and sees a window of opportunity that could get him off the hook, Who would with common sense pass it up? The Law clearly states to penalize the self-reporting companies for Blatent, Conspicuous, & Flagrant violations, & total disregard for human life. ""(Egregious)"", A company that dumped on public land in the cloak of darkness 244 pounds of pure chlorine gas on public lands near Grand Junction Colorado.

A key member of the Environmental Protection Agency disagreed with one of his key advisor's and his support for Norton, when he stated he was skeptical of Norton's support for the operation. He also said she only tolerated the idea but never really embraced it. Pete Michaelson, former district attorney for Summit County, and a strong advocate of pollution law enforcement. States "It was a real blend of skepticism, & reluctance, on the other hand a grant of authority". Some of the most contentious debate over Norton's record is tied to the Summitville case. The Sierra Club is said to contend Norton failed to enforce federal or state law.

Some thanks....and a little bit of blame...is credited to controversial.... Coloradan Gale Norton. Past Attorney General for Colorado, and now the new Interior Secretary. There are new battles for the environmentalists to deal with such as, Access to public lands, Oil drilling, mining and endangered species. But she for some reason energized environmentalists groups, with surges of fund raising, and new memberships and online activism.

Norton has been accused of being an extremist, and it was over fund raising and there are no facts she said. She is accusing some groups that attacking her might help fill their fund-raising coffers. Norton claims there is already evidence of a Norton effect showing up in the environmentalists fund raising efforts. In just the first 2 ½ months after Norton's confirmation the group has already reached 55% of last years fundraising. ½ million came in during her confirmation battle, and with a lot of people rolling up their sleeves.

A large brouhaha is still brewing over Summitville mine as it continues to haunt Norton. During her tenure in Colorado a lot of good for the State came right out of that office, and I'm sure as Interior Secretary, she will do her best. In her new position she is pulled to the left and to the right, The middle ground is when she is able to pull both sides together for the right decision for our environment.

EPA, ENVIRONMENTAL PROTECTION AGENCY, OMBUDSMAN BOB MARTIN, & INVESTIGATOR HUGH KAUFMAN.

Became heroes at least in a half dozen cities in the last decade as they weighed in on behalf of residents who were unhappy how EPA had acted or in fact failed to act….on the removal of toxic pollution. In Denver Martin and Kaufman championed residents in the neighborhood where EPA allowed the now defunct Shattock Chemical Co. to dispose of radioactive soil by burying it at their site near Santa-Fe Drive, a heavily populated area of west Denver. The waste is now being removed after a decade of dispute.

Now that it is being removed EPA is crying wolf to our heroes. Aggressive tactics pleased citizens but made a lot of enemies within EPA.

While winning a "cerimonial embrace" as heroes, from the citizens, the Washington based pair angered numerous EPA officials, whom they are accused of denouncing publicly as incompetent or even criminals. Larry has, "witnessed too many cover-ups by EPA for the Government and Corporate officials, and it's not always the people in the field, it is those working administration". Larry says, "if the shoe fits wear it". They accused Martin of presiding over rigged public hearings at which the outcome was all but decided in favor of community safety and protection. Jack McGraw, the acting director of EPA in Denver, said the hearings held by Martin & Kaufman, were demoralizing & downright humiliating the way agency employees were treated. Kaufman is accused of further humiliating & angered EPA Officials by unleashing barrages of allegations to the local media in advance of the hearings. Larry says, "The truth sometimes really hurts when it has been covered up for so many years".

Kaufman describes Martin as "upset by the controversy that is swirling around him, and says martin draws fire because he's doing a good job, and not just a lightning rod for EPA. Kaufman says, I'm an aggressive investigator, and that's why I'm good at it. EPA workers fear Kaufman and mostly decline to criticize him publicly.

Martin has been Ombudsman since 1992, and the move to rein in the Ombudsman's office only came after years of friction with EPA workers. Kaufman has been with EPA since it was founded in 1971. He has been accused of repeatedly being at the center of controversies over alleged EPA screw-ups. He is also accused of being a major player in the controversy that had surrounded Anne Burford who being President Reagans first EPA Administrator who stepped down from the position.

Carol Browner, in one of her last acts as EPA administrator under President Clinton, removed Kaufman as Bob Martin's investigator. Then last November, Christine Todd Whitman, who was President Bush's EPA administrator, gave orders for Martin to report to the agency' inspector general. Where it was decided he would no longer have authority over his own budget, or to hire his own staff, which virtually ended his partnership with Kaufman. A sweetheart deal was soon reached between EPA, & a Denver superfund site, it is said she crafted the deal for her husband's former employer. But critics of Whitman & the Environmental Protection Agency deal struck with Citigroup Inc. are saying that Whitman was involved in the decision because she was briefed on the company's Shattuck Chemical Co. site & has never formally explained the circumstances surrounding the Sweetheart deal that was made of Citigroup.

Larry says "Right is Right, & Wrong is clearly Wrong, Bob Martin as the Ombudsman, and Kaufman as the Investigator are heroe's to everyone of us. As they continue to keep Right, Right, and turn Wrong into Right for our safety, and the safety of our children & in order to clean-up our already polluted environment.

Hugh Kaufman is viewed by many as a tiger, He is a fireball not afraid of confrontation. An investigator who knew how to do his job.

Martin and Kaufman became heroe's to millions, for them legal maneuvers of their own had just begun. They sparked the world outcry of citizens, helping them to move an intransigent bureaucracy. Citizen lawsuits have also began in Florida, and Idaho. It had appeared for the first time, someone from the Federal Government was finally listening. The Ombudsman's office is the only arm of EPA that ever listened to the concerns of the people, in how to clean up Ultra Toxic, hazardous, and flammable materials form mostly now defunct and long gone chemical plants both government and corporation. Supporting Martin and Kaufman Senator Wayne Allard R-Colorado, even though doesn't believe in private legislation, says the bill is in part for the support for both Martin and Kaufman. Larry says, I guess it is all about who you are, since Allards office refused to assist the Land's for their part in closing the Rocky Mountain Arsenal, and Shell Oil Company, stopping over 40 years of polluting the countryside with the deadliest chemicals known to mankind. They are both now defunct but in the time of need by the Land's Allard was not available. Even with Congresswoman Diana DeGette, D-Colorado who backs Allards bill, her office refused to assist the Land's in getting a Bill before Congress that had already been approved by the Court of Claims who gave the Land's an award but it was their Congressman, and Senator's who needed to carry the ball. Representative DeGette, and Senator Allard were not there in time of need for the family that was insturmental in closing forever the Rocky Mountain Arsenal, and Shell Oil Company, well known as the most polluted piece of Land on earth. Senator Allard says citizens have a right to be heard by an independent arbiter. The Lands were heard by the Court of Claims who gave them an award that only the Congress, or Senate can put together. That award has been sitting in the files before clerks of Congress of the United States now for six years awaiting Land's Representative, or Senator to do their part prescribed by the Law of the United States.

EPA accuses Kaufman of browbeating, and definetly short of charm, Kaufman's behavior and his confrontational manner has not went down well. There is no one that is as knowledgeable, and educated as Martin and Kaufman. I guess it's all about knowing how to push your weight or position around, it 's called Provocateur, and that is exactly what it took to get the EPA to tell the truth. In this case as in the Shattuck pollution case, it was Senator Allard who threatened to hold up a key EPA confirmation, and that is exactly how Martin and Kaufman entered the case. Kaufman would remind you of a parent who sometimes has to get angry in order get a point across, sometimes you have to just stay stop it. It was also said a decade of conventional methods continuously failed to convince EPA, that radioactive waste does not belong in nearby downtown in Denver Colorado. Martin and Kaufman are still out there conducting investigations wherever they may be needed. Whitman EPA's new administrator's reason given for Martin's transfer was said it would give him more independence. Martin said, the transfer de-fanged his watchdog role at the agency. Martin resigned, and in the wake lawmakers are renewing their efforts to reverse the transfer. There is legislation to create independence for the Ombudsman, and Senator Wayne Allard R-Colorado, and other western Senators are pushing their bill in the Senate on similar legislation. Allard said, "If EPA thinks that Robert Martin's departure is going to make this issue desappear.... They are wrong....

MAN SENTENCED FOR DUMPING WASTE IN A 9 STATE DISPOSAL SCAM.

The disposal scam run out of Lakewood has ceased, and the mastermind has been ordered to spend 17 years in a Colorado Prison. It will match the longest sentence in the nation for an environmental crime. Just recently In a Jefferson County District Court sentenced Hormoz Pourat of California for over a 5 year period of time illegally storing, dumping, & shipping dry cleaning chemicals. He must also pay a fine of $100,000.00 & clean up costs yet to be figured, this was by far the harshest penalty, for an environmental lawbreaker & was prosecuted under organized crime, or racketeering, statutes.

In the sentencing the Court considered Pourat's length of time, & senior level responsibility with the company, the far reaching scope of illegal activity & destruction of the environment that had resulted. The primary damage was in Nevada, and Idaho to the landfills, where there were likely thousands of boxes with a protective lining, filled with dry cleaning solvents, that were crushed during the filling and will eventually eat through the protective lining and pollute local groundwater supplies.

The only case that comes close to being matched in an Idaho case approximately two years ago. The defendant was accused and tried for sending an employee to clean a 25,000 gallon storage tank that contained cyanide, & without providing him any safety protective gear. The 20 year old employee was soon overcome by the Cyanide gas & suffered permanent brain damage.

Pourat's brother Hormayoun whom they believe fled the Country, is considered a co-conspirator. Three other employee's of the two Companies had been sentenced earlier this year, and one is still awaiting trial. The Pourat brothers were Iranian-born had orchestrated this disposal scam covering up to nine states with more than 300 laundry shops, the scheme was ran with an iron fist. While in Lakewood AAD Disposal Inc. a dry cleaning disposal company, also had a sister operation in Vernon California, and was believed to be linked in this case. The Colorado operation collected roughly $80,000 a month for over 5 years from those who contracted with them to properly dispose of the contaminated waste that contained Perchloroethylene, (PERC), a common but very toxic solvent.

WE HAVE MANY CORPORATIONS, INDUSTRIES & GOVERNMENTAL AGENCIES, WHO NOW HAVE TO FACE UP TO WHATS HAPPENED, & WHATS NEEDED AS THEY FACE THE RULES OF CLEAN AIR, CLEAN WATER, CLEAN SOIL. & A SAFE CLEAN ENVIRONMENT FOR ALL.

LARRY D. LAND

There are tougher nitrate standards facing all sewage plants, Nitrate to Nitrite, laden water mainly caused by feces and fertilizers, and pose a deadly hazard to the young, and to pregnant women. Naturally occurring selenium....via irrigation water....into our rivers Mainly from the eastern plains and the western slopes. Selenium can be very toxic. Poplar trees are a natural for taking up selenium before it reaches our rivers. The Forest Service needs improvement, and a plan to eliminate old logging roads by means or revegetation as a way to cut down on the sediment pouring into, and clogging streams and creeks, the tributary's that feed the rivers flowing swift and decisive. This would a stop in boosting the population of the aquatic bugs which in turn are prey for the fish in the streams and rivers.

This is a broad approach to the control of pollution and is getting the attention of Corps. And Industries alike they must also worry or even fear the Clean Air, and the Clean Water Acts. Coal-fired power plants, have long been regulated under the Clean Air Act, and now the extent of the emissions into the air are polluting lakes, and rivers alike they could fall under the new water pollution rules also.

Industries huge fear is the controls where there have not been controls before. EPA has now announced it will begin regulating mercury emissions by the year 2007, although it will face many legal challenges. Mercury is Ultra Toxic chemical, the Department of Energy, a governmental agency has estimated mercury controls could cost the Industry $2 to $5 billion a year. Yet to be safe is better than being sorry years later for something that could have been made safe for millions suffering and dying today. This area is uncharted legal territory, and both State and Federal officials will be using and making new rules and laws in order to hold just the power plants responsible for water pollution 10 miles and even thousands of miles away. Just how tough can regulators be in allocating the responsibility, its almost as if it would be pitting polluters, against fellow polluters. Each one of them will be eager for the regulators to focus on a myriad of other sources that are harming the water quality, and have been skating while they've long been scrutinized. You don't allocate shares in a clean up that could be from many sources, it should be stopped, Controlled, at its source immediately. They are the polluters who run the risk.

How do you actually pin down the source of the pollution, Larry says, "it should be easy, every one has its particular entry and that's where the regulator should be testing, then add it up. Take 1, 3, 5, 16, = 25, the maximum is 10, then who gets pinched for polluting. Knowing the parameters of what can be done what is allowed, and what is not allowed then to purposefully violate the rules or law. If found in violation that government agency, Corporate Industry, or Company should then be written an immediate cease and desist order, and the doors locked, and business will cease, until the Attorney General of the State can commense legal action in the Court.

The unregulated industries are the polluters, they know exactly what to do and what not to do. What they have done was no accident, in most cases it was "intentional" for the whole purpose of greed. The general public not only pays for it with their lives, and the lives of their children, but they also pay for in taxes in the $billions to even attempt a clean up. Just like it was said by those involved in the big city sewage agencies, such as Denver, Chicago, Los Angeles New York, they have been controlled under the clean water act now for 25 years and still in business even when the requirements are becoming evermore restrictive, and that is exactly what should control industry. States have not and are not aggressive enough in citing polluted waters, and developing the necessary clean-up plan. Most State quality reports support the claim, the list of polluted waters is incomplete. Of all the lakes and rivers through out the United States only about one third have actually been assessed.

WATER AND AIR CLEAN-UP IS PROVING MURKY.

Power plants, farms, logging operations and Industries alike are coming under tougher scrutiny for polluting our lakes, streams and rivers. The sweeping water clean-up efforts are still in its infancy, with even bigger disputes coming, and regulators are looking deeper and well beyond the traditional culprits to really find out what is at fault. It's really the uncaring government that hires corporations to do the dirty work for them and Industry in general across the World. The teeth are now being placed in the Clean air Act, & the Clean Water Act. Not only getting locked into cleaning whats already been done we must look into stopping it now before it happens so our children aren't facing what we started and could not stop. Chemicals that fall after spewing from power plants and industry alike at the same time what is being washed into the streams, rivers and lakes. Where the vegetation and the trees have been stripped, and a real necessity of re-vegetation in the area necessary. Larry says, "The rules are there and need to be enforced "Today Not Tomorrow", or there may be no tomorrow".

TRI-STATE POWER PLANTS & OTHER OWNERS AGREED TO PAY $100 MILLION FOR AIR POLLUTION CONTROLS.

Tri-State Generation and other owners of a coal-fired power plant including Xcel Energy——agreed Wednesday to spent $100 million for air pollution controls, and six years in the courts. Those power companies will be passing it on to the consumers. These are just some of the Companies that are contributing significantly to acid rain, and acid snow. The Sierra Club has been involved heavily in making them clean it up, and making sure they follow the Clean Air and Water Acts.

EVEN STEEL MILLS GET CAUGHT POLLUTING THE AIR IN VIOLATION OF THE CLEAN AIR ACT. THE COST IS $1 MILLION TO $16 MILLION TO CLEAN IT UP.

Steel mill to pay a record $1 million dollar penalty to the State Health Department, for repeated air pollution violations. Formerly known as CF&I Steel. While they also agreed to spend $15 million on state-of-the-art electric arc furnace and other measures to reduce pollution. Then just under $1 million to pay keeping a State Health Inspector to monitor emissions on site for two years. The fine was the most significant civil penalty that this states air division ever assessed and collected. They must meet the clean air standards by January 1, 2003 or another $1 million fine will be levied.

SHELL'S ATTEMPT TO ONCE AGAIN ENTER BUSINESS IN COLORADO, IS OVER POWERED BY REJECTION.

Barrett a Denver oil and gas firm rejects Shell Oil Company offer of $2.2 billion. It was an unsolicited offer, but said they would open up the process to all qualified bidders. The board is in a position to maximize share holder value, while seeking proposals from other parties, and Shell will be invited. Shell was prepared to pursue a hostile takeover if spurned by the regional oil and gas independents. This would amount to $55 a share that was offered to signal major oil and gas companies are looking for ways to return to the Rocky Mountain area after exiting from a forced Closure of the Rocky Mountain Arsenal that Same cease and desist was issued to Shell Chemical Company at the Rocky Mountain Arsenal that is well known as the most contaminated piece of land on earth, notably near or over 100,000 acres, and Shell Oil played a large part in polluting the water, air, and soil during their 30 years in Denver manufacturing some of the deadliest pesticides in the world, that were eventually banned in the United States. They were very well known for the "SHELL NO PEST STRIPS" during the 1960's, that were later banned in the United States because they were so deadly. Every one had them hanging in closets, bedrooms, kitchens, dens, living rooms, bathrooms, and even the garages. It seemed like a good idea but it had not been tested before marketing. It was not only killed the bugs, spiders, and Fly's, it was also said to have killed the pets, birds, and fish in the fish tanks right through the water. As they were taken off the market, it was being said by our sources it was also very dangerous to humans, thought to be carcinogenic, and a potential mutant.

Our sources employees of Shell Oil Company at the Arsenal also told us about the times they had dumped pits full of the Vapona that was being returned to them by the train load from all over the United States. Along with the containers, the pallets, card board boxes, and other debris they and the Army scavenged and collected, fuel added and burned leaving huge clouds of different colored smoke to settle up against the Rocky Mountains, and as the air would get heavier it would settle down over the entire city of Denver like a black shroud. A nasty reddish and yellowish brown cloud would hang over the city for days at a time, the red alert, (or smog alert) would once again sound in Denver asking everyone to stay inside, close the windows, and those with breathing problems to dawn their respirators, for some they needed oxygen, others needed it available to them. The only thing people could hope for now would be a southwesterly wind blowing to the North East, other states across the country would get the remnants of the Arny, and Shell Oil Companies Ultra Toxic and hazardous pollution. This would happen after 5 p.m. so there was no one out to test the air for contaminants. Larry can remember turning Shell, and the Army into the Air Pollution Commission as to the times they were intentionally polluting the air to save money.

Shell Oil Company closed it's office's in Denver Colorado at the behest of our government, both State and Federal. Through Cease and Desist orders for their manufacturing and polluting the Country side as well as the City of Denver were over. Barrett is one of the Largest independent oil and gas producers in the region. Besides Shell Oil, Barrett has a number of interest shown by Anadarko, Exxon Mobile, & BP Amoco. Barrett described Shell Oil as imposing "artificial deadlines that only served Shell's Interests."

Barrett stated there is not a timetable for the negotiating process, but said they would promptly open its books to confidentiality agreements. "There were no assurances given a sale of the Company will ever occur, or on what terms". This statement was issued after the stock markets had closed. The stock closed up at $61.22 a share. A letter was sent by Barrett Chief Executive Peter Dea, to Shell Exploration President Walter Van de Vijver, encouraged Shell not to react in a hostile fashion. The letter went on to state "Because of your stated desire to negotiate a friendly acquisition of Barrett, we invite you to participate in the process".

For a foot back in the door Shell explained they would keep Barrett's office in Denver and retain all of its more than 200 employees. Barrett would not comment on what they would do if Shell were to force a hostile takeover. The word is now out that other Independent's in Denver area could become take over targets also. Mark Smith executive directory of the independent petroleum association of Mountain States, Rattled off a half dozen names; Forest Oil, Tom Brown, St Mary Land & Exploration, Basin Exploration, Prima Energy and Evergreen Resources, and of course Barrett. The Association of Mountain States Participants, some are not excited about the prospect of major oil companies returning to the area or Denver. However when Shell made that bid, it sent out a message, "as if to say the Gold rush is on.

I'm not sure what it is that Shell is thinking but they launched a hostile take over bid for Barrett on the 13th, day of March, staying tight with the 2.2 billion dollar figure giving Barrett a 24 percent margin above the stock price when the offer was made on March 1st, the stock soard past $60 a share shortly after the bid was announced. By the close of the day market on the 8th it closed at $61.22 many millions above Shells 24 percent plus margin, by the 12th it was still soaring above $62 after having one of their bad days and was off 52 cents a share. In order to have been successful, Had Shell offered a 36 to 40 percent margin, it probably would have been an immediate deal with the Stock holders. Had Shell presented that without the hostile attitude again with Barrett they would probably have been successful.

One anylist explains that he was not optimistic that Shell would be able to convince the Stockholders, most of them being large companies, and Institutions that $55 was the best they would get. The Anylist thought Barrett to be worth at least $67 per share or more. His thing the industry is flush with money (because of the high natural gas prices). There will be other's surface as potential bidders. Hostile takeovers are not usually successful especially in a rising commodity price environment.

With Barrett very much in control of it's shareholders they tried explaining to Shell they are urging the Shareholders to take no action. However Shell had decided on about the 15th of March to take its offer directly to the Barrett shareholders since it believed the price was fair and a bidding process would disrupt Barrett's operations. Shell Oil Company Official said we are concerned, as to Barretts shareholders as they also probably are. The prolonged process of bidding could be a destruction of the Barrett employees and have an adverse impact on their ability to effectively operate the business.

Six of the Barrett shareholders filed suit against Barrett at the time of the hostile takeover claiming the company's directors were obligated to get a better price for the company. The Rocky Mountain region is considered to have some of the most plentiful natural gas reserves in the country. In the Wake of Shell oil Companies bid, and attempted hostile take over of Barrett, will leave the door open for other major oil companies that will be looking at return to the Rocky Mountains.

RIVALS RATTLE SHELL OIL, SO THEY UPPED THE BID TO $2.4 BILLION.

Shell upped the anti, and extended the offer to Barrett to May 9 as the deadline. Barrett again urged shareholders not to accept the new bid until the Board can make a recommendation. At least Shell Oil is treating it more at negotiable than take over, even though the take over is still standing tough, even though they are beginning to recognize Barrett is really running a process where there will be bidding. Shell Oil, unit of the Anglo-Dutch giant Royal Dutch/Shell Group, made the bid for Barrett in March in order to boost its U. S. Holdings. Barrets Auction is already set in stone and will close on May 2nd prior to Shell Oils deadline of May 9th. Shell was very serious about the Takover of Barrett, and that seems to be the reason they raised its bid, Likened to a warning to others who may tread that Shell Oil is very serious. It is said by New York based Merger Insight's President Tom Burnett, "We've felt all along the top (bid) range is $65 (a share) and it could go even higher."

Shell is very serious and when they raised their own bid it was done to scare away any other bidders that may be tempted to jump in. All Shareholders have been advised to stand firm, and to hold their stock until they see further developments. Barrett on May 8th, stock climbed to $70.15, Shell Oil abruptly dropped out on the 7th of May, leaving Williams Companies, standing alone with an offer of $2.8 billion for the Denver based Oil and Gas firm. Williams Co. is the Country's second largest natural-gas pipeline operator. The selling price for an equivalent of $73 a share. Shell Oil Co said to pursue

Barrett no longer made economic sense. Williams is out of Tulsa, Oklahoma, based diversified –energy company with more than $10 billion in revenues last year. The Company has already had production in the Green River and San Juan basin in the Rockies, said the acquistion will more than double its natural-gas reserves and help fuel its growing gas-fired power-plant business. Williams said the Company plans to retain most of Barrentt's 237 employees. The Company also plans to maintain Barrett's Denver headquarters as Williams' principle office for Rocky Mountain exploration & production. Williams acquisition price represents A 60 percent premium over the $44.25 just before Shell made its initial bid in early March.

CHAPTER 21 FROM OUR BACK YARD, TO OUR SCHOOLS, TO THE WORK PLACE, & AROUND THE COUNTRY & OUR ENVIRONMENT. THE PAGES YOU'LL READ ARE FOR YOUR KNOWLEDGE OF THE REAL WORLD AND WHAT'S HAPPENED TO THE ENVIRONMENT AROUND US.

The eyes of America, and Governments from around the world have been opened to what has happened. "were all looking back at some of the most horrible tragedy's poisoning our Environment". It's now time to look ahead at science, & how it has changed our lives on earth. It has expanded our knowledge, our power, Let's not allow it to out-strip our Capacity. Together we can harness the power & educate our selves with the knowledge necessary to control the "Tremendous capacity" & use it with wisdom, and Not greed, and we can together leave our children a cleaner environment than ours' and with the same capacity or greater.

By Larry Land/Author of DEATH FACTORY U. S.A. PART I & II. PART III 2004

WE WILL 1ST DISCUSS LOVE CANAL, A STORY OF CORPORATE & GOVERNMENTAL GREED THAT CAUSED A "FATAL ERROR" IN GOOD JUDGMENT.

As we come to the end of the nineteenth century, America was totally unified. It had opened the book of entrepreneurial ideals, looking at shipping. The American waterways were unified to provide an efficient mechanism for shipping, many canals were opened across America, The Panama Canal, the C&O Canal, the Erie Canal, and Love Canal, and many more. The idea was to provide the efficiency of shipping in the American waterways.

William Love originally envisioned a "POWER CANAL" for the purpose in which a cheap supply of hydroelectric power in the Niagara Falls area of New York State. Construction began and was completed and was working as was envisioned by William Love. It wasn't long before things began to change. In 1942 Hooker's chemical and plastics corporation cut a deal with the current land owners and/or the Power Company, whereby the Corporation was then allowed to dump uncontrolled, any waste products into the Canal.

Hooker himself finally bought out the land, & all its surroundings by 1947. Hooker's defense was that the chemicals were dug into impermeable clay soil tombs and on top was clay soil and cement seals put in place. But indeed records show many tons of deadly and hazardous chemicals were dumped into Love Canal. Environmental concerns were not at a forefront of social consciousness, like they are now today, the local homeowners only tried to clean it up and were not apt to complain. Some of Hooker's power, was in fact Hooker Corporation that was a large employer in that area. What was once the Love Canal soon became a huge acreage in which children could play soccer.

During the 1950's dumping into Love Canal was completed. Hooker Corporation went to the greatest of lengths of the time to seal the chemicals forever. The Canal was dug into impermeable clay soil, filled with raw and toxic chemicals & then clay soil, & cement caps were placed on top to prevent any rain water from leaking into the tombs of toxic & deadly wastes. The precautions made were sufficient for the times, and as sufficient as to what they are now doing at the Rocky Mountain Arsenal and at Rocky Flats in Denver Colorado.

At this time the District School board was out looking for places to send all the baby boomer children that were entering school. Looking at this large field that was once Love Canal, they approached the owner's, Hooker Plastics and Chemical Corporation, and actually they were more that eager to rid themselves of this potential nightmare a virtual waste Land of death. A trap just waiting to happen. The Hooker Corporation gave this land and its' potential risks to the public. The Company even dug test holes into the ground to prove to the government executives there were toxic chemicals encapsulated in impermeable clay tombs. Despite all the warning, the school board prepared eminent domain cases. With somewhat of a reluctance the Corporation gave the land over nearly free, for a return to be loosened of any and all liability.

The first thing that was done when the government took over was to remove any safety mechanism, the cement and clay caps that were placed so as <u>not</u> to allow rain to seep in. This school would be the first without a basement for obvious reasons. But later the school had to construct a sewer line that actually and completely, penetrated the cement and clay caps and walls that had entombed the toxic and hazardous materials.

Surrounding the lines was permeable gravel. A storm drain was put in place that pierced the walls. The punctures allowed any and all chemicals to be swept into surrounding lakes, rivers, and wells. As the population and density increased pressure was now upon city government to sell the land for Development & thats what happened in the 1960's, and the trap was sprung.

It became an ever growing problem Besides breathing toxic fumes, people Were exposed to pools of chemicals actually bubbling up to the surface and that is not all that came up. The fingerprints also come popping up.

Slowly as the mass media began to draw attention toward the ever-growing problem, the United States soon got involved. Slowly at first homes were evacuated, and by 1980, everyone was allowed to evacuate.

Love Canal, is without doubt one of this Country's worst and most notorious hazardous waste sites in America, next to the Rocky Mountain Arsenal, the Rocky Flats and Martin Marietta had became. In 1972 when Larry blew the whistle on pollution, and it was heard around the World, & voiced at Love Canal. Larry was there in New York as it unraveled. In the 1960's and the 70's, and 80's having had a chance to speak first hand with some.

It wasn't the first, & wasn't the worst, but it did grab headlines, drew attention, & help to stimulate scientists, Industrial Corporations their leaders, government officials, & plain old grass root environmentalists. The Love Canal neighborhood is in the southeast section of the La Salle area of Niagara Falls, New York. William T Love envisioned and sought to develop an industrial community, a Model City. Waters from the Niagara River were to be routed around the Niagara escarpment (Niagara Falls) in order to produce almost free and cheap hydroelectric power. Below is a map clearly showing the landscape of Love Canal.

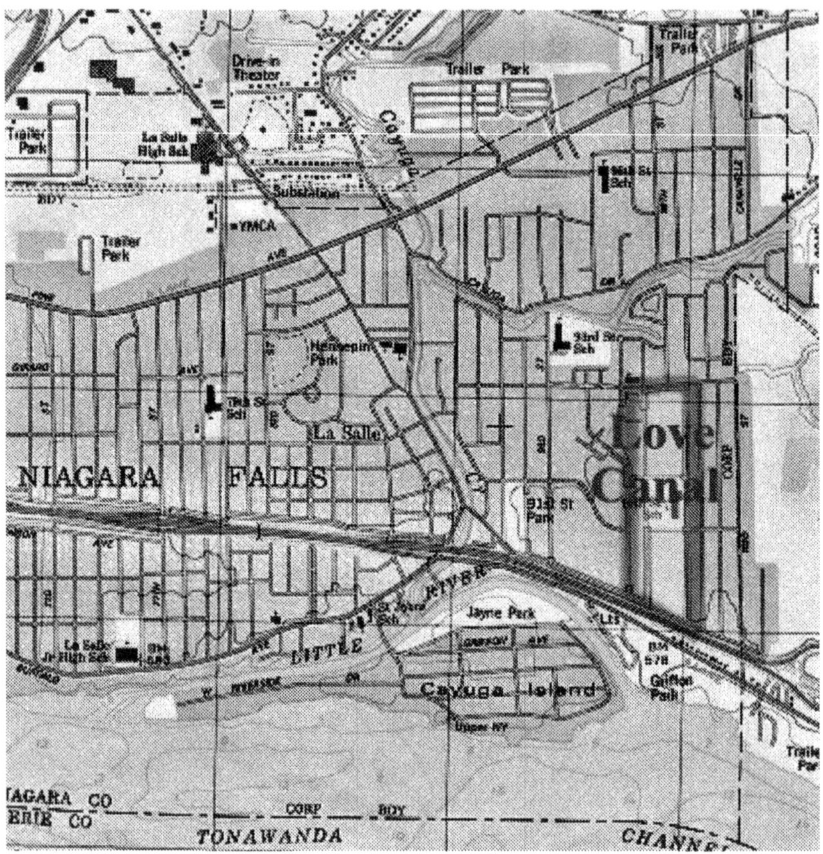

NEXT PAGE SEE one of a few 10,000 gallon containers that had been buried at Love Canal, then filled with deadly chemical waste being lifted and retrieved by a huge crane from its deadly resting place. One of how many.

The model city never happened, but work on the canal to transport waters from the Niagara River did. Hooker Chemicals and Plastics Corporation (now Occidental Chemical) purchased the site of the Love Canal in 1942. About 22,000 tons of mixed chemical waste was pumped into and buried at Love Canal. Shortly after Hooker Chemical had disposed of all this waste, & ceased use of the site, the land was sold to the Niagara Falls School Board for a price of $1.00. In 1955, an Elementary School was constructed on the Love Canal Property and opened its doors to students. Then subsequent development of what was again thought to be the model suburb, in which hundreds of families took up residence a blue-collar neighborhood of Love Canal.

The unusual heavy rain and snowfalls in 1975 and 1976 provided high ground-water levels in the Love Canal area. Portions of the Hooker landfill subsided, and 55-gallon drums began to surface, ponds and other surface water area became contaminated, oozing an oily like residue, along with noxious odors permeated the area. Corrosion of sump pumps & the infiltration of basement cinder block walls made it very apparent there was something wrong here. The Agency for Toxic Substances & Disease registry soon revealed A long laundry list of 418 chemical records for air, water, and soil samples in & around the Love Canal Area.

By April 1978 the New York Department of Health's Commissioner, declared the Love Canal area a threat to human health and ordered the fencing of the entire area that was even near the actual old landfill site. By August the Health Commissioner declared a health emergency at the Love Canal, and Subsequently closed the

99th street School. He then recommended temporary evacuation of pregnant women and young children, starting with the first two rings of houses around the site. The Governor at the time Hugh Carey announced his intentions of purchasing of all houses in the first ring, and later expanded it to ring two, 238 homes. Then simultaneously President Jimmy Carter announced allocation of federal funds & ordered the Federal Disaster Assistance Agency to assist the city of Niagara falls in finding a remedy for the Love Canal Site.

Officials of the area soon became concerned about the situation. Their primary concern was the actual extent and the impact the chemical contamination was having on the health of Love Canal residents. Second and perhaps even more important, was the total lack of readily available information to explain the science. "THERE WAS NONE" Levels of total uncertainty, Political and corporate agendas, and the manipulation of the media.... In general A complete and full paucity of any reliable information that could answer a very simple question, "IS IT SAFE".

......A CORPORATE, GOVERNMENT, AND THE FATAL ERROR......

The Hooker Company took great precautions, even when you compare it to the standards of today. The burying the containers into impermeable clay, cementing the side walls and cement caps. Comparing it to the standards of today it would be perfect. Even though sooner or later is was bound to come up, or come out. They did sell the Land to the Government for "$1.00" along with the consequences, but only when they expected or knew fully it to be developed. It is understood by our sources the use of the Land by the Government was no secret. Even after it was sold Hooker continued in an effort to stop the development of the land. It was the local government that went against all the warnings that had been repeated over and over again. As they tried to profit off of the contaminated land. They knew full well it was contaminated, it was no accident what they did when they dug away the walls and the caps off the tombs. The Government at the Rocky Mountain Arsenal, and Rocky Flats are doing the identical thing to hide the chemical and nuclear waste today. Like Larry says, it is nothing more than another fatal error that will come out in the future.

The government knew full well of the chemicals, but not only pierced the clay and cement walls and caps, they were obliterated by construction they knew full well was wrong at the time. This was done all for the sake of fill dirt, and to install the sewer lines. When they constructed the School for innocent children, and by selling the land for development, they virtually asked the attention to be brought to the site. The governments handling of the wasteland was shoddy, and their greed for the all mighty dollar, that caused the ultimate Love Canal disaster.

Who was at fault? And who will pay? It was the people who sent their children to the school, and the family's who bought homes their. It was the children, their parents, the grand parents, and their children's, children who suffered and inevitably paid the price, and still paying today, and will for decades to come. Lastly it was now the taxpayer's turn to pay the price to have it cleaned. But lastly but not least who was at fault? It was the fault of a few, in government and their greed, with full knowledge of what may happen & repeatedly warned of the eventual outcome.

Will it ever be clean, and just how clean is clean, If it is cleaned to the standard used by the United States Environment Protection Agency, it is not clean enough and will not fly, It must be the Department of Health State of New York, & independent Scientists, who just may set a safe standard for all.......

They say only good can come from bad things, and in a way that is the truth, the hard truth. I believe it has truly awakened America, and the World to the toxic waste problem. However they have millions of tons being manufactured today, and millions of tons that have already been produced and Corporate America, and its' government are just looking for a safe place to store it even temporarily, as the toxic waste problem grows. The Superfund is out their but it is totally inadequate, and very ineffective. With the growing interest in our environment especially since the 70's perhaps someday a real solution might come. Larry believes that places like the Rocky Mountain Arsenal's "Chemical and Biological Death Factory, and Rocky Flats nuclear weapons "Death Factory", and Love canals' Toxic Waste Dump Site has opened the eyes of America and the World to come up with the permanent solution.

"The Solution to the entire problem would have been forever solved had the government done their part in protecting the people.

By making a Law that which is put together must be made to take apart. All Chemical, Biological, Virulogical, and Nuclear weapons all of mass destruction. Until they can put it together like a puzzle, that can also be taken apart in the very same manner for which it was put together, could solve what has been made. In order for a product to be integrated into our system there must have been first a method to destroy it and any waste generated from it. Not just the haphazard methods, and promises of the past that have been devised and failed right up front they should not be allowed to continue, until they come up with a 100% workable solution with absolutely no room for errors.

By Larry D. Land

A permanent solution, and a real solution still might come if we worked on it as hard as greed has paid off for the government and giant corporations in the past at the cost of million of lives, and billions of dollars paid by the taxpayer, and not the polluter, because when our government pays for the corporate, and government mistakes, it is the taxpayer.

The Love Canal area today is starting to be re-inhabited. The chemicals there will not decompose for approximately 20,000 years, and the genetic mutations will survive indefinitely. Much the same for the Rocky Mountain Arsenal, and the Rocky Flats Nuclear Weapons Plant, will be made into huge National Wildlife refuges, the Hanford nuclear weapons plant, Jack Ass Flats nuclear proving ground in Nevada will not support life at all. The legend and stigma tied to them will live on in infamy. Lessons have been learned by some, and it will take that “some” to take control of those who don’t.

As we came out of the 19th century and into the 20th century William T. Love began to build His vision, of Love Canal. It worked & the transport of water from the Great Niagara river did, but then Love Canal was purchased by Hooker Chemical (is Occidental Chemical today.

MEDICAL TRAUMA STARTING WITH SIGNS & SYMPTOMS OF WHICH MEDICAL DOCTOR'S AREN'T TAUGHT OR EDUCATED IN. THEREFORE MIS-DIAGNOSING THE DEADLY SECRETS HAUNTING MEDICAL PRACTICE TODAY & WILL TOMORROW. THE UNKNOWN THAT DOCTOR'S ARE UNABLE TO & CANNOT TREAT, THEY'RE "TAUGHT TO LOOK AWAY" BECAUSE THE LAWS THAT ARE MADE TO PROTECT PEOPLE GO UN-ENFORCED BECAUSE OF INSURANCE COMPANIES & THEIR LOBBIESTS. INSURANCE INTERESTS ARE THE 2ND LARGEST POLLUTERS PROTECTING THE POLLUTERS FROM THE AXE.

The U. S. Army again is polluting the water people drink or have already drank. Installation of a filter to clean chemical contamination that seeped into water from the Pueblo Chemical Depot near Avondale Colorado. Containing 2.4 ppm. Trinitrotoluene, A byproduct of the explosive "TNT". Depot officials are still advising those affected to continue drinking bottled water provided by the Depot. Larry asks, Could this or could have this been absorbed through the skin while taking a hot bath that opens up the pores in the skin? "OR DID THEY FORGET ABOUT THAT?" Did the Doctor's even know or were they aware of this fact?

Anthrax cases began to spread like wildfire, and the United States was more aware of it than anyone else. That is exactly why they already had medication warehoused just in case. It was the United States who had sent the secrets, & the live virus screens directly to BAGDAD, personally to Saddam Hussein, all the risks, & the use of it as a biological & deadly potential as a weapon of mass destruction. Saddam Hussein then may have shared the secrets with the rest of the radical nations that are bound to destroy the United States & Israel. In one case there was a man begging for help with his Anthrax disease and was shunned by his Doctors, mainly because they could not identify & by the time it made national news, it was too late. At this point it mostly affected the people working with the mail or possibly something sent directly to a person. The Doctors were untaught, unaware and unable to identify it, just like many other cases of Bio/Chemical warfare items. The government has still kept the lid screwed on tight until it happens and when its too late for at least some.

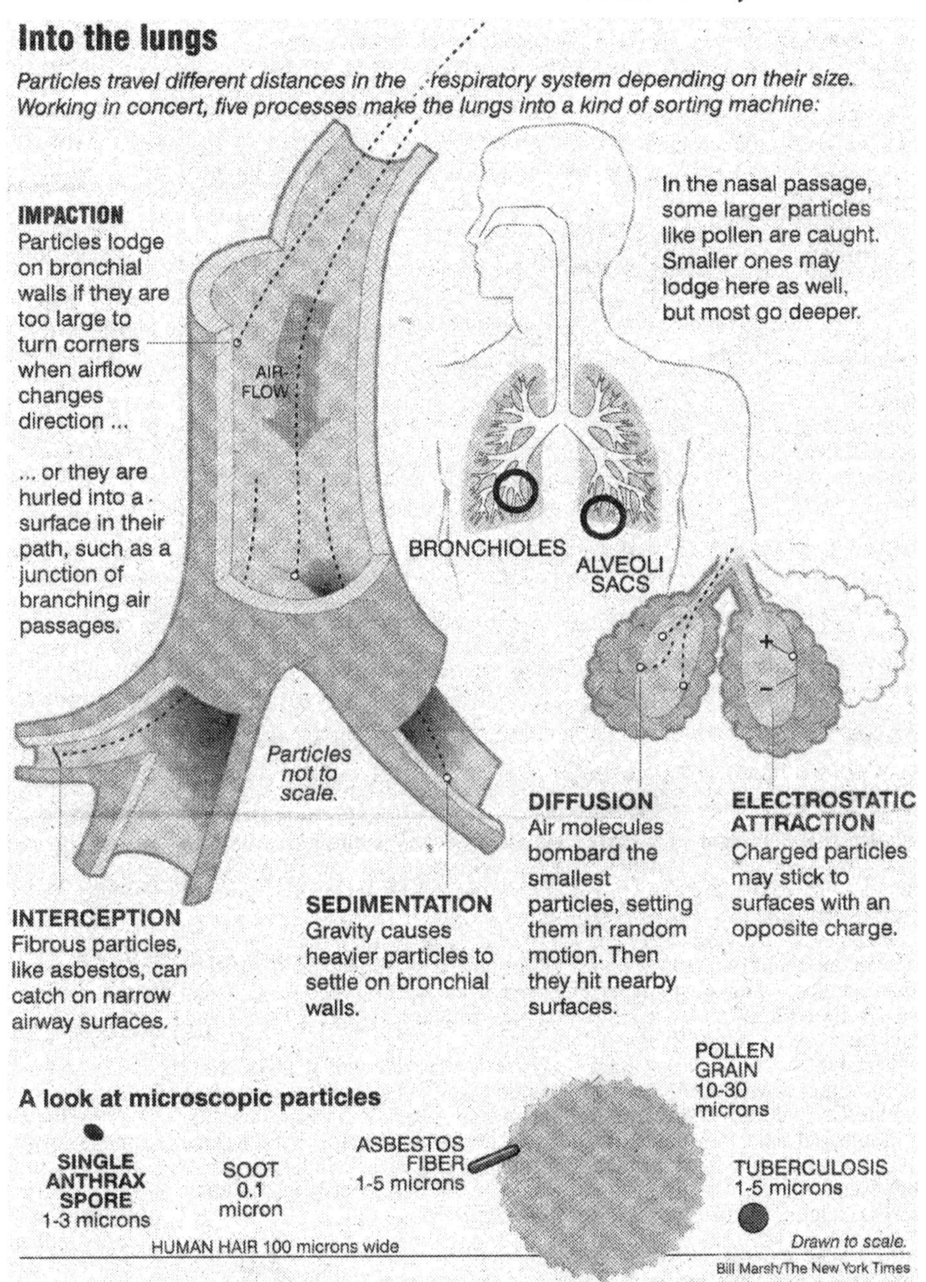

GULF ILLNESS, POINTS TO SARIN GAS, COULD BE ONE OF MANY IRAQI AMMUNITION DEPOTS THAT WERE DESTROYED.

Some of the illnesses Doctors were unable to immediately diagnose could have been caused by Iraqi Bio/Chemical warfare agents, of which Sarin at low levels has been named. However it's taken over 10 years, just to begin to diagnose the undiagnosed illnesses the Persian Gulf War Vets suffered and brought home to their families some of the same pain and suffering. Low levels is nothing more than a long painful death, instead of quick and easy. For the Vets this has been agonizing, painful & most heartbreaking because they were unable to get the help they should have had. There is first hand scientific literature available that would have told the so called experts, Sarin at lower levels is much the same as Mercury poisoning may take months even years after the overexposure to show what it had done, and the Doctors should be and are not aware of this data. I as a layman am totally aware of this and where it is and had been classified through 1989, other parts of it are still classified in the year 2003.

It is still not possible to eliminate Nerve agents at different levels. They did play a role in all the cases of Gulf War Syndrome. Where they were overexposed to not just Sarin, what about VX "The United States had also sent that to Iraq". Along with a multitude of other gases and Biological warfare agents, that could have been used and totally escaped it presence. See chapter 10 of DEATH FACTORY Part I for a better explanation for the Persian Gulf War Vets and their families. They continue even today only calling it exposure when in fact suffering the signs and symptoms, "they were Over. Exposed and the Army or the Doctors have not admitted it. If for once they did they would better understand the problems. Many of the problems in the Persian Gulf war have not yet even been discussed, & the government if fully aware of what happened hoping it will just go away.

After 12 years since the Persian Gulf War, they finally decided to research the long term health effects of exposure and over exposure. They do know for a fact that at least 100,000 American soldiers were exposed or even possibly over exposed to Bio Chemical agents at not only the Khamisiyah, but at every ammo storage depot that was blown-up & there were many, not realizing they contained at least Sarin but what else. Our source told us that Bio-Chemical agents were also released in the clouds of smoke as it was blowing over the Americans and the United States was aware the Iraqi's. may be using the wells as a method to spread death to the Americans. It was after the Gulf war was over the government became aware that the 122 mm rockets at Khamisiyah were loaded with Cyclo-Sarin, when blown up the evaporation is more deadly, it is completely vaporized, and is at least four times more effective, and covers an area of at least four times that of Sarin. Our source also said that the monitors used would be showing low levels Sarin when in fact it would be deadly levels. Then of course for over 10 years the pentagon was having the wrong soldiers treated, and didn't think it was so bad while in fact ignoring those that were deathly ill. They treated some for over 3 years and announced that 32,806 were not in the path, while then notifying 34,819 soldiers that were not being treated were thought to be in the plume of Cyclo Sarin, and were overexposed.

The pesticides used by American troops in the Persian Gulf War might well have contributed to the unexplained illnesses reported by many veterans, and cannot be dismissed. This was read between the lines of the governments poor explanation of what happened. Larry says, "of course they told their damned lies they can now make it just another statistic". There were about 13,000 troops who were given flea and tick collars while in the Gulf. They were unsafe and illegal. And not endorsed by the government, stating veterans were getting the collars from family members or friends. Then we have the troops assigned pesticide spraying & delousing duty who did not use masks, or other protection available. Doctor's are saying the pesticides are what caused the signs symptoms and illnesses developed by the troops, yet we have Mr. Bernard Rostker, who is not a Doctor of medicine but rather the under secretary of Defense for personnel states it could not have caused any health problems. Who do we believe the Under secretary of Defense who made an idiotic decision, and took it upon himself to make a diagnosis of the troops illnesses, or the Medical Doctors? Larry says "who is this idiot who called himself the Under secretary of Defense, and was able to make such an idiotic Statement. But then after making the idiotic state Bernard Rostker then had a change of mind and stated in his next breath, "Were not yet able to make a link, however the Pentagon cannot exclude the possibility that some pesticides [C]aused chronic health problems, and he urged more Research". Larry says, "I like this guy, he sure rolls with the punches". He also added his conclusion is much like that of the Pentagon when he suggested causes of the unexplained illnesses, including exposure to nerve agents during demolition operations in Iraq during and after the war, & the use of the nerve agent antidote known as pyridostigmine bromide, or PB., and the raw oil sending plumes of deadly gasses into and permiating the entire region every thing they ate, everything they drank, breathed, or absorbed through the skin.

DID THE PENTAGON, AND THE INVESTIGATIVE PANEL WORK DILIGENTLY? THEY DENIED A "COVER-UP" IN THE GULF WAR ILLNESSES, VETERANS GROUPS SAID IT WAS A WHITEWASH.

Yet a whole series of veterans' ailments are still unexplained 12 years after the war. The President said 'leave no stone unturned'. The illnesses include memory loss, nervous system disorders, headaches, joint pains and chronic fatigue. Allegations are going around among critics who do believe the Pentagon is hiding data about Iraqi chemical warfare agents or other toxins veterans may have been exposed to or "overexposed" too. To data researched has still not validated any specific cause of the Persian Gulf War Veterans or their families. Government claims scores of studies have been done as follows: (1) Iraqi Bio/Chem weapons, (2) Service members vaccinations, (3) Oil well fires, (4) The anti nerve agent tablets given to the troops, (5) desert sand and stress. The board appointed by the pentagon to complete the studies, was made up of largely.... Retired Military Brass......there was no independence from the pentagon. It was considered whitewash between the board that was doing the investigation......and the Pentagon......

Did the (DOD) Department of Defense, or the Pentagon, make it clear in the reports that were released the extent of the Bio/Chems. Saddam Hussein had at his fingertips. The defense of the United States, It's armed forces, and their families deserve answers instead of the runaround they have been getting now for over 12 years, and should not be undermined by our export policies that allow this government, Russia, France, Germany or any other government to assist any pariah nation, such as they did in Iraq, in the furtherance of their nuclear, chemical, & biological weapons programs, during the 1970's, 1980's & the 1990's, right into the Gulf war. Below is just one of the type culture collections and the technology that was exported to the Government of Iraq, issuances of export licensed by the United States Commerce Department.

1. Date: 08, February 1985,
   Sent to: Iraq Atomic Energy Agency,
   Material shipped: “”Ustilago nuda (Jensen Rostrup””
2. Date: 22, February 1985,
   Sent to: Minstry of Higher Education,
   Materials shipped: Histoplasma capsulatum var. farciminosum (ATCC 32136) Class III
   Pathogen
3. Date: 11, July 1985,
   Sent to: Middle and near East Regional A,
   Materials shipped: Histoplasma capsulatum var. farciminosum (ATCC 32136 Class III
   Pathogen.
4. Date: 2, May 1986,
   Sent to: Ministry of Higher Education,
   Materials shipped, Gacillus Anthracis Cohn (ATCC 10) Gatch # 08-20-82, 2 each.
   Class III pathogen.
5. Bacillus Subtilis (Ehrenberg) Cohn (ATCC 82)
6. Clostridium Botulinum Type A (ATCC 3502)
7. Clostridium porringers (Weillon and Zuber) Hauduroy, et al.
8. Bacillus subtilis (ATCC 6051)
9. Clostridium outline Type E (ATCC 9564)
10. Clostridium porringers (ATCC 12916)
11. Bacillus Anthracis (ATCC 14185) and (ATCC 14578)
12. Bacillus subtillus (ATCC 6633)
13. Hulambda 4x-8, clone: human hypoxathine

Much, Much more see Chapter 10 DEATH FACTORY U. S. A. PART I.

Several shipment of Escherichia-Coli (E. Coli and genetic materials (human and bacterial DNA) sent to: the Iraqi Atomic Energy Commission Shipped: 1989 just ahead of the war. Below another short list shipped to Iraq:

a. Saccharomyces cerevesiae
b. Bacillus Subtillus
c. Staphylococcus epidermis
d. Salmonella Cholerasuis
e. Klebsiella Pneumoniae

f. Bacillus Pumilus
g. Bacillus Megaterium
h. Clostridium Tetani

MATERIALS SHIPPED TO IRAQ AND CAPABLE OF FULL REPRODUCTION.

A. Bacillus Anthracis
B. Clostridium Botulinum
C. Clostridium Perfringens
D. Histoplasma Capsulatum
E. Brucella Abortus
F. Brucella Melitensis

IRAQ—ARMED WITH U. S. EXPORTS OF BIOLOGICAL WARFARE & RELATED MATERIALS, JUST IN TIME FOR SADDAM HUSSEIN TO PRODUCE EVERYTHING NEEDED FOR THE MOTHER OF ALL WARS. THE PERSIAN GULF WAR. ACTUAL SIGNS & SYMPTOMS & A LIFETIME OF PAIN & SUFFERING. "IF YOU WERE IN THE GULF WAR AND SUFFER ANY OF THESE SIGNS & SYMPTOMS OR ILLNESSES. EVERY CONGRESSMAN & SENATOR, HAS THIS LIST, & CAN TELL YOU EXACTLY WHAT IS WAS YOU WERE EXPOSED TO. THE PENTAGON AND THE DEPARTMENT OF DEFENSE HAS THIS SAME LIST OF HEALTH MALADIES."

a. Fever
b. Difficulty breathing
c. Chest pain
d. Septicemia
e. Often fatal
f. Vomiting
g. Constipation
h. Constant thirst
i. General weakness all over
j. Severe headaches
k. Pneumonia
l. enlarged spleen and liver
m. inflammatory skin disease with bleeding infected rashes
n. Chronic fatigue
o. loss of appetite

p. sweat at rest
q. muscle pain and aches
r. nausea and insomnia
s. Can cause major organ damage
t. Mortality
u. Kills muscle cells
v. produces other toxins, difficult to trace or treat, will lead to systemic illnesses.

SADDAM HUSSEIN PRODUCED & USED BIO/CHEMICAL WARFARE MATERIALS DURING THE GULF WAR, ALTHOUGH THEY DIDN'T WORK THE WAY HE HAD PLANNED, WHEN HE STATED IT WAS THE MOTHER OF ALL WARS, THAT IS EXACTLY WHAT HE MEANT HE THOUGHT HE WOULD WIN THE WAR BECAUSE OF THE BIO/CHEMICAL WARFARE HE USED. I HAVE GIVEN YOU THE LIST OF ILLNESSES ABOVE FOR ONLY SOME OF THE BIOLOGICAL WARFARE BACTERIA. BELOW I WILL LIST THE SIGNS, SYMPTOMS, AND ILLNESSES THAT WILL BE PRODUCED EVEN AT LOW CONCENTRATIONS, IMMEDIATE AND UNCONTROLED DEATH WILL OCCUR AT HIGHER CONCENTRATIONS.

1. Sarin nerve agents.
2. Cyclo Sarin nerve agents.
3. VX nerve agents
4. BZ nerve agents
5. DIMP. Di Iso Propyl Methyl Phos Phonate
6. IMPA. Iso Propyl Methyl Phos Phonate
7. Mustard agents
8. Lewisite

Iraq followed the Soviet doctrine; Based on this doctrine, a chemical warfare agent attack might carry a variety of nerve agents, vesicant and blood agents, blister agents, biotoxins, & Bio/Chemical agents.

SEVERE SIGNS AND SYMPTOMS OF BIOTOXINS.

a. bleeding from the rectum
b. blood in the urine

FAILURE OF THE GAS MASK OFTEN EXPLAINED BY THE USE OF "CYANOGEN".

*****The smell of amonia was common place amongst the troops, and the Center for Disease Control, Nerve Gas Division was totally aware of the use of Amonia. We believe that is exactly why the troops often talked about the smell of amonia. Could it be because the Department of Defense was totally aware of the low levels of Nerve agents, and were trying to cover it up, and lessen the affects on the troops with the use of Amonia which is non-toxic alkali they used to neutralize the exposure to nerve gas. But that did not stop the low level intoxication of the nerve agents that did the damage to some or most of the troops. Intoxication to low levels of Nerve agents will cause any of the follow signs, symptoms and illnesses, some will improve with time but some will remain or get worse.

1. Headaches, severe to life threatening	2. Ataxia	3. Confusion
4. Short term memory loss	5. Aphasia	6. Dizziness
7. Difficult retrieving thought	8. Dementia	9. Moodiness
10. Tight chest, difficulty breathing	11. Amnesia	12. Irritability
13. Burning at different areas on the body	14. Chronic fatigue	15. Sores/nose & mouth
16. Cold spots on different areas on the body	17. Overall weakness	18. Stumbling

19. Irritated eyes, nose, throat, and lungs	20. Nausea	21. Trembling/shaking
22. Diarreaha, and/or Colitis	23. Vomiting	24. Sleep disorder
25. Poor learning ability	26. Depression	27. Poor concentration
28. Multi-chemical-Sensitivity-Syndrome	29. Lungs burning	30. Burning stomach
31. Blisters over the body	32. Anxiety	33. Hypoglycemia
34. Inability to handle heat or cold	35. Red bruising	36. Gettiness
37. Toxic hypersensitivity	38. Euphoria	39. Poor concentration
40. Muscle, join, & bone hurting	41. Cancer	42. Mutations
43. Spontaneous Abortion	44. Convulsions	45. Slow death

Listed above are the normal signs, symptoms, illnesses, and health maladies a person will suffer from low level poisoning of nerve agents. Insecticide poisoning works much the same way. By ingestion, breathing, or absorbsing through the skin. An amount as small as the tip of a small pin can kill a person. In seconds or minutes, and that is what makes Cyclo Sarin much more toxic, it‘s not as volatile as Sarin, and can drift in a mist one cannot see with the naked eye. “The Mist of Silent Death“. The families as well could have been seriously exposed, and suffer the illnesses as explained above from anything that may have been brought home, including the affects it may have had on the Veteran. The only possibility for the veterans to avoid carrying it with them would have been not to have shown any of the illnesses that each item of BIO/CHEMICAL warfare items are known to produce, to have gone through an amonia bath, and to have not brought home anything that could have been contaminated, including the smoke from the oil well fires. Larry Says, "did the government, DOD Department of Defense do their job, did they lie and white wash what had happened? The answer would be yes". "Is it still being white washed? That answers is also believed to be yes". "Will the real truth ever be known? The answer is believed to be no, not in our lifetime". Most of it still remains classified. A specific General from the Pentagon was quoted as saying, "We can't let the public find this out, Classify it".

FACT: NERVE AGENTS, AND PESTICIDES PRODUCE SIMILAR ILLNESSES THE SAME SYMPTOMS AS PARKINSON'S DISEASE ALZHEIMER'S, HUNTINGTON'S DISEASE, LOU GEHRIG'S DISEASE, MOTOR NEURON DISEASE, VARIOUS CANCER'S MUTATIONS, KORSAKOFF'S SYNDROME.

1. Damage to the Central Nervous System	2. Slow speech
3. Damage to the bodie's nerve system	4. Ataxia
5. Muscle disorders	6. Aphasia
7. Bone disorders	8. Staring/unblinking eyes
9. Disorders of the eyes and ears	10. Runny nose
11. Disorder to every major organ in the body	12. drooling
13. Compromised Immune system	14. Bleeding rashes
15. Red bruising at a touch	16. Depression
17. Hormone disorders	18. Dementia

19. Skin blisters	20. Oily scaling skin
21. Loss of hair (in patches)	22. Difficulty walking
23. Stooped posture	24. Involuntary noises
25. Shrugging/ neck/head/shoulders	26. Growth retardation/ Dwarfism
27. Tic/habit spasms	28. Chronic Fatigue
29. Uncontrolled Shaking & Tremors	30. Slow speech

Larry says, "Everything you have seen is documented and being studied. The same affects & illnesses are caused by Mercury, Lead, & other heavy metals, including other chemicals, also exposures to radiation. This is meant as a reminder of experience, scientific studies, & knowledge stored now known and "Too Late for Many", through out the world. Sick Building Syndrome, and the illnesses related to Fiber Glass are also now know to cause many of the above, Plus destroys the DNA. From over exposure, and over exposure you may not even be able to see.

A. Korsakoff's Syndrome: Usually caused by a drug overdose, chemical overexposure, not ruled out are pesticides, or nerve agents especially so with BZ nerve agent. It is said to have been used in the Russian takeover of the theatre that killed many. The United States in my knowledge and from our sources the only country besides Iraq that ever had or used BZ nerve agents.

There is a lot of thoughts and talk out there, and from our sources that the United States had supplied Russia with BZ nerve agents that were used. It was used by the Russian police or soldiers that were clearly unfamiliar with what could happens when overexposed. There is exposure where you get the expected results, then there is overexposure when it may not have been used prudently, and with care. This nerve agent was much more than first explained as a simple sleeping agent. Larry says, "Our sources and experience, say It was a deadly nerve agent, a hallucinogenic type where a person goes into unconsciousness & loses control of the body & its functions just before death and very similar to a drug overdoses. It is said it also causes a totally Euphoric condition.

B. Motor Neuron Disease: a disease, of the nerve cells that carry signals from the brain to the muscles die, and eventually leads to paralysis or death. Something very common in the destruction in the brain done is by chemicals, nerve agents, heavy metals.

C. Huntington's Disease: People with this disease lose control of body movements. Violent irregular jerking motions, mental failure, paralysis and death. Genetic defect of the gene's. Nerve agents, Chemicals, Heavy metals, and by Fiber glass. & other hazardous materials, Fiberglass destroys DNA., have not been ruled out.

D. Endocrine High Blood Pressure: can cause damage to many other parts of the body, nerve agents, Heavey metals, Chemicals have not been ruled while "Lead" is a positive cause of hypertension.

E. Dementia: Mental impairment, Nerve agents, Chemicals, heavey metal poisoning, have not been ruled out as causation.

F. Alzheimer's disease: Nerve agents, heavey metal poisoning, Chemicals have not been ruled out as causation.

G. Multiple Sclerosis: MS: is a disease of the nervous system, mainly affecting the brain and spinal cord, Where messages from the brain are conveyed as electrical impulses traveling along nerve fibers. Theses impulses leak out or are not transferred along the nerve fiber, and lost to the point the impulse was being sent to. Nerve agents, Heavey metals, Chemicals, and other hazardous materials have not been ruled out as a causation.

H. Muscular Dystrophy: Is clearly caused by a genetic defect, and is inherited, that results in muscle wasting. Nerve agents, Chemicals Heavey metals, and other hazardous materials have not been ruled out.

I. Parkinson's disease: A chronic progressive disorder of the central nervous system. The underlying cause is believed to be deficiency of dopamine. A chemical that transmits nerve impulses in the brain. The area called the "basal ganglia", which helps regulate movement of the limbs and body, is most affected. Pesticides and Nerve agents, along with heavy metals, chemicals and other hazardous materials have not been ruled out, and pesticides are now being ruled in as producing Parkinson's disease. One pesticide they are zeroing in on is widely used on home grown fruits and vegetables and for killing unwanted fish in the nations lakes, rivers, and produces all the classic signs, symptoms and illnesses of Parkinson's disease in rats. Scientists are finding evidence showing that chemicals in the environment may be factors in this devastating disease.

At this point it may be linked to many different pesticides besides just rotenone. This was revealed at an annual meeting of the Society for Neuroscience, which is the largest meeting of brain researchers. Noting it was a major step in parkinson's disease research. This has been the best research we ever had for the disease linking it to an environmental agent. Since Rotenone is extracted from dried roots, seeds, and leaves of various tropical plants, Rotenone is found in 680 compounds marketed as organic garden pesticides and flea powders. Unlike most artificial pesticides, rotenone breaks down with-in days in the sunlight.

Parkinson's disease is one of the most common Neuro-degenerative diseases that affects nearly a million Americans over the age of 50…. A steady loss of cells, in a tiny region of the brain called the Substantianigra, that produced chemical dopamine, that is crucial for movement and cognition……Those affected develop jerky, tremulous movements that get worse with time.

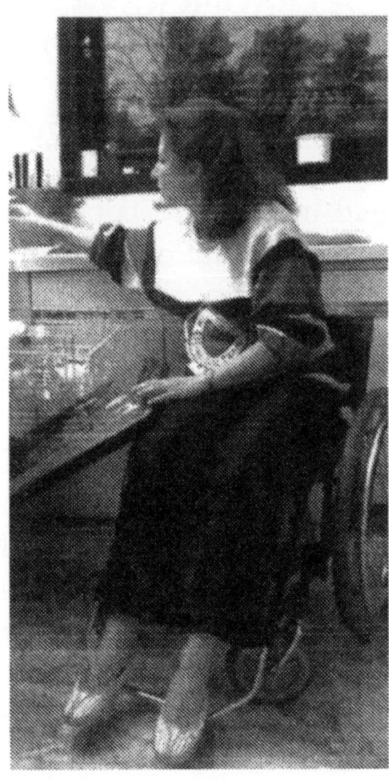

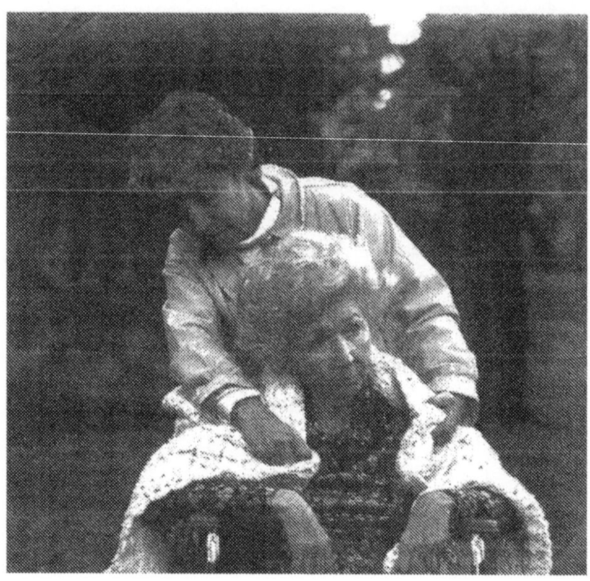

A realization that something is wrong. It could be caused by our polluted environment.

COLORADO, A LEADER REVAMPS PROCESS FOR POLLUTION REPORTS

The long delays in the case of former gun-sight factory prompt charges, and a knew system in place for document tracking, everything is now logged in and a response time is set up. Redfield Gun Sights Inc. reported the pollution problem to the State 3 years ago, and the State failed to act on it's letters. They say it would not have any impact, Even though 3 years have been lost. How would it not have an impact on how soon or in what manner, they would have been able to have addressed the problems?

It may not have delayed the remediation, but what about those polluted by plume of degreasing chemical, and how far did it travel in 3 years? How many people were potentially affected. To say it may not have delayed the remediation, is nothing more than an excuse for the States mishandling of a dangerous situation that had been self reported by Redfield. They have installed 122 homes with ventilation systems, But what about the three years of discovery that was lost? What about 3 years 100's of people breathed in this pollution of a deadly chemical DCE 1,1 dichloromethane? And additional discovery of 43 more homes that were above the safe threshold will receive ventilation systems.

Since the etiology, and immensity of this problem several years have passed where these people were being “Over Exposed” daily to a deadly cancer causing solvent. The illnesses, the diseases, and the cancers that will be seen here are just now in the metamorphosis stage, and soon the Monster will make his unwanted presence, those responsible will deny any responsibility, and those people will most likely suffer from the unknown. Medical monitoring should be set up immediately, because if they wait 5, 10, 15, years when things begin to happen No One Will Know or will be Familiar with What Took Place, or the consequences of the deadly pollution. Your expert witnesses (DOCTORS) will be blaming it on everything but what actually happened to cause the problems.

LOOK AT THE MAP BELOW, SEE HOW FAR & WIDE AN AREA THAT HAS ALREADY BEEN AFFECTED. IF MEDIUM GRAY SHADE IS CONSIDERED HAZARDOUS, THE BLACK AREAS MUST BE CONSIDERED TOXIC & DEADLY.

It wasn’t until 1997, when the State found airborne DCE contamination at a nearby Colorado Department / transportation maintenance facility that Redfield was asked to check for possible airborne contamination.

No one would have thought of doing the airborne testing on the Redfield contamination until after the discovery at the CDOT facility in 1997. This is what they are using As an excuse. “The misfiling by the Department of Health of Redfields original self-reporting didn’t delay the remediation now under way. But they did not discuss the Jeopardy placed on the lives of those who were living with the deadly toxic fumes for several years because of that “Deadly Delay, Mistake, or Oversight”

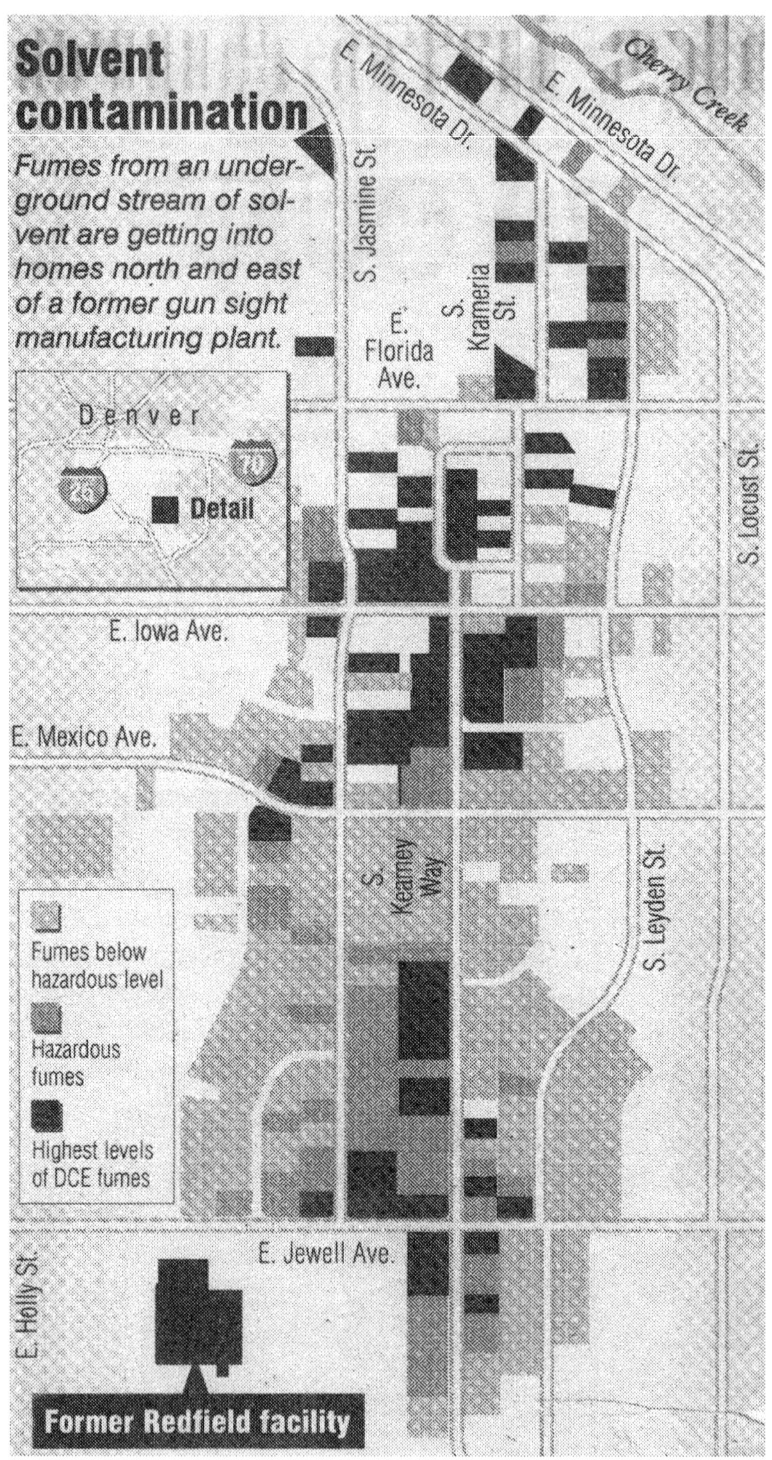
Solvent contamination
Fumes from an underground stream of solvent are getting into homes north and east of a former gun sight manufacturing plant.
Denver
25
70
Detail
E. Minnesota Dr.
E. Minnesota Dr.
Cherry Creek
S. Jasmine St.
E. Florida Ave.
S. Krameria St.
S. Locust St.
E. Iowa Ave.
E. Mexico Ave.
S. Kearney Way
S. Leyden St.
Fumes below hazardous level
Hazardous fumes
Highest levels of DCE fumes
E. Jewell Ave.
E. Holly St.
Former Redfield facility

WHY MORE AMERICANS ARE DYING FROM NERVE DISORDER

Afflictions called motor neuron disease, Steep increases in mortality from a mysterious Disease have occurred in the United States, and this is all since the 1960's. A new study says Motor Neuron Disease (MND) is referring to a group of neurological conditions that damage nerve cells in the brain and spinal cord, causing muscle weakness, wasting, and many other symptoms. A close and most known condition is amyotrophic lateral sclerosis or Low Gehrig's disease.

Researchers are now beginning to suspect some unknown agent that damages nerve cells over periods of years or decades. Larry says if 2 + 2 were put together these are the very signs and symptoms of Mercury Poisoning, a material used in Paints, Pesticides Herbicides, and fungicides, like in showers, near and around pools, restaurants, public toilets, sold for use in homes, and was banned in the United States of America by EPA in the late 1980's because of it's deadly toxicity. Once a person reaches the stage of overexposure to low levels of Mercury, the deadly and crippling outcome takes years of illness and disability. It is cancer causing, & it can cause mutations, deadly to a fetus. It has been totally banned in the United States. The entry into the body is different than most other toxic materials and that is why it had fooled even EPA for so many, many years. It enters the body via the nerves, by ingestion, breathing it, or even is absorbed through the skin, eyes, ears, mouth and throat. It attacks the Central Nervous System almost immediately via the nerve route into the brain. The Neuron transmitters in the brain are immediately affected, much like a nerve agent or pesticide which contain or use mercury in its manufacturing.

Researcher's are closer than ever now and will be able to put more names behind the silent killers in our environment. It causes immediate damage to all or at least some of the nerves in the brain & spinal cord, depending on the exposure or over exposure, repeatedly and the duration of the exposures.

Researcher's now believe the rise in incidents of MND, suggests that the population has experienced a much greater exposure to causal factors. Larry says, "again if they would just put 2 + 2 together they would have the answers to the puzzle". It was in the late 50's and early 60's that the Chemical Industry began using Mercury in just about everything from solvents, to pesticides, to herbicides, to Nerve agents, to Paints, to Fungicides, an array of household items and others. For over 40 years we have not only been exposed but "OVEREXPOSED" to mercury in our environment, there are many other deadly toxic, and Ultra hazardous materials that are used everyday in our environment. Larry says, "They need to clean up the Chemical Industry in our nation and our environment."

The hazardous chemical agents in our environment are considered a great possibility if not a certainly Alzheimer's disease, and other degenerative nerve disorders. The steepest increases have been occurring amongst older people, and steady increases in people over 40. MND's has shown increases as the American population ages. But it also increased in younger age groups. With Mercury now banned in the United States since 1988 and Lead since 1967 in the United States Larry believes within the next 30 to 40 years we will see a steep decrease in Alzheimer's, and many other Motor Neuron Diseases in the United States.

The increases we have seen are very real, & all statistical indications cite environmental Factors are playing a very aggressive role. Researchers have speculated that MND is a sequel of polio that occurred earlier in the century. That theory emerged when researchers took note MND and Polio do involve damage to the same nerve cells in the spinal cord.

"IS THE UNITED STATES SEEKING LOOPHOLES TO AVOID CUTTING POLLUTION? ENVIRONMENTAL GROUPS SAY YES"!

The United States stands accused at the HAGUE, as the U. N. Conference began to set rules for reducing harmful gasses released into our environment, and into the Atmosphere. The major issue seemed to revolve around how much countries could trade off against their commitments to roll back their own pollution by helping other countries reduce theirs, or by planting new forests to absorb the greenhouse gasses.

The biggest polluters are those who need to be forced to change their ways. The United States, which advocates unlimited emissions trading, and the small developing countries who fear it will defeat the effort to force Large Industrial polluters to stop or clean it up. The Oil industry of course is exploiting every loophole, which is the noose around nature's neck. The United States continues to hide in a cloak of darkness, calling it a most cost effective way to meet the global target of reducing greenhouse gasses while they use others to cover up their polluting ways. All the developing countries want is a guarantee to technology, and a fund to help them in adapting to climate changes, that will soon threaten their economic and physical survival.

The imperative banner means "work it out" so the figures add up, even though there are still loopholes to avoid cutting pollution. The target date of 2012 the rolling back emissions of carbon based gases by an average of 7 percent, by Japan, Europe, and the United States. The Kyoto Protocol has been ratified by 30 Countries, Mexico is the most industrialized country on the list, and none are among the heaviest polluting countries. 55 countries need to ratify the agreement in order to come into force.

Methamphetamine is the cheap man's cocaine, also known as speed, Ice, Crystal or better known as crank. The use has tripled now in 2003 reaching toward 15 million users. The Hardware stores, have huge increases in sales of thinner, & specifically Xylene, Toulene, Benzene, and then the Fertilizer plants sales are up, but also they are being targeted by thieves, stealing Anhydrous Amonia close to pure nitrogen, and the sale and theft of cold medicines is on the rise, all natural chemicals in the manufacturing of Methamphetomine. Then the empty or near empty containers, and all the waste pollution from it are being buried and leaving small but ultra toxic chemical dumps everywhere.

The U. S. Forest service has agreed to reimburse a small Colorado water district, for contaminating the wells with a suspected cancer causing chemical. The Forest service will pay $288,000 Teller County for all past cost of treating water, and will establish a process for all future treatment costs. And of course the Forest service did not admit responsibility for the contamination.

The chemical "EDB" ethylene di-bromide, was banned in 1983 by the EPA, the first discovery of it in the water was in 1996. The district near Woodland Park, has and still serves about 225 households. The Forest service was using the chemical long before it was banned in 1983, what about all the people who have drank the water back then and even now, "Are they on the slow road to death". When cancer takes them as victims, no one will ever know why they now have cancer, one to ten years or even decades from now.

The EPA's office of the Ombudsman is still under fire, and there is an effort in the Senate by Senator Wayne Allard who introduced legislation making the Ombudsman's office independent from EPA who is attempting to set greater controls over it. It all started after several Ombudsman's decisions were critical of the Agency. For instance it was the Ombudsman's office who finally convinced EPA to reverse a decision allowing the defunct Shattuck Chemical Co. to bury low-level radioactive waste in a Denver neighborhood instead of shipping it to a toxic waste facility.

LANDFILL OPERATOR FINED $100,000 FOR ILLEGAL LANDFILLS.

Phil J. Spano Sr. settled with the State of Colorado Department of Public Health and Environment. He must also monitor the sites for escaping methane and any leakage of contaminants to the groundwater. These sites were operated by Browning-Ferris Industries under a lease from Spano. His permits were only to grade the sites, with rubble, then he accepted the fly ash from power generators. As far as it is known he did not accept any toxic materials, anyway were not aware of any says spokesman for the Department of Health. Spano's operation was finally closed down by citizen complaints, It had created a major problem getting him into compliance, and was going on for a long time.

COORS BREWERY IN GOLDEN COLORADO SPILLED 77,000 GALLONS OF BEER INTO CLEAR CREEK KILLING OVER 50,000 FISH, YEARS EARLIER A SPILL INTO THE SAME CREEK KILLED NEARLY 15,000 FISH.

The chief environmental regulator for the Colorado Department of Health and the Environment, said a punitive penalty was Coors donation to a group cleaning up mine waste on clear creek was sufficient, and included years of sewage waste violations. This did not include the 1991 fish kill or air and toxic waste violations. This was an accident & not willful when a Coors employee accidentally sent 77,000 gallons of beer into Clear Creek. Yet EPA questions the punishment of Coors, of course it as a tax-deductable donation and EPA that was a benefit to the State and not EPA.

Brighton Colorado's South Platte river, once an industrial waste dump is turned into an Oasis. The South Platte River between Henderson road and Colorado 7 is now a wildlife corridor right in the heart of the metro area.

For more than 40 years this area has gone from a toxic Industrial waste dump to a beautiful wildlife corridor. It is the known by Adams County Officials as the Crown jewel in their open space program. Near completion is the hiking trail along the river links Adams, Denver, & Arapahoe Counties. The serenity of the river & its once pristine environment has been returned with modern day improvements.

There is still some concrete on the banks, and maybe an old tire peering through the sandy bottom of the river. It will get even better. The river is home to at least a dozen kind of minnows, and a home to carp fish. The fish support the raptors, and herons, which we notice take flight around every bend in the river. Eagles make their secondary homes along the South Platte during the winter months, then fly back North for nesting & raising their young. It's not clear whether the river will ever support game fish, since they were never natural on the South Platte River even before cities & farms forever altered its course.

The original Platte river was a wide flood plain, that would fill when the water come pouring out of the Mountains, then once again by July would be no more than a trickle. Compared to the pioneer days the river still remains channeled, to control such a wide flood plain, and to make full use of the water, through out Colorado, and Nebraska, out into Iowa.

ENVIRONMENTAL PROTECTION AGENCY STILL PLAYING POLITICAL FOOTBALL WITH REGULATIONS FOR ARSENIC IN DRINKING WATER.

First of all they make it much higher up to 50 ppb before the Clinton Clinton Administration, then during the closing days of his administration it was lowered to 10ppb. Then EPA rescinded the rule saying it needed to be reassessed, because of the impact of the change on the health of the residents and the economic burdens the tougher standards would impose. President Bush felt to be defensible to everyone, it needed to be reviewed in order to give consideration to stakeholder groups. Larry Says, "or is it the stockholders of large Industries that would be forced to clean up their polluting ways with a deadly chemical, called arsenic". The only ones to benefit would be the Chemical Industries allowing them to be less responsible for what they have done and are still doing. Larry says, "Bush has swung too far in protecting those who pollute our nation walk away and let the public pay for what they were allowed to do. I say clean it up, it never should have ever been allowed to 50ppb. 10ppb is even toxic but livable 50ppb would be toxic to humans, especially pregnant women & children“. I hope Bob Martin the ombudsman for EPA wins his battle in the Senate, takes his job back he an his investigator Mr. Kaufman who's hands are tied right now by EPA, I believe they would never allow something like this to take place. They always seemed to be the peoples voice within the EPA. This should be the job of the regulators from all the Departments of Health across the Country.

LOWRY LANDFILL, ONE OF THE NATIONS MOST CONTROVERSIAL TOXIC CHEMICAL WASTE LANDFILLS NEXT TO DENVER COLORADO. OUR SOURCES TELL US IT CONTAINS ULTRA TOXIC CHEMICAL WASTE FROM THE (RMA) ROCKY MOUNTAIN ARSENAL & SHELL CHEMICAL COMPANY. OUR SOURCES ALSO TELL US IT ALSO MAY CONTAIN RADIOACTIVE PLUTONIUM WASTE FROM ROCKY FLATS NUCLEAR WEAPONS PLANT IN JEFFERSON COUNTY.

The Landfill toxins are leaching toward more groundwater, EPA tells cleanup officials to contain the wayward pollutants coming from the Lowary Landfill. It is one of the largest Superfund sites in the United States. Several cancer causing solvents have soared hundreds of times beyond the Environmental Protection agency's Standards in just the last two years. One toxin 1,2-Dichloroethane...has already been recorded in groundwater at 4,000 times the safe standard. EPA Officials who thought they had it contained back in 1994, are admitting there are pathways moving water contrary to regional groundwater flow charts. It's got us disturbed and may cost upwards of $100 million just to start in our efforts to contain this monstrous problem. They now have to refigure what they thought they had figured out back in 1994 which called for underground walls forcing the contaminated water toward another wall on the North end, and into the treatment center.

EPA is now blaming it on the City of Denver since the Lowry Superfund clean-up is not protecting human health, and the environment, and is requiring the cleanup officials to take even more aggressive steps to control it. EPA is now talking of levying fines against the city of Denver and it's waste management, which are responsible for over-seeing the EPA-supervised clean-up. Regulators are now fearing the solvent-laden groundwater is somehow making a wrong turn…toward northwest Denver with nothing in its way, instead of to the northeast where it could be captured and treated. "WATCH OUT, CITY OF DENVER AND BEYOND". Then on another hand TCE and PCE, two ultra toxic solvents are escaping to the east beyond the reach of the underground containment wall.

There are several areas now where contamination is turning up just outside of the Landfill's underground walls…that had been completed in 1998, and never surrounded the pollution in the first place as workers were trying to avoid gas and phone lines. The general public in the area are "chompin at the bit". So to say, they feel there's not enough scientific work being done to get that stuff out of here. Larry says its too late, it is there to stay and the only thing that can be done is try to contain it and possibly treat it. The life span on these toxins run into the thousands of years.

Regulator's are asking the question, could it have been present before the construction of the containment walls, are the walls really containing anything? Has the toxins found new avenues of escape. They know it is moving offsite to the Northwest and into the city of Denver. Workers are drilling several more test wells to track the flow of the contaminants. The contaminants have already moved beyond the EPA's compliance boundaries within the site.

EPA is sidestepping the questions as to the dumping of Radioactive waste at the Lowry Landfill, stating there is no credible evidence to support the claim. The agency was scolded for inconsistent and inaccurate responses to the claims of the presence of Radioactive contamination at the site. The new fact sheet in an effort to stop "hemming and hawing" over the matter, and characterized the report as "calling a space a space".

EPA Officials continue to insist that the low levels of plutonium at the Lowry hazardous site pose no public health threat. However one wants to remember it only takes about 8 pounds of Plutonium, to create an atomic bomb. It would be a dirty bomb that could contaminate the entire state of Colorado, and depending on the wind drift could pollute every state in the United States of America. they don't even know the amount of plutonium, possibly CESIUM, BERYLLIUM, RADIO ACTIVE IODINE, amongst all the deadly ultra toxic solvents…. "BEING FOUND IN THE WATER"…..that were used to manufacture the bomb parts at Rocky Flats Nuclear Weapons Plant are leaking out & into the under ground water for Denver residents……(LACED WITH RADIO ACTIVE MATERIALS……Ultra Toxic Solvents that leaked from the Redfield manufacturing plant for up to 10 to 15 years is also (LACED WITH TRITIUM)……..What your reading is documented facts that has already affected Denver residents……The Brutal vulnerability and long term ramifications are yet to come……

Officials are now saying "whoa it's true Several Ultra Toxic Solvents, Laced with highly radioactive materials are there"……"in the "WATER AND THE AIR", at and around the Lowry Landfill……and the old Redfield manufacturing plant….Hundreds of homes in Denver have been supplied ventilation systems for their homes….."10 years too late"…..and long after Redfield had self reported the Ultra Toxic Leak to EPA……

EPA 's fact sheet represents the unraveling of a complex web of collusions, corruption & conflicts of interest between EPA, it's contractor's & the major Industry's Polluter's & the dumpers of Ultra Toxic, and radioactive Waste without reporting what it was……Many residents have already paid the price & don't even know it, and they probably will for years or even decades later. "THIS IS AN ALL OUT EFFORT TO DENY LIABILITY IN ORDER TO COVER UP NEGLIGENT, WILLFUL, & RECKLESS ENDANGER MENT TO THE PUBLIC". EPA's public outreach was cited in several areas, (1). EPA project manager in 1998 told the audience at a public meeting there was "ABSOLUTELY NO PLUTONIUM AT THE 480 ACRE SITE"……(2). "JUNE 1999 EPA BROCHURE STATED, """THERE IS NO CREDIBLE EVIDENCE OF PLUTONIUM CONAMINATION AT LOWRY LANDFILL"…. "WHEN IN FACT THEY WERE AWARE OF RADIOACTIVE CONTAMINATION WHEN THEY ALLOWED ROCKY FLATS NUCLEAR WEAPONS PLANT DUMP SOLVENTS LACED WITH PLUTONIUM AT THE LOWRY LANDFILL.

EPA admitted they are running the Lowry Ultra Toxic Chemical and Radio Active waste through the Denver sewage treatment process…. They are treating the toxic waste by diluting it along with 10's of millions of gallons of Denver sewage water that flow into the plant from households and other industries in Denver. They call this treating the waste water into the South Platte River. This method of treatment is not at all treatment it is a "scam by EPA" on how they get rid of Ultra toxic Solvents and Radioactive materials and calling it treated when in fact these materials cannot be removed from the water in what they are calling treatment......and spreads it across the landscapes and rivers.

The Lowry Landfill began as a bombing range used by the Air Force in the 1930's. In 1966 the Land was turned over to Denver & was used as a Municipal & Industrial & Toxic waste Landfill. Millions of gallons of Industrial wastes, Sludges, Oils, Caustic liquids, Solvents laced with Radioactive materials & Plutonium, were dumped into unlined pits which were described by our source as being some of the most permeable soils. This was done without any soil tests, perk tests, or Hydrological, or Geological engineering done. With development moving closer and closer to the Ultra Toxic and Radioactive Landfill, housing is now with in a quarter of a mile, with Ultra toxic waste already in the water.

COLORADO SUPERFUND SITES, ESTIMATED DATE FOR CLEAN-UP

All of or large portions of the Swansea, Elyria, Clayton, Cole, and South Globeville, neighborhoods in North East Denver are sitting on a powder keg of contaminants. Metals through ingestion of soil particles & inhalation of airborne particles are considered ultra toxic, The possible ground water contamination has not yet been assessed, yet nor are they saying it is or is not safe, and people are still using it. The full clean up of the area is just getting started, and the estimated date of completion is 2004, The water contamination has not yet been considered.

There were three smelters operated from the 1870's through the 1950's, refining gold, silver, copper, lead, zinc, arsenic, and cadmium. There were no rules for controlling pollution, contaminating the air, water, and soil. Now there are people living in their new homes they have purchased. Since most of these materials may take years even decades to show it's monstrous face, in the form of Cancer, Mutant capabilities, Central Nervous System Injuries that begin showing up in children as Learning disabilities, Short term memory loss. Failure to thrive, Lung disease, Asthma, Multi Chemical Sensitivity Syndrome, Chronic fatigue, just to name a few things one might watch for.

Denver Radium, where it came from and how will it be cleaned up.

There are many sites in Denver, estimated, clean up possibley sometime in 2003. There are 65 properties with radioactive soils & debris from radim processing in the early 1900's. Contamination, is Radium, Thorium, and Uranium, with decay will also produce Radon Gas, the groundwater was polluted at some of the sites. Another of the Superfund Sites.

PJKS AIR FORCE BASE JEFFERSON COUNTY, 20 MILES SOUTH WEST OF DENVER, ESTIMATED CLEAN UP DATE 2005. A 460 acre complex near Waterton Canyon, It was used for Missile & Rocket research Development and manufacturing. Contaminations: Volatile Organic Compounds, Rocket Fuel, Hydrocarbons, Organic & Inorganic Compounds, also radionuclides in the soil, and groundwater, and probably the air. (PROBABLY THE HONEST JOHN ROCKET).

NEVADA'S YUCCA MOUNTAIN PLAN IS UNPREDICTABLE IT'S BEEN ON AGAIN, OFF AGAIN NOW FOR OVER 12 YEARS. SEE BELOW 2002, CONSTRUCTION CONTINUES, IT'S COMPLETION DATE WAS 2000, THEN 2002, 2204, 2006, AND IT'S NOW BEEN MOVED AHEAD TO 2010.

Yucca mountain is an ancient volcano, as most volcanic mountains are, it would be totally unpredictable, on or about Sept. 2002, earth quakes reported in the region.

JOE CAVARETTA/ASSOCIATED PRESS

Workers at Yucca Mountain in Nevada get off a train that takes them through they mountain. The Senate on Tuesday approved the plan to bury radioactive waste deep inside Yucca Mountain.

As understood around 4.5 on the Richter scale, recorded in Boulder Colorado & were all within the Yucca Mountain Region.

COULD THIS CONSTRUCTION BE THE BEGINNING OF THE WORLDS WORST NUCLEAR DISASTER OR "DIRTY BOMB" THAT WOULD BE CAPABLE OF POLLULTING THE ENTIRE WORLD WITH DEADLY NUCLEAR VOLCANIC CINDERS, NOT ONLY SMOTHERING OUR WORLD WITH ASH, BUT BURNING EVERYTHING AT THE SAME TIME. ONE THAT WOULD MAKE HISTORY, IF HISTORY WERE TO SOMEHOW SURVIVE.

Stop and think for a moment of what the Department of Energy "DOE", is forcing on the World, "realizing at the same time DOE is a cabinet post in our government. Plans to store over 770,000 tons of Radio active waste, these were figures back in the early 1990's. Those figures have by now in 2003, and with the Construction still ongoing Those figures have probably quadrupled. We are talking of the most hazardous and deadly components known to mankind. Radio-Active Uranium, Plutonium, Beryllium, Cesium, Ytrium, Radio Active Iodine, millions of gallons of volatile radioactive solvents amongst others to be permanently stored inside a known volcano that has generated earthquakes as a warning, "NOT TO DO IT"...........Were talking of a Nuclear Volcano that would be capable of destroying the entire world for now and forever.

Your Congressperson's, and Your Senator's working hand in hand with the DOE, & our government in making such a decision for us today that may destroy our tomorrow, our children's tomorrow, and the future of us all. Please pick up a pen write your Representatives from your State since this will affect us all. "ASK THEM TO STOP NOW", any consideration now & forever of using Yucca Mountain to store any nuclear waste short term or permanent. We can make our voices be heard together around the World, There are reasons why the Carlsbad New Mexico permanent storage depot failed, for a similar reason & same reasons, """ YUCCA MOUNTAIN MUST FAIL"".

Author of Death Factory U. S. A. Larry D. Land

Larry will document in his book, and to you all the reasons Yucca Mountain Must Fail. Even though it may sound like fiction, it is not, the information your about to read are the brutal facts of what our representatives have to work with in Congress and the Senate. The deadly ramifications of Yucca Mountain, if allowed to become a permanent storage facility for nuclear waste. What happened at CherNobyl was only a fly in the ointment.

THE SHOW DOWN AT YUCCA MOUNTAIN NEVADA.

Even the Republicans lobbyist's who are against depositing nuclear waste at Yucca Mountain are restrained by President Bush's wrath, and may find the doors to the White House locked to them. 770,000 tons was the original figure, it could now be more at 3.8 millions tons of nuclear waste.

The Democrats in Congress were blocking Bush's desire to open up Yucca Mountain immediately, while the Democratic leadership were making good on its assurances that Yucca Mountain Project was dead. 2002 we have had a Republican change of the guard. It's like having the "Fox guarding the henhouse". The House and the Senate are now ready to override that protection we all had from the Democrats. The great State of Nevada has rejected the Yucca Mountain Project now for nearly 15 years, and may lose their battle to the Republican Guard. (THE FOX)

While it is only 90 miles from Las Vegas, the gambling interests are afraid it will deter the high rollers being caught in Las Vegas.

Larry says, "If Yucca Mountain blows it's top with millions of tons of nuclear waste stored in side the whole world may only be given seconds to day" ""God, or maybe Allah, I wish I had spoke up & stopped this insanity"". This may be the dirty bomb that has been promised to the World, by the Evil doer's in this world. I hope President Bush takes time just to read this part of "Death Factory U. S. A".

The once Majority leader of the Senate Tom Daschle (D) promised the World the Waste Repository, "Yucca Mountain" would die in the Senate. Larry says, "even though I'm not a Democrat, I am not a Republican, one swings to far to the left, and the other swings too far to the right, and I am caught up in the middle of this like most of you". DEATH FACTORY U. S. A. will be my letter to the World and needs your support with letters to your Representatives in the Congress and the Senate, rejecting the use of Yucca Mountain, a Volcano ready to blow it's top, just waiting till the trap has been set as a Waste repository that could blow the World's Top".

Prominent Republicans in the Government, warned the White House that any Pro Yucca decission might yeild a new Democratic sweep in the elections, and a negative decision from the White House would be a death sentence for nuclear power. DOE, and President Bush did approve the Yucca Mountain Project. The Nevada interests have set out recruiting the best lobbyists money can buy for the battle in 2003, 04, and beyond.

If your relatives, your friends, children, your grandchildren need a good explanation of what's happening in our world, and what they can do to protect themselves, & to protect our future together, Get them a gift, Ask them to get a copy of DEATH FACTORY U. S. A. Part I, JAN. 2003, PART II, BY MID. 2003, & Part III, BY THE YEAR 2004. Part III will bring everyone up to date as to what's happeing, & what has happened and the World Clean up of Toxic and radioactive wastes that has been achieved in our world. Larry says, "Word of mouth is the best tool for knowledge of what we all can do together". Not since Rachel Carson's "Silent Spring", nor John Regenstein's, "American the Poisoned" has a book so "Dynamically & Boldly Documented the Hazardous Biological, Chemical, & Nuclear Warfare this nations Government, The "DOD" The "DOE" The "USCD" & the Pentagon, & Industry Giants from the United States & through out the World. "They have waged war on their own people", with DEATH FACTORY'S amongst us all. MAN HAS MOUNTED SCIENCE, THAT IS NOW RUNNING AWAY, WE MUST TAKE BACK CONTROL OR SCIENCE WILL BE THE DEATH OF MAN....THE CHEMICAL, BIOLOGICAL & NUCLEAR WARFARE MATERIALS ARE WELL BEYOND HIS STRENGTH AND CONTROL...SCIENCE NOW HAS THE EXISTENCE OF MANKIND IN ITS POWER.... THE HUMAN RACE MAY COMMIT SUICIDE, BY POISONING IT & ALL OF US TO DEATH, AND/OR BLOWING UP THE WORLD & ITS HISTORY AT THE SAME TIME....... "YUCCA MOUNTAIN NEVADA"

LARRY D. LAND

Remember nuclear waste will be coming from all parts of the United States by train, or by truck, it could be passing through or near your town or city...... Union Pacific Railway is looking to the Government for long term contracts transporting nuclear waste right through our back yards, and next to the schools, as well as the trucking industry trying to grab a part of it..........The Bush team arrayed for battle to approve everything from the shipping to the Storage at Yucca Mountain. "BEWARE"

Larry says, "Even though disabled, he is exercising his rights under the Constitution of the United States Amendment I, giving him the right to 'Freedom of Speech', 'Freedom of the Press', & 'Amendment XIV, giving him the right to Equal Protection under the Law, & Due Process of the Law' (without fear of reprisal), as it would not have been available in any other part of the World, He says he feels much like a lobbyist, being able to bring DEATH FACTORY U. S. A. Part I, Part II, & Part III. A book outlining our dilemma to the people in the United States, the world, and to our Government as well. The books will be available in book stores everywhere in the World, Just ask for it if they are temporarily out they will order it for you.

A PANEL OF SCIETISTS SAY NO, IT WILL BE FRAUGHT WITH UNCERTAINTIES 10'S OF THOUSANDS OF YEARS FROM NOW.

The go ahead from the White House was to be given in just weeks, of receiving this information. Although the Scientist's emphasized they were making no judgment for or against long term burial of millions of tons of nuclear waste, just delivering the facts, that need to be considered before making the final decision.

The fact finding panel of Scientists were created by Congress as a technical watchdog in the search for a nuclear waste site that would be safe. Larry says, "Yucca Mountain is not safe and again the Fox is left guarding the hen house". The scientists were not about to put their neck on the chopping block of Republicans or Democrats, that is why they stated it would be impossible to avoid unexpected problems. "What would that say to you" "It would be just like loading a known (bunker) Volcano with Nuclear explosives that would destroy the World. It seems like playing the odds, one chance in a trillion that it would not blow it's top off." Larry says, "It's in the right State Nevada, he wonders if the odds would even be a trillion to one & if President bush is ready to play the odds against our lives, His hope is that Congress and the Senate are not ready to play those odds, if they are we need to change their minds, saving their lives, our lives and the Worlds future as well".

The Panel has limited confidence even with DOE's prediction of hundreds of years, what about after? The life time of these material is forever. Larry says, "We will have set the timer for our own nuclear destruction". The panel said "eliminating all uncertainty associated with future performance of the repository would never be possible". How much uncertainty is acceptable? "NONE". The panel also said if DOE were to present computer models that would attempt to predict the future performance, thousands of years into the future would pose many scientific uncertainties. There would remain gaps in data Gaps in the data the basic understanding of how the volcanic rock, Hydrology, Geology as well as future volcanic conditions, including the man made barriers would contain the Ultra Toxic, and very explosive nuclear capabilities of the chemical and nuclear waste....will perform during tens of thousands of years and beyond the panel said in it's report....? Larry says, "This World relys on ""BEYOND A SHADOW OF DOUBT".

WHAT HAPPENS, IF THE SAFETY CONCERNS OF OUR NATION ARE WAIVED & WE SAY WHO DONE IT? WAS IT BUSH & DOE, OUR REPRESENTATIVES BE THEY REPUBLICAN OR DEMOCRAT & THOSE OF US CAUGHT IN THE MIDDLE THAT WILL BE USED AS FOTTER.... TOGETHER WE CAN CHANGE THE MIND OF DOE, AND

PRESIDENT GEORGE BUSH. WE AS A NATION MUST STAND UP FOR RIGHT, ALWAYS REMAINING POSITIVE SAYING NO TO YUCCA MOUNTAIN. LARRY SAYS, "LET'S ALL JOIN NEVADA IN ARGUING THE GOVERNMENT IS IGNORING THE SAFETY CONCERNS FOR US OUR CHILDREN & THE FUTURE". WRITE YOUR REPERESENTATIVES & BE HEARD OF YOUR FUTURE CONCERNS.

The scientific debate over Yucca Mountain may be terribly complicated, but the politics playing this out are very simple at best. I would invite a national poll after everyone has had an opportunity to read DEATH FACTORY U. S. A. Part I, II, III, realizing there are 103 nuclear power plants across this nation who plan to ship spent nuclear waste across our country for storage at Yucca Mountain. Larry says, "he would put up the odds the poll would show 99 to 1 against Yucca Mountain as a Nuclear repository".

Those of you who do care about this issue care deeply, and I know you want to help. I know what it is like when we feel our hands are bound, I would not let that happen & hope you won't either, my hands are not bound, & by our Constitution yours are not bound. Pick up a pen or a phone, talk with your representative, their a lot more willing to talk with you think, they want to carry your feelings your thoughts & your message with them all the way to Washington DC., to the desk of President George W. Bush.

In Nevada John Sununu, President Bush's first chief of staff, proclaimed that, "If Nevada is not willing to do its part in what is part of a national plan for homeland security? Maybe Americans ought to vacation somewhere else".... However Nevada's Las Vegas Mayor Oscar Goodman, described appropriately John Sununu as, "that piece of Garbage" Larry says," if you take a moment to think of what Sununu was doing by pushing the Yucca Mountain repository right down the throat of every citizen in the State of Nevada, in fact every citizen in the United States we could probably agree".

The Republican leadership in Nevada, Gov. Kenny Guinn, Sen. John Ensign, and Rep. Jim Gibbons all restated their opposition to the Yucca Mountain project. Reid said it is only because President Bush pledged not to approve Yucca Mountain project unless "good Science" supported it. If Bush accepts DOE, administrator Abraham's recommendation to use Yucca Mountain as a permanent repository, Gov. Guinn then has 60 days to veto it, then can be overridden by

Congress & that power of persuasion has already begun on both sides. Abraham's recommendation is not based on information from the general accounting office who concluded DOE "Abraham" is not ready to make a site recommendation because "Abraham, &/or DOE does not yet have all the technical information necessary to make that decision. Larry says "this is not unusual for DOE, to go off half cocked unloaded and unprepared". Larry says, he feels he will make the right decision."

DOE's Secretary, states that Yucca Mountain is a significant milestone, in finding a final resting place for all the spent nuclear waste, nuclear fuel, and all the solvents that were used in the manufacturing process, that was a legacy of our Cold War victory. Larry says, "He failed to discuss the scientific future and the legacy we are leaving our children, their children and the children's children." "Were handing them a sick or deadly future, that future has already been handed to many including our Parents ourselves, our children, who have or will suffer from Cancer, Birth defects, Pain and suffering from a multitude of unknown diseases, CNS injury, learning difficulties, Short term memory deficit just to name a few, caused from NUCLEAR, BIOLOGICAL, VIROLOGICAL, & CHEMICAL, Warfare weapons & its waste". "THIS IS NOT NORMAL, NOR WAS IT AN ACCIDENT, IT WAS INTENTIONALLY INFLICTED BY OUR GOVERNMENT, CORPORATIONS, & INDUSTRY ALIKE WHEN THEY WAGED WAR ON THEIR OWN PEOPLE, USING THEM AS GUINEA PIGS JUST TO SEE THE AFFECTS". "DOCUMENTED PROOF"

Right out of the Senate of the United States. "Nuclear waste is being shipped into the United States from at least 40 countries from around the World for storage. Tons of Nuclear waste has been crossing our highways for decades Idaho alone has just completed 4,600 shipments of high level nuclear waste. They are saying in not one instance has a shipment cask released radioactive material. Nuclear transportation casks have been subjected to every imaginable test to ensure their safety. Casks have been dropped onto steel spikes, hit by locomotives, crashed into walls at 70 mph".

But Larry says, "They were never stored the shipping casks inside a volcano. It could erupt with enough heat and gas to destroy every shipping cask immediately. As for the millions of tons of nuclear waste. In a split second our world would disintegrate. It could become an unknown comet in the galaxy". Larry says, "It sound like they are trying to compare a volcano to a locomotive at 70 mph. This would be like comparing several nuclear bombs to the same locomotive at 70 mph".

The Senate goes on to say a sites suitability is based on sound and thorough scientific investigations that would leave not doubt in the scientists minds. But Larry "says all the scientific reports to date 2003, do not support using Yucca Mountain as a repository". and from sound scientific data from the panel of scientists appointed by Congress and the Senate. They have clearly stated Yucca Mountain repository would be too unpredictable. They stated it is fraught with uncertainties, and the unexpected is impossible to predict over 10's of thousands even 100's of thousands of years. "Where your not allowed even one mistake, not even unexpected impossibility, would destroy the entire world. What does a common sense person do when? "There is a shadow of doubt? We Say No to Yucca Mountain as a repository for any waste nuclear or other wise.

Governor of Nevada "R" Kenny Guinn says, "Yucca Mountain is flawed". When it comes to Abraham's idea Kenny said "THIS DECISION STINKS". Larry says, "Yucca Mountain is first of all a damn volcano, and the Scientific panel appointed by the Congress of the United States call it unpredictable, & have indicated "NO" to avoid the un-unexpected from now to 100's of thousands of years from now. Earthquakes just recently at the site, should be taken as a warning to say "NO" to what could be a "Nuclear Volcano" capable of destroying the entire World. Governor Kenny Guinn says, "This is America's fight not just Nevada's, It's something that should concern every citizen of this nation." Is it right for Abraham from DOE, and President Bush to make that final decision without considering the safety that could affect everyone? "NO" I hope you feel the same way. As Larry does, who is lobbying the people of the United States, so say "NO" to Yucca Mountain From his book "DEATH FACTORY U. S. A. Part I, II, III,.......

Here is another test the Government has not talked about or even done, Governor Kenny Guinn ® of Nevada says, "Let's examine the rail accident of July 18, 2001 in the Baltimore's Howard Street Tunnel, that ignited a fire that burned for five days. A report assessing this accident in particular had the train been carrying <u>spent nuclear fuel.</u> It concludes that a deadly amount of radiation would have been relesed from the containers. This could be in your community, maybe close to your community meaning OverExposure to radiation and death. Who will provide the protection that would ensure its safety from Terrorist attacks? Yucca Mountain will create a massive new target for terrorists. I will take President Bush's own statement, "We will not fail"," Then why is Abraham, so eagerly pushing through a questionable project to President Bush. Larry says, "he questions the credentials of Abraham, Secretary of DOE Department of Energy, & his eagerness by using massive pressure on the White House, & his lobbyist's in forcing this project into service without considering the safety of our people", Larry is asking the people of the United States to also question his credentials, by acknowledging your safety concerns of our nation to your Congress-person, or your Senator's. Your immediate response is necessary letting them know you are concerned about your safety, the safety of your children, and our future involving the plan for Yucca Mountain.

"Fortunately, this is a nation of laws, and one of considerable common sense. With our backing, of a few good scientists, & a few good lawyers, together we can maybe unravel the house of cards created by years of ineptitude and political maneuvering at Yucca Mountain. I have every confidence in President Bush, who promised me that the project would not be mindlessly advanced in the face of bad science or any Political meneuvering".

ROCKY ROAD FOR THE NUKE WASTE PLAN.

Thanks to Molly Ivins, Rocky Mountain News 2002, W/modifications.

Another bad idea. What are they, cheaper by the dozen? With the Bush administration deciding to dump all the light-level nuclear waste in America into some yet-to-be excavated tunnels at Yucca Mountain in Nevada. Most people prefer to avoid nuclear waste topics on the grounds that it is depressing and scarey......like in Science Fiction coming true.

Most people are thinking well good it's Nevada at least it won't be here, "WRONG" on both counts. Yes Nevadan's are predictably unhappy and seriously irate...but it also brings nuclear garbage right to your front door and to your schools, by the closest rail traveling through your town, city, State, or closest Interstate by Truck.

Putting many Millions of tons of nuclear waste from at least 131 nuclear plants and more, then we have at least 100 or more military facilities operating under the guise of DOE Department of Energy……Using Yucca Mountain, as a repository a known volcano that has volcanic action within "considerable earthquakes". It is our's, every man, woman, and child, in the World. We are allowing them to create a monster that could destroy the entire world. Yucca Mountain could fast become the Worlds first & last Thermo-Nuclear-Volcano, that could destroy us all, the World, leaving it as a comet made up of burning cinders in our galaxy. Before that happens it will insanely escalate the chances for terrorist attacks…on the trucking industry, our rail systems & tunnels, along with Yucca Mountain.

******** @@@@@@@@ Future aliens, as the comet goes by will say

***** ******* that was earth, until science became their

******** @@@@@@@@ master, and destroyed their entire world.

SAY "NO" TO YUCCA MOUNTAIN, IT IS WHERE OUR FUTURE LIES.

P. HALF LIFE OF PLUTONIUM 100'S OF THOUSANDS OF YEARS.

— Creators Syndicate

THE UGLY BEAST RAISES IT'S HEAD ONCE MORE, as the government now limits nuclear power Industry's liability to the public in case of an accident.... No other industry enjoys this type Federal protection. The Limited Liability is now only $200 million for each power plant, with a cap for the entire industry at $9.1 billion. "BUT UNFORTUNATELY" SANDIA National Labs has estimated the cost of one big reactor accident at more than $550 billion......This was the Price Anderson Bill passed by Congress. Worse yet, it created a law that any new nuclear power plants will be built by Merchant Generator's.... Much like Enron.... Who will compete in the new deregulated markets... This gives all of them an added incentive to build, & operate cheap, & as usual safety at the bottom of the list....

The Senate version of the Bill does not authorize Price Anderson for Commercial reactors....If the bill were to pass as it stands, any new reactors would not be covered under the Senate Bill. At the very least we can hope the Senate version will become law, so the risk of doing business will actually fall on commercial reactors, instead of taxpayers. "Like everything else there are no guarantees".

HOUSE PANEL CONFIRMS YUCCA MOUNTAIN FOR NUCLEAR WASTE, IN A 41-6 VOTE WHICH COULD DEFEAT NEVADA'S VETO POWER

This is a victory for President Bush as the House overwhelmingly confirmed Yucca Mountain 90 miles northwest of Las Vegas will be the nations permanent storage site for millions of tons of nuclear waste. The American people are caught up in a pathway of destruction that may destroy the World. First by truck, then by rail, and barge, building America's first and probably last Thermo-Nuclear-Volcano. Final licensing has not yet taken place for the site to open, as the Governor of Nevada Kenny Guinn has veto'd President Bush's designation of Yucca Mountain. This set into motion the law's special procedure allowing Congress to override the Governor's Veto in a vote by both Chambers of Congress.

THE WAY LARRY SEEN IT, & HEARD IT, "COLORADO OFFICIALS ARE SPLIT, MAINLY CONGRESSMAN SCHAFFER WHO HAS ALWAYS BEEN FOR GOVERNMENT, AND INDUSTRY ALIKE WHILE IGNORING THE PEOPLE WHO ELECTED HIM INTO OFFICE. GOOD-BYE TO MR. SCHAFFER 2003 HE IS NO LONGER IN THAT RESPONSIBLE POSITION. HE ALSO IGNORED THOSE WHO POLLUTED THE NATION. AS WELL AS CONGRESSMAN MCGUINNIS, & CONGRESSPERSON DEGETTE, THEIR GUILTY OF SUPPORTING POLLUTION IN COLORADO, WHILE IGNORING THEIR CONSTITUANTS INSTEAD OF CLEANING IT UP.

"Even though Senator Ben Nighthorse Campbell ®. CO. has ignored some Government and Industy pollultion, he does oppose the shipments of nuclear waste to Yucca Mountain Nevada, And for the safety of the American people and our world"……COLORADO SHOULD BE PROUD OF SENATOR CAMPBELL…….HE PROBABLY HELPED SAVE YOUR LIVES AND THE WORLD……….

As for Senator Wayne Allard he has always ignored the safety of the American people by supporting Government and Industry polluters in Colorado and abroad.

As Senator's waiver & bow to the pressures of President Bush, the DOE & Secretary Abraham, & all the Lobbyist's who don't want to be locked out of the White House. They all want to be there when Secretary Abraham hands out the 9 pieces of silver for turning their backs on the World.......

Republican Senator's Bob Bennett & Orrin G. Hatch of Utah after being wined and dined at the White House by President Bush, Secretary of DOE, Spencer Abraham & all it's Loyal Lobbyists, announced Monday their support for the disposal of nuclear waste inside a known volcano Yucca Mountain. "THE VOLCANO" A PANEL OF SCIENTISTS, THE WORLD'S BEST WHO WERE SELECTED BY OUR PRESIDENT, BY OUR CONGRESS, & BY OUR SENATE, TO GIVE THEIR EXPERT OPINIONS ON THE SAFETY OF YUCCA MOUNTAIN AS A REPOSITROY FOR NUCLEAR WASTE.

Yucca Mountain to be used for a repository for millions of tons of nuclear waste was described as being fraught with uncertainties, & as being very unpredictable for now or into the future. They stated with this volcanic mountain, it would be impossible to avoid the unexpected. Larry says, "they are saying, we are building our own possible destruction & the destruction of the entire world. Being fraught with the unavoidable, unpredictable uncertainties now & into the future". This is what you call beating around the ""BUSH"" So "Bush" took the bull by the horns & said "READY OR NOT" Nevada here we come.

It really sounds more like Utah Senator's were "blackmailed by the White House, as President Bush, Spencer Abraham and their loyal lobbyists. Said there will by no more beating the "BUSH". When they informed the Utah Senators, they have been negotiating with a reservation 70 miles west of Salt Lake City for permission to store nuclear waste there of coarse on an intrim basis that could be forever, since the government is known for their lies, and making promises they never keep. They also roped and hog-tied, Senator Richard J. Durbin, (D)-Ill. Who will also announce his support for the repository in Yucca Mountain.

THE SENATE OF THE UNITED STATES OF AMERICA IS SENDING THE NUCLEAR WASTE FROM ACROSS OUR NATION TO NEVADA.

Giving out a message to the American people, the World & Terrorists every where. The Senate approval came in a 60-39 vote for Yucca Mountain to be the permanent Nuclear waste site for millions of tons from around every corner of the United States. Senator Wayne Allard of Loveland Colorado said, “Oh Well most of the waste will be transported through other states anyway“ Senator Harry Reid from Nevada lashed out at the nuclear lobbyists & their unending source of money for perpetuating “the big lie“ that the Nevada nuclear waste permanent storage repository was urgently needed. Sen.® John Ensign, Nev. Replied Not in my back yard. There were 15 Democratic Senators who voted with most all the GOP senators for the waste site. Only 3 (R) Sens. Ensign, Nev., Chafee, R.I. & Campbell, CO. opposed the Dump repository.

Senator Trent Lott Minority leader (R)-Miss. Said the Senate was forced to act or be “squeezed” by the deadline. Majority leader Tom Daschele brought up the matter, forcing Republican Senators to use a provision of a 1982 waste law, that allows any Senator to demand a vote, and the same law required the Senate to act by July 26, or the Yucca site would be shut down automatically. Larry says, “We now have the fox, who by outwitting with trickery, ingenuity and cunning, is not only guarding the Hen house, but rules the roost as well.® With the Coyote, and Wolves held at bay.

Be sure to pick up copies of DEATH FACTORY U. S. A. Part I, Part II, & Part III for yourself, for a relative, or for a friend. Part III will follow year 2004 and will complete Larry’s documented account with “DEATH FACTORY’S IN OUR MIDST“. With a hands on tribute to “SILENT SPRING” by RACHEL CARSON and AMERICAN THE POISONED by LEWIS REGENSTEIN.

WITH BEST WISHES LARRY D. LAND

About the Author:

It was Larry's dream to one day own the very place he was born and raised. He was born one mile north of the Rocky Mountain Arsenal that had begun manufacturing death gas at the very same time. Little did Larry realize he was on a path that would one day again cross that with the Rocky Mountain Arsenal, but in 1972 it did. Larry felt he had fulfilled his dream when he purchased the home place. It didn't dawn on him that his parents were really run down and had no reason why. Until he drilled a couple of new wells and within just days he had nearly 700 cattle lying dead or near dead, and they suffered from toxic poisons in their water. Larry's parents died long before their time. Larry was broke, he had 1.5 million dollars net worth in livestock, and they were all dead or destroyed by GB and Sarin nerve agents that which had found it's way to the land wells. Larry began an investigation that would last over thirty years. During the investigation Larry discovered there were over two hundred burial dumps for the toxic waste.

Over the thirty-year time that Larry worked to investigate the events surrounding his place of birth, he would encounter many trials and difficulties. Following Larry's legal attempts to 'cease and desist' the state of Colorado files similar actions against the United States Army and Shell Oil Company. After years of effort, teams of lawyers, chemists, geologists, and hydrologists, in 1983 the Shell Oil Company was removed from Colorado and the US Army ordered to clean up the mess. Estimated time to complete the clean up project range as far into the future as 2060.

In 1983 Larry moved to Minnesota to begin anew. Toxic poisons would continue to haunt his future. In 1984, just days before his chance to testify before congress, Larry was a victim of a hit and run by a car swerving over from the opposite side of the road and struck him on his bicycle throwing him to the ground, and then, after coming to a grinding stop. The driver first tried to strangle him, and was over powered by Larry, the man went to his car cursing, Larry thought OH" OH" it's over now. He could invision the man coming back with a gun to end it. Instead he got into his car hit reverse with the tires smoking tried backing over top of him, he was just barely able to clear the wheels as he rolled over and over. As the driver sped out to the street his tires howling, I believe he thought he had ran over Larry. To put an ending to this story, four years later in 1988 when Shell oil was trying to make their insurance companies pay the price for clean up. One insurance company had came across some documents that could shed some light on the hit and run may have been orchastrated by Shell Oil Company, then another memo dated a few days after the hit and run stating “Let the sleeping dog lay.” To say they had accomplished all that was necessary for now.

The Washington County Sheriff Department, in Stillwater Minnesota Mr. Land’s Employer used a very toxic and hazardous industrial Lead based Bridge Paint that required an oxygenated airline respirators at all times. This was used to paint the interior of our work area without ventilation, using high pressure sprayers. Then in 1990 they completed painting his work area again with a Mercury based paint. The Sheriff’s department in which Larry worked poisoned Larry twice more, enacted steps to reduce his physical abilities to perform his job, were affected. Larry over exposed with high levels of fiberglass particulate and then compounded the problem by refusing to work with the now disabled Larry. The two toxic poisonings involved using deadly lead based paint and an alkyd, alkyd, epoxy paint containing high levels Mercury Biocide banned for use in the United States by the EPA. The paint was so toxic it continued to affect Larry for an additional few years. Larry was totally disabled by his employer Washington County, by negligent use of Toxic and Hazardous Materials used in his work place. Then for 2 ½ months while cutting, wrapping and installing fiberglass on all plumbing pipes in the new building, without ventilation. Air samples taken for examination by a laboratory specializing in indoor pollution in the work areas discovered Larry had been over exposed and a Toxicologist from the University of Minnesota testified there was a minimum of 12 to 15% fiberglass for every breath taken.

www.ingramcontent.com/pod-product-compliance
Ingram Content Group UK Ltd.
Pitfield, Milton Keynes, MK11 3LW, UK
UKHW040603210726
13854UKWH00008B/1876